T0199855

Toxicokinetics and Risk Assessment

Edited by

John C. Lipscomb
National Center for Environmental Assessment
Office of Research and Development
U.S. Environmental Protection Agency
Cincinnati, Ohio, U.S.A.

Edward V. Ohanian
Office of Science and Technology
Office of Water
U.S. Environmental Protection Agency
Washington, D.C., U.S.A.

CRC Press
Taylor & Francis Group
Boca Raton London New York

CRC Press is an imprint of the
Taylor & Francis Group, an **informa** business

CRC Press
Taylor & Francis Group
6000 Broken Sound Parkway NW, Suite 300
Boca Raton, FL 33487-2742

© 2007 by Taylor & Francis Group, LLC
CRC Press is an imprint of Taylor & Francis Group, an Informa business

First issued in paperback 2019

No claim to original U.S. Government works

ISBN 13: 978-0-367-45345-9 (pbk)
ISBN 13: 978-0-8493-3722-2 (hbk)

Version Date: 20130312

Preface

Risk assessment is sometimes erroneously referred to as an art. The scientific basis for risk assessment is rather objective, though, at times, not clearly or well annunciated to the general public and to those in laboratory research organizations. The goals of this text are to develop and communicate a clear description of the noncancer risk assessment process and its reliance on uncertainty factors (those addressing extrapolations of toxicity from animals to humans and within the human species); to present a synopsis of the limited available guidance on replacing default values for uncertainty factors with data-derived values; to distill, communicate, and discuss considerations that must be given to the generalization of findings of toxicity from animals to humans; to demonstrate proper methods for the development and interpretation of toxicokinetic data; to address and reinforce statistical considerations; and to provide a historical perspective and recent examples of the use of toxicokinetic information to inform the selection of uncertainty factors in risk assessments developed by the U.S. Environmental Protection Agency.

This text is aimed at several audiences: (i) the cadre of risk assessors who are undecided about the potential application of quantitative toxicokinetic findings for uncertainty factors; (ii) organizations, including those tasked with developing bench top and/or whole-animal studies of chemical metabolism and tissue dosimetry for risk assessment purposes; and (iii) those who make decisions about which areas of risk science to advance. At present, the International Programme on Chemical Safety has finalized a document that guides consideration of the quantitative value of chemical-specific toxicokinetic data for use in defining values for uncertainty factors, and in 1994, the U.S. Environmental Protection Agency finalized its guidance on the application of chemical-specific and animal species–specific measures of biochemical and physiologic differences that can lead to "data-derived" values for toxicokinetic variability between animals and humans. While the U.S. Environmental Protection Agency's Integrated Risk Information System contains examples of inhalation risk assessments where species differences in toxicokinetics have been used for uncertainty factor values, very few oral risk assessments have been conducted at such a level of detail. It is hoped that unambiguous communication of methods, demonstrations of feasibility, and completed examples will stimulate additional research and policy work in this arena, which could culminate in the development of advanced risk assessment methods addressing this topic.

Regardless of policy, the availability of data may limit success if researchers are not fully cognizant of the implications and potential use of their data in risk assessment. This text will clearly demonstrate the application of physiologically based pharmacokinetic modeling in human health risk assessment and will separately demonstrate methods for data generation, extrapolation, and inclusion of laboratory results for chemical metabolism and computer-based predictions of biochemical constants to address animal-to-human and human interindividual variability; specific attention will be devoted to Monte Carlo sampling techniques and the statistical considerations given to data demonstrating or predicting population variability, including that variability based on genetic polymorphisms. By communicating these topics, those investigators tasked with developing laboratory studies to address human risk and safety concerns will see the inclusion of such data in the desired end product—the completed risk assessment.

This text comprises the various topics of history and bases for human health risk assessment and its uncertainty factors; presents the available guidance on the inclusion of toxicokinetic data in risk assessment; communicates techniques and considerations to develop laboratory and whole-animal data of sufficient quality upon which to base estimates of variability and uncertainty factor derivation; presents, in some detail, the field of physiologically based pharmacokinetic modeling and computer-based predictions of biochemical values for inclusion in toxicokinetic models; demonstrates the value of characterizing human variability in chemical metabolism in large banks of human tissue samples; presents statistical considerations for the development and interpretation of study results; demonstrates considerations given in the translation of external exposures to tissue concentrations for application in mixtures and cumulative risk assessment; and demonstrates success stories where toxicokinetic information has been used to replace default values for uncertainty factors with data-derived uncertainty factors. In that regard, this text is unique—it addresses concerns of multiple organizations and divergent staff levels, and it is self-contained. While several, but not each, of these individual topics have been addressed at the "text" level, there is no other single reference available that addresses all these topics. The development of this text will prove to be a boon to the development of additional studies to address toxicokinetics in animals and humans and to guide their evaluation as a means to replace default uncertainty factors in risk assessment, representing an overall improvement in their scientific basis.

Readers should be able to readily recognize the impact that their own fields of study—whether biochemical analyses, physiological measurement, computer simulation, in vivo studies, or pharmacokinetic analysis—can have on improving quantitative estimates of risk. Those scientists conducting risk assessment activities involving dose extrapolations should have a heightened awareness of the value, as well as the complexity, of toxicokinetic information. Through consideration of the information contained in this book, researchers and risk assessors should find common ground upon which to plan, conduct, and interpret new studies to continue the refinement of the scientific basis for health risk assessment.

John C. Lipscomb, PhD, DABT
Edward V. Ohanian, PhD

Acknowledgments

The editors are deeply indebted to the chapter authors who devoted their time to developing the materials, and to their employers. We are thankful to the managers who were gracious (but cautious) in providing latitude for all involved to embark on this opportunity. We also appreciate the several scientific communities and societies that are solidly behind efforts like these.

We are grateful to Ms. Bette Zwayer and Ms. Lana Wood for document preparation, without whose caring attitude and technical expertise this book would not have moved forward.

While this text is edited, and to some extent authored, by employees of the U.S. EPA, the views and opinions expressed herein are those of the respective authors and editors and do not represent those of the U.S. EPA.

Mention of trade names or commercial products does not constitute endorsement or recommendations for use.

Contents

Preface . . . . *iii*
Acknowledgments . . . . *vii*
Contributors . . . . *xvii*

**1. Overview of Approach to Noncancer
Risk Assessment** *1*
Lynne Haber
Introduction to the National Academy of
    Sciences Risk Assessment Paradigm 2
Use of Uncertainty Factors in
    Risk Assessment 6
Interindividual Variability 9
Experimental Animal to Human
    Variability 10
Subchronic to Chronic Extrapolation 10
LOAEL to NOAEL Extrapolation 11
Database Uncertainty 11
Use of TK Data to Modify or
    Replace Uncertainty Factors 13
Outstanding Issues, Research Areas,
    and Uncertainties 21
Summary 21
References 22

2. Guidance for the Development of Chemical-Specific Adjustment Factors: Integration with Mode of Action Frameworks *27*
Bette Meek and Andrew Renwick
Introduction 27
Development of Chemical-Specific/Compound-Related
  Adjustments Factors: The Construct 29
Guidance for Development of CSAF 36
Integrating Development of CSAF with MOA
  Frameworks—An Example 39
Conclusions 44
References 44

3. Derivation and Modeling of Mechanistic Data for Use in Risk Assessment *47*
Ronald L. Melnick
Introduction 47
Evaluating Human Risk from Animal
  and Mechanistic Data 48
Derivation of Pharmacokinetic and
  Pharmacodynamic Measures from
  Toxicology and Mechanistic Information 50
The Necessity to Test Mechanistic Hypotheses 56
Conclusion: Pharmacokinetic and Pharmacodynamic
  Modeling Is an Iterative Process 61
References 63

4. Empirically Observed Distributions of Pharmacokinetic and Pharmacodynamic Variability in Humans—Implications for the Derivation of Single-Point Component Uncertainty Factors Providing Equivalent Protection as Existing Reference Doses *69*
Dale Hattis and Meghan Keaney Lynch
Introduction 69
IPCS Assumptions of Single-Point Factors
  Representing Overall Interindividual Variability
  and PK/PD Components 72

Description and Update of Previous Work to Construct
a Straw Man Proposal for a Quantitative
Distributional Goal for Defining Reference
Doses Protection Objectives 73
Expansion of the Estimated Pharmacokinetics Variability
for Some Age-Related Differences 79
Basic Inferences for the Split of Lognormal Variance in
Interindividual Variability Log(GSD) Between
PK and PD Portions of the Pathway from Oral
Administration to Biological Responses 80
Uncertainties in Current Estimates of PK
and PD Variability for an Untested Chemical 81
Effects of Perfect Knowledge of Overall and
Component Interindividual Variability on Doses
Needed to Achieve the Straw Man Protection Goal 84
Lessons for the Derivation of Single-Point Factors
to Represent Overall Interindividual Variability
and PK/PD Components 89
References 92

5. **Use of Classical Pharmacokinetic Evaluations
in Drug Development and Safety Assessment** *95*
Rakesh Dixit and Peter Ward
Introduction 95
Classical Pharmacokinetics and Toxicokinetics 96
Tissue Drug Toxicokinetics 102
Qualitative and Quantitative Assessment
of Metabolites 103
Drug–Drug Interactions 105
Importance of Understanding Drug Metabolism
in Discovery and Early Development 113
References 119

6. **Considerations for Applying Physiologically Based
Pharmacokinetic Models in Risk Assessment** *123*
*Chadwick Thompson, Babasaheb Sonawane,
Andy Nong, and Kannan Krishnan*
Introduction 123
Pharmacokinetic Data Required for Risk
Assessments 125

Choosing PBPK Models Appropriate for Use
in Risk Assessment 126
Evaluation of Dose Metrics for PBPK
Model-Based Assessments 128
Generic Examples of the Use of PBPK Models
in Risk Assessment 130
Perspectives and Concluding Remarks 135
References 137

7. **Considerations of Design and Data When Developing
Physiologically Based Pharmacokinetic Models** *141*
Peter J. Robinson, Jeffery M. Gearhart, Deirdre A. Mahle,
Elaine A. Merrill, Teresa R. Sterner, Kyung O. Yu,
and John C. Lipscomb
Introduction 141
Physiological Parameters 146
Chemical-Specific Parameters 149
Metabolism: Gas Uptake and In Vitro Methods 152
Model Development and Validation 155
Advanced PBPK Models 161
PBPK/PD 162
Conclusions: Application to Risk Assessment 163
References 164

8. **In Silico Predictions of Partition Coefficients for Physiologically
Based Pharmacokinetic Models** *167*
Kannan Krishnan, Andy Nong, and Sami Haddad
Introduction 167
In Silico Approaches for Estimating PCs 168
Mechanistic In Silico Approaches for
Predicting Interindividual Differences
in PCs 169
Interindividual Differences in PCs, Tissue Dose,
and Risk Assessment 178
References 181

9. **In Vitro to In Vivo Extrapolation of Metabolic Rate Constants
for Physiologically Based Pharmacokinetic Models** *185*
Gregory L. Kedderis
Introduction 185
The Structure and Function of Enzymes 186

Fundamentals of Enzyme Kinetics 187
Enzymes Involved in Biotransformation 194
In Vitro Systems to Study Biotransformation 199
Design of In Vitro Experiments to Predict
    Biotransformation Kinetics 200
Extrapolation of Kinetic Parameters to
    Whole Organisms 201
References 206

10. **Use of Physiologically Based Pharmacokinetic Modeling to
    Evaluate Implications of Human Variability** *211*
    P. Robinan Gentry and Harvey J. Clewell III
    Introduction 211
    Determinants of Impact 212
    Pharmacokinetic Variability Between Individuals
        of Different Ages 213
    Pharmacokinetic Variability Among Individuals
        at the Same Age 217
    Pharmacokinetic Variability Resulting from
        Differences Between Subpopulations 218
    Conclusions 225
    References 227

11. **Developmental Aspects of Children's
    Pharmacokinetics** *231*
    Gary Ginsberg, Dale Hattis, and Babasaheb Sonawane
    Children's Pharmacokinetics: Overview 231
    Absorption, Distribution, Metabolism, and
        Excretion in Children 233
    Development of Phase II Conjugation Systems 238
    Development of Renal Elimination 239
    Variability in Pediatric Pharmacokinetic Data 240
    Development of Children's PBPK Models 240
    Summary 244
    References 245

12. **Sensitive Populations and Risk Assessment** *251*
    Michael Dourson and Daniel Drinan
    Introduction 251
    General Principles 251

Uncertainty Factors that Account for Sensitivity 254
Mechanistic Information and Sensitive Populations 259
Summary 264
References 264

13. **Statistical Issues in Physiologically Based
    Pharmacokinetic Modeling** *269*
    Weihsueh A. Chiu
    Introduction 269
    Sensitivity, Identifiability, and the Role of Statistics 270
    Case Studies 274
    Conclusions 281
    References 281

14. **Drug Development and the Use of Pharmacokinetics/
    Toxicokinetics in Selecting the First Dose of
    Systemically Administered Drugs in Humans—A
    Nonclinical Perspective** *285*
    Abigail Jacobs
    The Overall Drug Development/Approval Process 285
    Selection of the First Dose in Humans 290
    Pharmacology and Toxicology Data 291
    Within Species Evaluations 292
    Cross Species Comparisons 293
    Use of PK in Picking the First Dose in
        Humans and Uncertainties 296
    Use of PK in Picking the Maximum Dose to
        Be Used in Human Volunteers 298
    Conclusions 298
    References 299

15. **Pharmacokinetic/Physiologically Based
    Pharmacokinetic Models in Integrated Risk
    Information System Assessments** *301*
    *Robert S. DeWoskin, John C. Lipscomb, Chadwick Thompson,
    Weihsueh A. Chiu, Paul Schlosser, Carolyn Smallwood,
    Jeff Swartout, Linda Teuschler, and Allan Marcus*
    Introduction—Increasing Use of PK/PBPK
        Models in IRIS 302

The Use of PK and PBPK Models in the
IRIS Program 302
Evaluation of PBPK Models Used in
IRIS Assessments 308
Specific Examples of Use of PK/PBPK
Models in IRIS System 310
PBPK Models or PK Data Used in the Derivation
of Reference Values 314
Vinyl Chloride 319
Dichloromethane 323
Xylene 328
Boron and Compounds 332
The IEUBK Model 335
References 341

Index *349*

Contributors

Weihsueh A. Chiu National Center for Environmental Assessment, Office of Research and Development, U.S. Environmental Protection Agency, Washington, D.C., U.S.A.

Harvey J. Clewell III CIIT Centers for Health Research, Research Triangle Park, North Carolina, U.S.A.

Robert S. DeWoskin National Center for Environmental Assessment, Office of Research and Development, U.S. Environmental Protection Agency, Research Triangle Park, North Carolina, U.S.A.

Rakesh Dixit Toxicology Department, Johnson & Johnson, PRD, San Diego, California, U.S.A.

Michael Dourson Toxicology Excellence for Risk Assessment (TERA), Cincinnati, Ohio, U.S.A.

Daniel Drinan Toxicology Excellence for Risk Assessment (TERA), Cincinnati, Ohio, U.S.A.

Jeffery M. Gearhart Air Force Research Laboratory, Wright-Patterson AFB, Dayton, Ohio, U.S.A.

P. Robinan Gentry ENVIRON International Corporation, Ruston, Louisiana, U.S.A.

Gary Ginsberg   Connecticut Department of Public Health, Hartford, Connecticut, U.S.A.

Lynne Haber   Toxicology Excellence for Risk Assessment (TERA), Cincinnati, Ohio, U.S.A.

Sami Haddad   Department of Biological Sciences, University of Quebec at Montreal, Montreal, Quebec, Canada

Dale Hattis   George Perkins Marsh Institute, Clark University, Worcester, Massachusetts, U.S.A.

Abigail Jacobs   Center for Drug Evaluation and Research, Food and Drug Administration, Silver Spring, Maryland, U.S.A.

Gregory L. Kedderis   Independent Investigator, Chapel Hill, North Carolina, U.S.A.

Kannan Krishnan   Groupe de Recherche Interdisciplinaire en Santé et Département de Santé Environnementale et Santé au Travail, University of Montreal, Montreal, Quebec, Canada

John C. Lipscomb   National Center for Environmental Assessment, Office of Research and Development, U.S. Environmental Protection Agency, Cincinnati, Ohio, U.S.A.

Meghan Keaney Lynch   Boston University School of Public Health, Boston, Massachusetts, U.S.A.

Deirdre A. Mahle   Air Force Research Laboratory, Wright-Patterson AFB, Dayton, Ohio, U.S.A.

Allan Marcus   National Center for Environmental Assessment, Office of Research and Development, U.S. Environmental Protection Agency, Research Triangle Park, North Carolina, U.S.A.

Bette Meek   Existing Substances Division, Health Canada, Ottawa, Ontario, Canada

Ronald L. Melnick   National Institute of Environmental Health Sciences, National Institutes of Health, Research Triangle Park, North Carolina, U.S.A.

Elaine A. Merrill Air Force Research Laboratory, Wright-Patterson AFB, Dayton, Ohio, U.S.A.

Andy Nong Groupe de Recherche Interdisciplinaire en Santé et Département de Santé Environnementale et Santé au Travail, University of Montreal, Montreal, Quebec, Canada

Andrew Renwick School of Medicine, University of Southampton, Southampton, U.K.

Peter J. Robinson Air Force Research Laboratory, Wright-Patterson AFB, Dayton, Ohio, U.S.A.

Paul Schlosser National Center for Environmental Assessment, Office of Research and Development, U.S. Environmental Protection Agency, Washington, D.C., U.S.A.

Carolyn Smallwood National Center for Environmental Assessment, Office of Research and Development, U.S. Environmental Protection Agency, Cincinnati, Ohio, U.S.A.

Babasaheb Sonawane National Center for Environmental Assessment, Office of Research and Development, U.S. Environmental Protection Agency, Washington, D.C., U.S.A.

Teresa R. Sterner Air Force Research Laboratory, Wright-Patterson AFB, Dayton, Ohio, U.S.A.

Jeff Swartout National Center for Environmental Assessment, Office of Research and Development, U.S. Environmental Protection Agency, Cincinnati, Ohio, U.S.A.

Linda Teuschler National Center for Environmental Assessment, Office of Research and Development, U.S. Environmental Protection Agency, Cincinnati, Ohio, U.S.A.

Chadwick Thompson National Center for Environmental Assessment, Office of Research and Development, U.S. Environmental Protection Agency, Washington, D.C., U.S.A.

Peter Ward Drug Metabolism Department, Johnson & Johnson, PRD, San Diego, California, U.S.A.

Kyung O. Yu Air Force Research Laboratory, Wright-Patterson AFB, Dayton, Ohio, U.S.A.

1

Overview of Approach to Noncancer Risk Assessment

Lynne Haber

Toxicology Excellence for Risk Assessment (TERA), Cincinnati, Ohio, U.S.A.

Uncertainty factors (UFs) are used to extrapolate from the available data to estimate doses anticipated to be without increased risk for adverse effects. The U.S. Environmental Protection Agency (EPA) specifically considers five areas of uncertainty: interspecies differences, intraspecies variability (including sensitive populations), the lack of a no observed adverse effect level (NOAEL), the lack of lifetime studies, and completeness of the database. Other organizations use a generally similar approach, but differ in the specific areas considered. Toxicokinetic (also called pharmacokinetic) considerations are relevant to many considerations for uncertainty factors. In particular, the interspecies and intraspecies uncertainty factors can be divided, respectively, into toxicokinetic and toxicodynamic components. Uncertainty factors may be considered to fall on a continuum of increasing incorporation of chemical-specific data, ranging from qualitative defaults to a fully data-derived approach using a biologically based mechanistic model. Intermediate approaches include database-derived uncertainty factors, categorical approaches, and chemical-specific adjustment factors (CSAFs). CSAFs incorporate data on the chemical of interest to quantify interspecies or intraspecies variability in toxicokinetics (TK) or toxicodynamics (TD), without requiring the mathematical sophistication of a physiologically based pharmacokinetic (PBPK) model, although PBPK model output can also be used to develop CSAFs.

INTRODUCTION TO THE NATIONAL ACADEMY OF SCIENCES RISK ASSESSMENT PARADIGM

As described by the National Research Council (1–3) of the National Academy of Sciences (NAS), there are four components to human health risk assessment: hazard identification, dose–response assessment, exposure assessment, and risk characterization. This chapter presents an overview of the four components, with particular attention to applications of toxicokinetic data to inform the risk assessment process. While the focus is primarily on noncancer risk assessment, cancer risk assessment is also addressed, particularly in the context of noting similarities and differences between the approaches for noncancer and cancer endpoints.

In the hazard characterization step, scientific studies are reviewed to determine what effects a chemical can cause (noncancer or cancer), including identification of the target organ(s) and identification of the "critical" effect(s) (i.e., the first adverse effect or its known precursor that occurs in the most sensitive species as the dose rate increases) (4). In some cases, one needs to determine whether an effect is adverse or not, and the relevance to humans. Other considerations in the hazard characterization step include the characteristics and relevance of the experimental routes of exposure.

Consideration of weight of evidence is important for all hazard characterization, but is of particular importance in consideration of the varied types of data available for cancer assessment. For cancer toxicity, hazard characterization considers the available epidemiological information, chronic animal bioassays, other short-term screening bioassays, mechanistic studies, genotoxicity tests, structure–activity relationships, metabolic and pharmacokinetic properties, and physical and chemical properties. Based on the overall evaluation, the chemical is placed into a cancer category (5,6) or a weight-of-evidence narrative and a descriptor is developed (7). Moreover, the hazard characterization can provide specific information about the conditions under which a chemical is likely to be carcinogenic, which may depend in turn on the chemical's mode of action (MOA), a general description of the sequence of events by which a chemical causes cancer. For example, it may be "likely to be carcinogenic by the route of inhalation but not by ingestion," or "likely to be carcinogenic under conditions that lead to cytotoxicity and regenerative hyperplasia in susceptible tissues, and not likely to be carcinogenic under exposure conditions that do not cause cytotoxicity and cell regeneration."

Dose–response assessment is the second step in the risk assessment process. In the absence of ethically obtained human data (the scientifically preferred data for risk assessment), the dose–response assessment for either cancer or noncancer toxicity is determined from animal toxicity studies. Doses that induce adverse effects are identified based on effects in experimental animal species that are relevant to humans, and a critical study is identified that shows an adverse effect at the lowest administered dose. The default assumption is that humans may be at least as sensitive as the most sensitive experimental species. Traditionally, noncancer assessments have assumed that the adverse effect exhibits a threshold in the dose–response curve, while cancer assessments have assumed that no threshold exists. Using this approach, safe doses [e.g., Reference dose (RfD)[a] or concentration (RfC), tolerable daily intake (TDI),[b] minimum risk level (MRL)[c]] are calculated by dividing a point of departure by uncertainty factors (described in more detail below). Dosimetric adjustments are often applied to inhalation exposure concentrations. The point of departure may be a NOAEL, a lowest-observed-adverse-effect level (LOAEL), or a statistically derived value corresponding to a specified level of risk [e.g., a benchmark dose (BMD)].

In contrast to this approach for noncancer effects, cancer assessments by many organizations have traditionally calculated a risk per unit dose, rather than a safe dose level, because of the assumption that no dose is without risk. This risk per unit dose can be used to calculate the dose corresponding to a specific level of risk, or a de minimis risk (e.g., the dose that is estimated to correspond to an increased cancer risk of one in a million). Such extrapolations far below the range of the data introduce significant uncertainties into the risk assessment process. The risk per unit dose continues to be used for chemicals that are believed to act via processes that would be expected to exhibit a linear dose–response at low doses. However, the differences in dose–response assessment methods for noncancer and cancer effects are diminishing as the focus on MOA increases. For example, when a chemical is shown to cause cancer via an MOA other than DNA reactivity (or other biological process expected to show a linear dose–response), the U.S. EPA derives an RfD or RfC for that chemical using an uncertainty factor approach (7).

A parallel step in the process to the hazard identification and dose–response assessment is exposure assessment. In exposure assessment, the

---

[a] The RfD is defined by the U.S. EPA as "an estimate (with uncertainty spanning perhaps an order of magnitude) of a daily oral exposure to the human population (including sensitive subgroups) that is likely to be without an appreciable risk of deleterious effects during a lifetime. It can be derived from a NOAEL, LOAEL, or BMD, with uncertainty factors generally applied to reflect limitations of the data used." The RfC is the parallel inhalation concentration. Other agencies use generally similar definitions.

[b] Health Canada defines the TDI as follows: "The tolerable daily intake (or tolerable intake) expressed on a body weight basis (e.g., mg/kg b.w./day) are the total intakes by ingestion, to which it is believed that a person can be exposed daily over a lifetime without deleterious effect."

[c] Agency for Toxic Substances and Disease Registry (ATSDR) defines a chronic MRL as "an estimate of daily human exposure to a dose of a chemical that is likely to be without an appreciable risk of adverse noncancerous effects over a lifetime of exposure (based on studies of 365 or more days)."

intake of a toxic agent from the environment is quantified and may consider any combination of oral, inhalation, and dermal routes of exposure, depending on the use of the risk assessment. Factors considered include the source, type, magnitude, and duration of contact. In a quantitative risk assessment, these factors are typically combined to estimate a potential human dose rate (and concentrations to which organisms in the environment are exposed). An exposure assessment may also be focused on one particular medium and one route of exposure, such as the oral intake of a drinking water disinfectant byproduct, or it may include exposure from all sources (aggregate exposure). Exposure can be quantified by direct measurement at the point of contact using personal monitoring devices. This method gives the most accurate exposure value for the period of time over which the measurement was taken. Exposure can also be estimated through environmental fate and transport modeling. An increasing number of models are available for use in developing exposure assessments; many of these are described in the U.S. EPA's exposure assessment guidelines (8).

Risk characterization is the most important and final part of a risk assessment. It summarizes and interprets the information from hazard identification, dose–response, and exposure steps, often by quantitatively comparing exposures with doses that are associated with potential health effects. The risk characterization identifies the assumptions, limitations, and level of confidence in risk estimates and communicates the actual likelihood of risk to exposed populations (9–11). The uncertainties identified in each step in the risk assessment process are analyzed and the overall impact on the risk estimate(s) is evaluated quantitatively and/or qualitatively. A good risk characterization provides transparency in decision making, clarity in communication, consistency (i.e., in harmony with other actions by the same decision-making body), and reasonableness (based on sound judgment).

Several different approaches are used in risk characterizations for comparing the dose–response and exposure assessments, depending on the approach used for the dose–response assessment and the available data. A hazard quotient (HQ) is the ratio between the exposure and the RfD, RfC, or similar value typically derived in noncancer risk assessments. HQ values below 1 indicate that an effect is unlikely, while the probability of an effect, the percentage of people affected, and the severity of the risk usually increases as the HQ increases above 1. However, noncancer risk values such as RfDs or ADIs are generally imprecise, and small differences in the HQ in the region of 1 cannot be distinguished scientifically from a value of 1. The use of a bright-line cutoff is a policy decision and should not be interpreted as meaning that any HQ above 1 carries a risk while any HQ below 1 is safe for every individual. Any risk from small exceedances of the RfD or RfC will generally apply only to the most sensitive individuals in the population, and larger exceedances are generally required before the fraction of the population affected may increase. Conversely, while exposures at or below the RfD or RfC are considered protective for sensitive people for most chemicals, one cannot rule out the possibility that exposure at these levels may result in some small risk for highly sensitive individuals for some chemicals.

The margin of safety (MOS) or margin of exposure (MOE) approach is similar to the HQ. However, while the HQ is calculated by dividing the exposure (estimated daily intake) by a derived risk value (e.g., RfD), the MOS or MOE is calculated by dividing the point of departure in an animal study (NOAEL, LOAEL, or BMD) by the exposure. The larger the MOS, the greater the presumed safety. The MOS must be interpreted by experts depending, in part, on the completeness of the toxicity database from which the NOAEL of the critical effect is derived (12). Factors that may be considered in the MOS or MOE approach include:

1. The uncertainty arising, among other factors, from the variability in the experimental data and intra- and interspecies variation,
2. The nature and severity of the effect,
3. The human population to which the quantitative and/or qualitative information on exposure applies,
4. Differences in exposure (route, duration, frequency, and pattern),
5. The dose–response relationship observed, and
6. The overall confidence in the database.

These assessment factors are very similar to those covered by the uncertainty factors of the tolerable intake or RfD approach. However, in contrast to the RfD approach, where expert judgments about the appropriate factors are considered in the determination of a "safe" dose that is then used to estimate a guidance value, the MOS approach relies on expert judgments to reach conclusions about given exposures on a case-by-case basis.

One advantage of the MOS approach over the use of RfD-based guidance values is that not all toxicity databases are strong enough to develop an RfD, yet these databases may be used to determine a point of departure for deriving a MOS. However, additional care is needed in the interpretation of this MOS, because the database is correspondingly weaker. An additional advantage is that one can identify a single NOAEL/point of departure (POD), compare it to different exposure scenarios (e.g., subchronic, chronic), and interpret the MOS based on what is of concern for that scenario, without needing to derive a separate safe dose estimate for each scenario.

For DNA-reactive carcinogens (and for carcinogens for which the MOA is not known), the risk characterization for cancer is also conducted by comparing a toxicity benchmark with exposure. In this case, however, the toxicity benchmark (expressed as risk per unit dose) is multiplied by total exposure, resulting in an estimate of risk (e.g., 5×10^{-5}). Note that the toxicity benchmark is often a statistical upper bound on the risk per unit dose, so that the overall estimate of risk is also a bound, and the actual risk may be lower. For carcinogens that act via MOAs that generate nonlinear responses, risk characterization is done as for noncancer endpoints.

Toxicokinetic considerations can enter into the risk assessment paradigm at a number of points. In the hazard characterization phase, information on the degree of absorption and distribution of the chemical can provide information about likely and plausible targets. All aspects of TK (absorption, distribution, metabolism, and excretion) contribute to the determination of tissue dose, which ultimately determines the dose–response. Consideration of interspecies differences in TK can also provide information on the relevance to humans of effects observed in experimental animals. Finally, TK are relevant in relating external doses calculated as part of an exposure assessment to internal dose. It is important to use consistent approaches and assumptions for calculating internal dose in the dose–response and exposure phases of the risk assessment. Toxicokinetics can also be important in evaluating interactions between chemicals in mixtures. For example, certain solvents can increase the bioavailability of other chemicals. Chemicals that induce enzymes may increase the metabolism of other chemicals, leading to lower tissue dose if the parent chemical is the toxic form, or higher tissue dose if a metabolite is the toxic form. Conversely, inhibition of metabolic enzymes can increase a chemical's activity. For example, piperonyl butoxide inhibits the activity of the enzyme cytochrome P450, and so can increase the biological half-life of P450 substrates, thereby increasing their potency. Because of this activity, piperonyl butoxide is often used as an insecticide synergist.

USE OF UNCERTAINTY FACTORS IN RISK ASSESSMENT

Uncertainty factors are used to extrapolate from the available data to estimates of safe doses. The U.S. EPA considers specifically five areas of uncertainty in deriving chronic noncancer risk values: (i) differences between experimental animals and humans (interspecies variation, UF_A), (ii) differences between average humans and sensitive humans (intraspecies variability, UF_H), (iii) the lack of a bioassay (UF_L), (iv) the lack of lifetime studies (subchronic to chronic, UF_S), and (v) the lack of bioassay that tests a variety of endpoints such as in young experimental animals (database, UF_D). Consideration of an additional UF to ensure protection of children is mandated by the U.S. Food Quality Protection Act (FQPA) for pesticide safety evaluations. Issues addressed by this additional factor are also considered for UF_D and UF_H, and the EPA recommends that the use of the FQPA factor should be modified when UF_D has already been used (13,14).

Thus, the RfD or other noncancer risk values are calculated using the following equation:

$$RfD = \frac{POD}{UF} \tag{1}$$

where POD is the point of departure, which may be a NOAEL, LOAEL, BMD, or other similar value, and UF the product of all relevant uncertainty factors:

$$UF_A \times UF_H \times UF_L \times UF_S \times UF_D \qquad (2)$$

where UF may be 1 or other values, as described below.

Other organizations use generally similar approaches, although they may differ in some details. For example, Health Canada includes factors to account for intraspecies variation and interspecies variation, and a combined factor of 1–100 to account for inadequacies of the database (15). These inadequacies may include lack of adequate data on developmental, chronic, or reproductive toxicity, use of a LOAEL versus a NOAEL, and inadequacies of the critical study. Other considerations and possible adjustments might be made for essential substances; severe irreversible effects; or interaction with other chemical substances commonly present in the general environment. As summarized by the International Programme on Chemical Safety (IPCS) (16), the World Health Organization (WHO) and other organizations also use an additional factor in cases where the NOAEL is derived from a severe irreversible effect, such as teratogenicity or non-genotoxic carcinogenicity.

A brief description of commonly used UFs and their justification is shown in Table 1. These factors are based on scientific judgments of available toxicity, toxicodynamic, and toxicokinetic data and inherent uncertainty. The six specific categories shown follow the approach of the U.S. EPA. Although not all health organizations apply these factors as discrete divisors, most groups consider the uncertainties defined in Table 1 in at least three areas: inter-individual variability within the human population, interspecies extrapolation, and less than ideal data. Moreover, while the UF can be derived on a chemical-specific basis, a default factor of 10 is often used to account for each area, with some allowance for overlap. These individual factors are multiplied to develop the total UF, although there is some allowance for overlap where multiple factors are applied. Due to the application of multiple health protective factors inherent in the multiplicative UF approach, the acquisition of additional data often results in a higher RfD (i.e., a conclusion that higher exposures are "safe"). In addition, if the chemical is an essential nutrient at low concentrations, the daily requirement may be taken into consideration in developing the overall UF. For example, the manganese RfD is based on essentiality data together with toxicity data from an exposed population, and the RfD for zinc took into account the recommended daily allowance for this essential element (4). Each of the areas of uncertainty is discussed briefly below. Detailed reviews of the bases for these factors, including post hoc analyses of the scientific and empirical basis for the magnitude of these factors, are available (24–26).

As described in the following sections, chemical-specific data can be used as the basis for UFs other than the default. In the absence of data allowing more precision, the U.S. EPA uses factors of 1, 3, or 10. A value

Table 1 Description of Typical Uncertainty and Modifying Factors in Deriving Reference Doses or Reference Concentrations[a]

Standard UFs	General guidelines
H (interhuman)	Use a 10-fold factor as a default when extrapolating from valid experimental results from studies using prolonged exposure to average healthy humans; this factor is intended to account for the variation in sensitivity among the members of the human population.
A (laboratory animal to human)	For RfDs, use a 10-fold factor as a default when extrapolating from valid results of long-term studies on experimental animals when results of studies of human exposure are not available or are inadequate; for RfCs, this factor is reduced to threefold when a NOAEL (HEC) is used as the basis of the estimate; in either case this factor is intended to account for the uncertainty in extrapolating animal data to humans.
S (subchronic to chronic)	Use a 10-fold factor as a default when extrapolating from less than chronic results on experimental animals or humans; this factor is intended to account for the uncertainty in extrapolating from less than chronic NOAELs to chronic NOAELs.
L (LOAEL to NOAEL)	Use a 10-fold factor as a default when deriving an RfD or RfC from a LOAEL, instead of a NOAEL; this factor is intended to account for the uncertainty in extrapolating from a LOAEL instead of a NOAEL.
D (incomplete database to complete)	Use a 10-fold factor as a default when extrapolating from valid results in experimental animals when the data are "incomplete"; this factor is intended to account for the inability of any single study to adequately address all possible adverse outcomes.
Modifying factor (MF)	The MF was used in some early assessments, but was phased out by the U.S. EPA (17). Use professional judgment to determine an additional uncertainty factor termed an MF that is greater than 0 and less than or equal to 10; the magnitude of the MF depends upon the professional assessment of scientific uncertainties of the study and database not explicitly treated above (e.g., the number of animals tested); the default value for the MF is 1.

[a]Professional judgment is required to determine the appropriate value to use for any given UF; see text. The values listed in this table are default values that are frequently used by the EPA in the absence of data. The maximum uncertainty factor used with the minimum confidence database has historically been 10,000 for an RfD and 3000 for an RfC. However, recent guidance for the RfD recommended limiting the total UF to no more than 3000, and avoiding the derivation of a risk value when there are four or more full areas of uncertainty.

Abbreviations: RfDs, reference doses; RfCs, reference concentrations; UFs, uncertainty factors; NOAEL, no observed adverse effect level; HEC, human equivalent concentration; LOAEL, lowest-observed-adverse-effect level, MF, modifying factor.

Source: From Refs. 19–23.

of 3 is used as an intermediate value, as it represents the square root of 10, rounded to one significant figure. This meaning behind the uncertainty factor of 3 is the reason for the potentially confusing practice that the product of two UFs of 3 is 10. Other organizations use a linear scale for intermediate values. For example, Health Canada often uses an intermediate value of 5, and the WHO often uses values of 2 or 5.

INTERINDIVIDUAL VARIABILITY

This UF (also called the UF for intraspecies variability, or the UF for protection of sensitive subpopulations) assumes that humans vary in the response to a given dose of chemical and that available human studies may not have adequately characterized the response of sensitive individuals, usually due to small sample size. This factor may also assume that subpopulations of humans exist that are more sensitive to the toxicity of the chemical than the average population. The default value of this UF is 10, but it may be reduced if data are available from the sensitive population in humans. For example, if the risk value is developed based on an effect in a sensitive population, at least some, if not a substantial portion, of human variability, is taken into account. The RfD for nitrate is based on methemoglobinemia in infants, a highly sensitive population, and provides an example of an RfD based on effects in a sensitive population, and the use of a human variability UF of 1. Logically, this UF cannot be less than 1, as the variability in the total population can only be greater than or equal to the variability in a test population; total population variability cannot be less than that in a population sample. Information on human variability in TK and TD can also be used more quantitatively to modify this UF, as described below. As shown in Figure 1, UF_H moves down the dose–response curve, as the responders at the lower end of the dose–response curve are the more sensitive individuals.

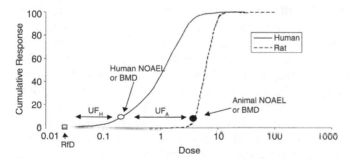

Figure 1 Intraspecies and interspecies uncertainty factors relative to the dose–response curve. *Abbreviations*: NOAEL, no observed adverse effect level; UF_H, uncertainty factor for inter-individual variability; UF_A, uncertainty factor for interspecies variability; RfD, reference dose; BMD, benchmark dose.

EXPERIMENTAL ANIMAL TO HUMAN VARIABILITY

If adequate toxicity data on humans do not exist, then experimental animal data can be used as the basis of the assessment. This UF (also called the UF for interspecies variability) addresses differences in TK and in tissue response between laboratory animals and humans. The basic assumptions for this factor are that the results seen in experimental animals are relevant to humans, that toxicokinetic and toxicodynamic differences exist among species, and that humans are assumed to be more sensitive than animals at a given mg/kg/day dose or mg/m^3 concentration. The default value for UF_A is 10, but several organizations use a default factor of 3 for extrapolating from animals to humans for inhalation data when dosimetric adjustments are used to account for some of the toxicokinetic differences between animals and humans (19,27,28). These dosimetric adjustments, and the calculation of the human equivalent concentration (HEC) for inhalation exposure, are described further below. Work is also ongoing at the U.S. EPA to develop guidance for deriving a human equivalent dose (HED), including use of default body weight$^{3/4}$ scaling to harmonize the oral noncancer and cancer assessments (28). Similar to the situation for UF_H, information on interspecies differences in TK and TD can be used to modify UF_A. As shown in Figure 1, UF_A extrapolates from the animal dose–response curve to the human dose–response curve.

SUBCHRONIC TO CHRONIC EXTRAPOLATION

As noted in their definitions, risk values such as RfDs or ADIs are generally intended to apply to lifetime exposures. While a full lifetime exposure is generally considered to be 2 years for rats and mice and 70 years for humans, exposures of greater than approximately 10% of the lifetime (i.e., more than 90 days for rodent studies) are often considered to be chronic exposures. Other agencies (e.g., ATSDR) define chronic exposure as a year (365 days) or longer. However, often only subchronic toxicity data are available. In such cases, risk assessors often extrapolate between durations using the assumption that an effect seen at shorter durations will also be seen after a lifetime of exposure, but may occur at lower doses. They may also assume that some effects may only be seen after an experimental group is exposed chronically. This may be either due to accumulation of damage, or due to accumulation of the toxicant, resulting in ongoing damage. This UF addresses the question, "If data in subchronic studies exist on which to base the estimation of a subthreshold dose, would data in chronic studies yield lower NOAELs?" Note that consideration of the appropriate value for UF_S requires consideration of the chemical's mode of action (MOA). For example, if the chemical causes toxicity via a reactive metabolite (which therefore does not accumulate in the body) with no evidence of progression of damage, a reduced UF_S may be appropriate. For organizations that use this factor, the default value is 10 when extrapolating from a subchronic study, while a value of 1 is used when using a chronic study. A reduced factor of 3 is typically used when the study is longer than subchronic duration (e.g., a 6–7 months rat study, as used for the cumene RfD), or if there is evidence based on other routes or on the MOA that toxicity does not progress with additional exposure (e.g., hexachlorocyclopentadiene RfD) (4).

LOAEL TO NOAEL EXTRAPOLATION

This UF addresses extrapolation from a LOAEL when a NOAEL or NOAEL surrogate is not available for the critical effect. More recent assessments often use a BMD (29–31) as the basis for the assessments.[d] While using a BMD may remove the need for UF_L, it is not always possible to use a BMD in the assessment.[e] The available dose–response data may not be sufficient for estimating a BMD, or the assessment may have been conducted prior to the widespread adoption of the BMD approach. A default factor of 10 is used in extrapolating from a LOAEL, with a factor of 3 typically used for LOAELs of minimal severity. A decreased factor for UF_L for mild severity of the critical effect has been used in a number of assessments on the EPA's Integrated Information System (IRIS), such as the bromomethane RfC and the aroclor 1254 RfD. As for all UFs, the choice of the factor should take into account the understanding of the underlying biology.

DATABASE UNCERTAINTY

It is often the case that the available toxicity tests in experimental animals (or available information in humans) do not cover all endpoints or life stages. In such cases, scientists might judge that the critical effect may be missed because the database is not complete and, further, that an additional factor is needed to account for this uncertainty (32). For the purpose of determining a critical effect and calculating a chronic RfD/RfC for

---

[d]BMD modeling involves fitting a flexible mathematical model to the dose-response data and then determining the dose associated with a specified incidence of the adverse effects, the BMD. The BMD, or a lower confidence limit on the benchmark dose (the BMDL), can then be used as the point of departure for deriving an RfD, in place of a NOAEL or LOAEL. Advantages over the NOAEL/LOAEL approach include (i) the BMD is not limited to the tested doses; (ii) a BMD can be calculated even when the study does not identify a NOAEL; and (iii) unlike the NOAEL approach, the BMD approach accounts for the statistical power of the study. Numerous examples of BMD use in the dose-response assessment part of the risk assessment process are available on IRIS (4).

[e]Some assessments have used an uncertainty factor or "extrapolation factor" with a BMDL, in recognition of the projected response at the BMDL. However, this is generally not standard practice by the EPA or many other government agencies.

noncancer health effects, the EPA generally considers a "complete" database to be composed of

- two adequate[f] mammalian chronic toxicity studies by the appropriate route in different species,
- one adequate mammalian multi-generation reproductive toxicity study by an appropriate route, and
- two adequate mammalian developmental toxicity studies by an appropriate route in different species.

These studies test the experimental animal at all life stages, including in utero and perinatal exposures. Generally, a database with well-conducted studies of all these study types would not need further studies because new data would not likely lower the NOAEL of the critical effect or identify a new critical effect, and thus the RfD would not likely change. However, sometimes chemicals have studies in all these areas and scientists judge that additional data are still needed to identify dose thresholds for sensitive endpoints (e.g., a developmental neurotoxicity study or immunotoxicity study). In such cases, a value other than 1 may be used for UF_D.

Based in part on the results of Dourson et al. (32), a UF of 10 is generally used if both the second chronic toxicity study and the reproductive toxicity study are missing. A factor of 3 is generally used if the database is missing one of these two studies. The developmental toxicity studies are less likely to identify the critical effect, and so have a smaller impact on this UF. Mechanistic considerations are also taken into account in development of this value. For example, if a chemical is highly reactive and causes most of its effects at the portal of entry, the potential for reproductive and developmental effects is lower, the absence of such studies may be of less concern, and a lower UF_D may be chosen. This is because the chemical would react at the portal of entry and would not be available systemically to cause systemic, reproductive, or developmental effects. Conversely, if a chemical is structurally related to known reproductive or developmental toxicants, a higher UF than indicated by the rules of thumb described above may be chosen, although the maximum would be 10, except under very unusual circumstances.

While UF_D compensates for data gaps relative to a "complete" database, there is also a minimum database; only screening-level assessments, not comprehensive assessments, are appropriate for chemicals that do not meet the minimal database criteria. Risk assessors consider a minimum database for estimating an RfD as a single, well-conducted, subchronic mammalian bioassay by the appropriate route. However, there is a relatively high likelihood for

---

[f] As determined by professional judgment. Typically, studies should have been adequately conducted and published in refereed journals, or be unpublished reports that adhered to good laboratory practice (GLP) guidelines and have undergone final QA/QC (27). The U.S. EPA and others have published guidelines in this area. For example, see Refs. 33,34.

such a database that additional toxicity data may change the RfD, and the associated confidence in the risk value is lower. Examples of confidence statements for RfDs and RfCs can be found on the EPA's IRIS (4).

In considering the areas of uncertainty addressed by UFs, it is useful to distinguish between uncertainty and variability. Variability describes a characteristic inherent in the parameter being measured. For example, variability exists in human body weight, no matter how large the sample size or how precisely weight is measured. In contrast, uncertainty addresses a lack of information. Additional and better experiments can reduce or remove the uncertainty. For example, uncertainty regarding whether a chemical is a developmental toxicant can be removed by well-conducted developmental toxicity studies. More rarely, the effects of the chemical may be well characterized, but uncertainty in the underlying biology remains. The first two UFs described in the previous paragraphs (UF_H and UF_A) primarily address variability, while the remaining three uncertainty factors (UF_L, UF_S, and UF_D) address uncertainty. Thus, UF_S can be removed by filling a data gap, while UF_A is modified by better characterizing the TK and TD differences between animals and humans. The use of TK data to modify uncertainty factors is described in the following section.

USE OF TK DATA TO MODIFY OR REPLACE UNCERTAINTY FACTORS

There is significant historic precedent for the use of data to move away from default values for uncertainty factors. A classical example of the use of toxicokinetic data is the development of methods to account for interspecies differences in tissue dosimetry via the inhalation route (19,27,28). These methods describe dosimetric adjustments for converting chemical concentrations inhaled by experimental animals to HECs that would result in the same tissue dose. The calculation of HECs takes into account whether the material is a particle (aerosol) or a gas (or vapor), the anatomic location of the target for the endpoint(s) of interest (systemic, or region of the respiratory tract), and differences in breathing rates and respiratory tract regional surface areas. As shown in Figure 2, the respiratory tract can be divided into the extrathoracic, tracheobronchial, and pulmonary regions. The HEC is the product of the dosimetric adjustment factor for the region of interest (DAF_r) and the exposure concentration. For particles, the DAF_r is the regional deposited dose ratio ($RDDR_r$), while for gases the DAF_r is the regional gas dose ratio ($RGDR_r$). Because use of these adjustments accounts for much of the toxicokinetic differences between animals and humans, a reduced uncertainty factor of 3 (to account for toxicodynamic differences) is used for interspecies extrapolation when inhalation dosimetric adjustments are used (27).

A similar concept can be applied to oral exposures by adjusting intakes using allometric relationships based on ratios of human to animal body

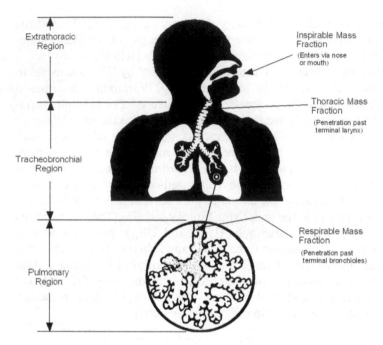

Figure 2 Respiratory tract regions. *Source*: From Ref. 27.

weights or surface areas. This is routinely done for cancer, but historically has not been done for noncancer risk assessment under current U.S. EPA methods. However, recent draft U.S. EPA guidance aims to harmonize these approaches by use of a hierarchy of approaches to derive the HED, with body weight $(BW)^{3/4}$ as the default scaling approach (35). These approaches (the RfC dosimetry and allometric scaling for oral exposure) can be considered as categorical default approaches (Fig. 3), which are based on common characteristics of compounds, such as their chemical/physical properties.

It should be noted that the $BW^{3/4}$ scaling means that the daily applied doses (in mg/day) are scaled in proportion to body weight raised to the 3/4 power, *not* that the scaling factor itself is the ratio of $BW^{3/4}$. This approach is exemplified below. In this example, the point of departure (NOAEL or BMDL) is 15 mg/kg-day in rats; the rat body weight is 0.38 kg and the anticipated human body weight is 70 kg. The human dose is scaled as follows:

$$\text{Dose human} = \frac{\text{dose} \times \text{scaling factor}}{\text{body weight}_{\text{human}}} \tag{3}$$

The animal dose is expressed in absolute terms:

$$15\,\text{mg/kg} \times 0.38\,\text{kg} = 5.7\,\text{mg} \tag{4}$$

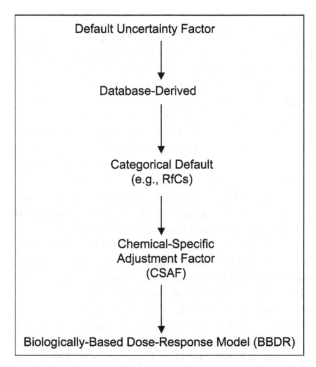

Figure 3 Continuum of use of data. *Abbreviation*: RfCs, reference concentrations.

The scaling factor is calculated:

$$\left(\frac{BW_H}{BW_A}\right)^{0.75} = \left(\frac{70}{0.38}\right)^{0.75} = (184)^{0.75} = 50 \tag{5}$$

$$(5.7 \text{ mg/day} \times 50)/70 \text{ kg} = 4.1 \text{ mg/kg day} \tag{6}$$

Note that, as illustrated in this example, the daily applied dose (in mg/day) is scaled in proportion to body weight raised to the 3/4 power. The scaling factor applied to the point of departure is *not* the ratio of $BW^{3/4}$.

There are many other cases in which the available toxicokinetic and mechanistic data were used as the basis for moving away from default UFs. For example, a 1996 review by Dourson et al. (24) evaluated the published information for 24 TDIs and tolerable concentrations (TCs) developed by Health Canada, 346 IRIS RfDs (RfDs on the EPA's Integrated Risk Information System) developed by the U.S. EPA, and 46 IRIS RfCs. The authors found that data were used to modify UF_H for 3.6% of the reviewed values, primarily for RfDs, because human epidemiology or clinical data (e.g., for essential trace elements) were available. In general, modifications to UF_H

have historically been semi-quantitative, considering both TK and TD holistically as part of an overall consideration of the MOA. For example, a reduced factor of 3 was used for UF_H for the methyl methacrylate RfC (4), because both human and animal studies show that the critical effect is nasal irritation, and human variability for this endpoint is reduced. A reduced factor of 3 was used for the molybdenum RfD based on the large and diverse human population in the principal study (because such diversity would take into account a significant portion of the total human variability), as well as considering molybdenum essentiality. More recently, separate consideration of TK and TD has been used to modify UF_H, as for the boron RfD (4), and as discussed further below in the context of CSAFs.

Dourson et al. (24) also reported that data were used to modify the interspecies uncertainty factor for 10% of the values overall, and for almost all of the RfCs, due to the use of inhalation dosimetry and categorical default uncertainty factors (19,27,28). Data were used even more frequently to modify the other three uncertainty factors. The subchronic to chronic factor was modified in 26% of the assessments where the factor was used. This is typically done when data are available that indicate minimal increased response or severity of response with increasing exposure duration, or when the principal study falls between durations of 90 days and 2 years (for rodent studies). Evidence that a chemical is rapidly metabolized, does not accumulate in the body, and that tissue damage does not accumulate would also support a lower UF in this area. The LOAEL to NOAEL factor was less than 10 in 17% of the assessments. The choice of this factor depends on a judgment of how close the LOAEL is to the (unknown) biological threshold for response. If the response incidence or severity is very low, a reduced factor is often used because the risk assessor judges that the LOAEL is relatively close to what the NOAEL would have been if lower doses were tested. A database uncertainty factor of less than 10 was used in 47% of the cases where this factor was used.

In light of the growing understanding of the mechanistic basis of toxic effects, there also has been a shift in the thinking about defaults. While early assessments considered the default as the starting point, with chemical-specific data being used to "move away" from defaults, more recent guidance (7) describes the need to start with consideration of all the data and to invoke defaults if needed.

Uncertainty factors may be considered to fall on a continuum (Fig. 3). At one extreme are default uncertainty factors, which are used in the absence of sufficient data to inform the choice of factor. Next are database-derived uncertainty factors. These are based on the overall database of information, not group data or rigorous chemical-specific data. Database-derived uncertainty factors were described above in the description of the individual uncertainty factors. The next level is categorical default, for which the use of a default factor of 3 for interspecies extrapolation for RfCs developed using inhalation dosimetry, or the use of $BW^{3/4}$ scaling are examples.

The next step on the continuum is the use of CSAFs. The IPCS is leading an international initiative on CSAFs, an approach that provides a framework for incorporation of adequate chemical-specific data on TK and TD to replace default uncertainty factors in dose–response assessment, based on an understanding of MOA.[g] Under the IPCS framework (16,36–38), the default uncertainty factor of 10 for *interspecies* differences is divided into a default subfactor of 2.5 for TD and a default subfactor of 4.0 for TK. The default uncertainty factor of 10 for *intraspecies* differences is divided into default subfactors of 3.2 for TD and 3.2 for TK. The composite factor is the product of these four subfactors, including both CSAFs and the residual default subfactors that were not replaced due to lack of relevant data. IPCS defines toxicokinetics (or pharmacokinetics) as "the process of the uptake of potentially toxic substances by the body, the biotransformation they undergo, the distribution of the substances and their metabolites in the tissues, and their elimination from the body. Both the amounts and the concentrations of the substances and their metabolites are studied." Toxicodynamics (or pharmacodynamics) is defined as "the process of interaction of chemical substances with target sites and the subsequent reactions leading to adverse effects."

Under the IPCS framework, any or all of the toxicokinetic and toxicodynamic subfactors can be replaced by CSAFs. To do so, one must first identify the active chemical species (e.g., parent or a metabolite). Without knowledge of the active species, one cannot appropriately address the implications of kinetic differences, and a CSAF cannot be developed. Similarly, knowledge of the active species is essential for evaluating variability in TD. Data on toxicokinetic variability are usually derived from in vivo kinetic studies or in vitro studies with suitable interpretation. Note that simply applying allometric scaling does not result in a CSAF, because such scaling may not fully account for interspecies differences in chemical-specific metabolism. Development of the CSAF for interspecies or intraspecies differences in TK is based on an appropriate dose metric. In the absence of sufficient information to determine the best dose metric, the default is to use either area under the curve (AUC) or clearance rate for the active entity. This is considered a health-protective default, because there is likely to be greater variability in these parameters than other common metrics, such as peak concentration (C_{max}) (38). Dose metrics may either be measured directly or estimated using a validated PBPK model. IPCS recognizes that the calculation of tissue dose in a PBPK model may include both TK and some aspects of TD, but the degree of overlap cannot currently be quantified.

---

[g] While the MOA understanding incorporated into a toxicokinetic subfactor may be limited to the definition of the active species, MOA plays a larger role for the toxicodynamic subfactor, in ensuring that the endpoint measured is relevant to the critical effect.

This is because the PBPK model may include local metabolism in the target tissue, while the IPCS guidance (38) considers local metabolism to be the first step in the toxicodynamic cascade. For TD, the data are generally derived from in vitro studies of human and animal tissues in which the critical effect(s) occur. Based on either the critical effect or a key event related to the critical effect, one then makes a quantitative interspecies comparison of the concentrations that cause an effect of defined magnitude (e.g., EC_{10}, the effective concentration resulting in a 10% response for the endpoint of interest) in the respective human and animal tissues, rather than a comparison of response attained at a defined dose.

Particularly for CSAFs addressing human variability, a key issue is having a large enough sample size to adequately capture the range of population variability. For this reason, information on human kinetic variability is generally derived from approaches that include measured or modeled human variability, especially variability in the relevant physiological or biochemical parameter(s). Alternatively, generic information on rate-limiting physiological parameters may sometimes be used, as was done for the boron RfD (4). For TD, data could be derived from in vitro studies in target tissues obtained from "average" versus sensitive humans. However, there are rarely sufficient data to replace the default UF for intraspecies differences in TD, because in vivo human data tend to address both kinetics and dynamics without clearly distinguishing between them, and it is difficult to obtain a sufficient sample size to address TD using in vitro analyses. (The issue of sample size is less of a concern for evaluation of interspecies TD, because that involves comparison of means, rather than characterization of human variability.) Further, the shape of the population curve for a particular biological parameter is usually not known: it could be unimodal, with the sensitive sub-population comprising the lower "tail" of the distribution for the healthy population of the dose causing a specified effect (or the upper "tail" of sensitivity). Alternatively, the distribution could be bimodal, with the sensitive sub-population having its own distribution that overlaps only slightly with the "tail" of the healthy population distribution (Fig. 4).

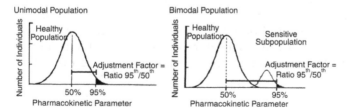

Figure 4 Development of chemical-specific adjustment factors for human variability in toxicokinetics/toxicodynamics.

A limited number of assessments are available that have used CSAFs. The recent boron RfD (4) used data on mean boron clearance in pregnant rats and in pregnant women for the interspecies toxicokinetic adjustment factor. Because the sample size in the study on boron clearance in pregnant women was too small to capture the population variability, the intraspecies toxicokinetic adjustment factor was based on the variability in the glomerular filtration rate (GFR), including consideration of preeclamptic women (a potentially susceptible population).[h]

Another example of a risk value developed using a CSAF is the Health Canada tolerable daily concentration (TDC)[i] for 2-butoxyethanol (also called ethylene glycol butyl ether, EGBE) (39,40) developed for effects on the blood (hemolytic anemia). Health Canada determined that a simple comparison of the AUCs in blood for humans and rats for the active metabolite is as informative for interspecies scaling as the available PBPK model, and therefore calculated a CSAF for interspecies kinetics based on this dose metric. Health Canada also developed a CSAF for interspecies TD, based on in vitro data on the sensitivity of erythrocytes to hemolysis caused by the active metabolite. The usual approach for calculating the CSAF for interspecies TD is based on a specified response level, such as the EC_{10} (16). However, in this case, the quantitative data were sufficient to show that humans are at least 10 times less sensitive than rats, but the human EC_{10} was not reached. Therefore, the CSAF was not quantitatively derived from the ratio of EC_{10} values. Further details on calculating CSAFs are provided in Chapter 2.

PBPK models provide a specific example of the enhanced use of data in the development of CSAFs, or in calculating HEDs to replace adjustment factors for interspecies kinetic differences. PBPK models describe the flow and transformation of the chemical by the body (TK) using species-relevant organ and tissue volumes, blood flows, and kinetic transformation parameters (41). This allows the estimation of the biologically important dose delivered to the target organ(s). PBPK modeling can be used to calculate tissue dose, to improve interspecies extrapolation, allow route-to-route extrapolation, and evaluate mechanistic questions. In addressing interspecies differences, PBPK models can be used to calculate the HED or HEC, eliminating the need for an uncertainty factor for the toxicokinetic component of interspecies differences. [In such cases, a partial uncertainty factor (default value) of 3.2 (for interspecies extrapolations) or 2.5 (for intraspecies variability) would still be used for the toxicodynamic component.] PBPK

---

[h] Preeclampsia means the development of hypertension with proteinuria and/or edema as a result of pregnancy.

[i] The tolerable daily concentration (TDC or TC) is an airborne concentration to which it is believed that a person can be exposed continuously over a lifetime without deleterious effect (15).

models, together with Monte Carlo techniques, can also be used to quantify human variability in TK (42,43) as part of the development of a CSAF, but no completed governmental assessments using this approach were located. Validated models that address the critical effect both in the experimental animal species and in humans are needed in order to use a PBPK model to develop a risk value. Adoption of this approach has been relatively slow, due to such issues as the time investment required to develop models, the need for transparency, and the need for regulators to become comfortable with evaluating the science and data used as the basis for model development. However, recent assessments in the IRIS database (4) based on PBPK models include the assessments for vinyl chloride and EGBE; others are in development.

One limitation in the development of PBPK models has been the determination of the appropriate chemical-specific parameters. As described further in Chapter 8, in silico methods have been developed to calculate partition coefficients using quantitative structure–property relationships (QSPRs) or biologically based algorithms. Chapter 9 discusses the use of in vitro studies using tissue preparations or genetically expressed enzymes in order to determine in vivo metabolic rate constants. The in vitro approach can be particularly useful for deriving human metabolic rate constants, where it may not be ethical or practical to obtain adequate in vivo data.

While PBPK models only describe a chemical's kinetics, BBDR models also include descriptions of a chemical's interactions with the body (TD), including a mathematical description of how the toxic agent interacts with and damages biological targets. While a number of PBPK models have been developed and used in risk assessments, the number of BBDR models is much lower, due to the very high data demands. Perhaps two of the most well-known are the model of formaldehyde (44) and the Moolgavkar–Venzon–Knudson (MVK) model of multistage carcinogenesis (45), a general toxicodynamic model that can be combined with a specific chemical kinetic model for a full BBDR. BBDR models for noncancer endpoints have also been published. These include a model for chloroform hepatotoxicity (46), domoic acid neurotoxicity (47), and developmental toxicity of 5-fluorouracil (48). BBDR models can be used to estimate dose–response at low doses not amenable to experimental evaluation. Due to the intensive effort required for development of BBDRs, relatively few have been used in risk assessment to date. However, Health Canada used a biologically motivated case-specific model for development of its formaldehyde inhalation cancer assessment (49). The ground-breaking effort expended in development of the formaldehyde model has contributed significantly to the development and improvement of the methods and models, as well as model parameters for BBDR modeling, and future models are likely to require less effort. Because a BBDR takes into account both kinetic and dynamic aspects, it could be used to replace both the kinetic and dynamic portions of the interspecies uncertainty factor.

OUTSTANDING ISSUES, RESEARCH AREAS, AND UNCERTAINTIES

While there has been considerable improvement in the scientific basis of uncertainty factors over the past 20 years and, more importantly, in the use of data in their development, a number of issues remain. For example, an improved approach to interspecies dosimetry including a rigorous and codified approach for the interspecies extrapolation for the oral route would be beneficial. In a manner analogous to the RfC dosimetry, this approach could be based on readily evaluated aspects of a chemical's characteristics and MOA, and would allow one to use categorical default uncertainty factors for TK. Such a categorical default approach could consider issues such as whether the toxicity is due to a parent, a stable metabolite, or a reactive metabolite, using a framework such as that suggested by Clewell et al. (50). Further improvements could also be made to the inhalation dosimetry, based on knowledge of respiratory tract physiology gained over the past 10 years (51,52). In addition, it would be useful to include improved consideration of the impact of metabolism on systemic dose from inhaled chemicals.

Improved evaluation of dosimetry would also be useful in enhancing the evaluation of risk to children, by helping to identify the conditions under which children would receive tissue doses significantly different from that of adults. Two recent papers (53,54) evaluated the relative dose to children and adults under several different scenarios. Development of a broader framework that addresses relative tissue doses for oral and inhalation exposure to a variety of chemicals and chemical forms would be useful (e.g., see Ref. 18 for some initial steps toward development of such a framework).

SUMMARY

Toxicokinetics plays a role in the hazard characterization, dose–response assessment, and exposure assessment portions of the risk assessment paradigm. Toxicokinetics can inform the choice of UFs for noncancer assessments, and TK is explicitly considered in the UFs for interspecies and intraspecies variability. Uncertainty factors may be considered to fall on a continuum of increasing incorporation of chemical-specific data. The incorporation of toxicokinetic data ranges from qualitative considerations, through categorical defaults, to CSAFs, and sophisticated modeling approaches such as PBPK and BBDR. These latter methods allow the incorporation of chemical-specific data to refine the accuracy of the assessment. Further work on categorical defaults for the oral and inhalation routes, as well as consideration of dosimetry for children, will continue to improve assessments.

ACKNOWLEDGMENT

John Lipscomb is acknowledged for his useful comments on this chapter.

REFERENCES

1. NRC (National Research Council). Risk Assessment in the Federal Government: Managing the Process. Washington, DC: National Academy Press, 1983.
2. NRC (National Research Council). Science and Judgment in Risk Assessment. Washington, DC: National Academy Press, 1994.
3. NRC (National Research Council). Understanding Risk: Informing Decisions in a Democratic Society. Washington, DC: National Academy Press, 1996.
4. U.S. EPA. Integrated Risk Information System (IRIS). U.S. Environmental Protection Agency. Washington, DC: National Center for Environmental Assessment, 2005. Available at http://www.epa.gov/iris.
5. Health Canada. Human Health Risk Assessment for Priority Substances, Environmental Health Directorate, Canadian Environmental Protection Act. Ottawa, Ontario, Canada: Health Canada, 1994. Available at http://www. hc-sc.gc.ca/hecs-sesc/exsd/psl2.htm.
6. IARC (International Agency for Research on Cancer). IARC Monographs on the Evaluation of Carcinogenic Risks to Humans: Preamble. Lyon, France: World Health Organization, 2004. Available at http://monographs.iarc.fr/monoeval/preamble.html.
7. U.S. EPA. Guidelines for Carcinogen Risk Assessment, EPA/630/P-03/001B. Washington, DC: U.S. Environmental Protection Agency, 2005. Available at http://www.epa.gov/iris/cancer032505.pdf.
8. U.S. EPA. Guidelines for Exposure Assessment, EPA/600/Z-92/001. U.S. Environmental Protection Agency, Office of Health and Environmental Assessment. Washington, DC: Exposure Assessment Group, 1992.
9. AIHC (American Industrial Health Council). Improving Risk Characterization. Washington, DC: American Industrial Health Council, 1992.
10. AIHC (American Industrial Health Council). Advances in Risk Characterization. Washington, DC: American Industrial Health Council, 1995.
11. U.S. EPA. Science Policy Handbook: Risk Characterization, U.S. Environmental Protection Agency. Washington, DC: Office of Science Policy, 2000.
12. van Leeuwen CJ, Hermens JL. Risk Assessment of Chemicals: an Introduction. Dordrecht: Kluwer Academic Publishers, 1995.
13. U.S. EPA. Determination of the Appropriate FQPA Safety Factor(s) in Tolerance Assessment, U.S. Environmental Protection Agency. Washington, DC: Office of Pesticide Programs, 2002. Available at http://www.epa.gov/pesticides/trac/science/determ.pdf.
14. Fenner-Crisp P. The FQPA 10x safety factor: How much is science? How much is sociology? Hum Ecol Risk Assess 2001; 7:107–116.
15. Meek ME, Newhook R, Liteplo RG, et al. Approach to assessment of risk to human health for priority substances under the Canadian Environmental Protection Act. Environ Carcinogen Ecotoxicol Rev 1994; C12:105–134.
16. IPCS (International Programme on Chemical Safety). Assessing Human Health Risks of Chemicals: Derivation of Guidance Values for Health-based Exposure Limits, Environmental Health Criteria 170, International Programme on Chemical Safety. Geneva, Switzerland: World Health Organization, 1994. Available at www.inchem.org/documents/ehc/ehc/ehc170.htm.

17. U.S. EPA. A Review of the Reference Dose and Reference Concentration Processes, EPA/630/P-02/002F. U.S. Environmental Protection Agency. Washington, DC: Risk Assessment Forum, 2002. Available at http://cfpub.epa.gov/ncea/cfm/recordisplay.cfm?deid=55365.
18. Haber LT, Hack CE, Zhao Q. Meeting Materials for a Peer Consultation on a Draft Framework to Evaluate Whether the Default Uncertainty Factor for Human Kinetic Variability is Adequate for Protecting Children. Cincinnati, OH: Toxicology Excellence for Risk Assessment, 2005. Available at http://www.tera.org/peer/adultchildtk/actkwelcome.htm.
19. Jarabek AM. Inhalation RfC methodology: dosimetric adjustments and dose–response estimation of non-cancer toxicity in the upper respiratory tract. Inhal Toxicol 1994; 6:301–325.
20. Barnes DG, Dourson ML. Reference dose (RfD): description and use in health risk assessments. Regul Toxicol Pharmacol 1988; 8:471–486.
21. Dourson ML, Stara JF. Regulatory history and experimental support of uncertainty (safety) factors. Regul Toxicol Pharmacol 1983; 3:224–238.
22. Jarabek AM. Interspecies extrapolation based on mechanistic determinants of chemical disposition. J Hum Ecol Risk Assess 1995; 1:641–662.
23. Dourson ML. Methods for establishing oral reference doses (RfDs). In: Mertz W, Abernathy CO, Olin SS, eds. Risk Assessment of Essential Elements. Washington, DC: ILSI Press, 1994:51–61.
24. Dourson ML, Felter SP, Robinson D. Evolution of science-based uncertainty factors in noncancer risk assessment. Regul Toxicol Pharmacol 1996; 24:108–120.
25. Kalberlah F, Schneider K. Quantification of Extrapolation Factors, Fb 797. Final report of the research Project No. 1116 06 113 of the Federal Environmental Agency. Schriftenreihe der Bundesanstalt für Arbeitsschutz und Arbeitsmedizin Dortmund, Fb 797. Bremerhaven: Wirtschaftsverlag NW, 1998.
26. Haber LT, Dollarhide JS, Maier Λ, et al. Noncancer risk assessment: principles and practice in environmental and occupational settings. In: Bingham E, Cohrssen B, Powell CH, eds. Patty's Toxicology. Vol. 1, 5th ed. New York, NY: Wiley, 2001:169.
27. U.S. EPA. Methods for Derivation of Inhalation Reference Concentrations and Application of Inhalation Dosimetry, EPA/600/8-90/066F. U.S. Environmental Protection Agency. Washington, DC: Office of Research and Development, 1994. Available at http://nepis.epa.gov/pubtitleORD.htm.
28. Jarabek AM. The application of dosimetry models to identify key processes and parameters for default dose–response assessment approaches. Toxicol Lett 1995; 79:171–184.
29. Crump KS. A new method for determining allowable daily intakes. Fund Appl Toxicol 1984; 4:854–871.
30. U.S. EPA. The Use of Benchmark Dose Approach in Health Risk Assessment, EPA/630/R-94/007. U.S. Environmental Protection Agency, Office of Research and Development. Washington, DC: Risk Assessment Forum, 1995.
31. U.S. EPA. Benchmark Dose Technical Guidance Document. External Review Draft, EPA/630/R-00/001. U.S. Environmental Protection Agency. Washington, DC: Risk Assessment Forum, 2000. Available at http://cfpub.epa.gov/ncea/cfm/recordisplay.cfm?deid=20871.

32. Dourson ML, Knauf LA, Swartout JC. On reference dose (RfD) and its underlying toxicity database. Toxicol Ind Health 1992; 8:171–189.
33. U.S. EPA. Health Effects Test Guidelines, EPA/100/B-00/002. U.S. Environmental Protection Agency. Washington, DC: Office of Science Policy for the Office of Prevention, Pesticides, and Toxic Substances, 1998.
34. U.S. FDA. Redbook: Toxicological Principles for the Safety Assessment of Food Ingredients. U.S. Food and Drug Administration, Center for Food Safety and Applied Nutrition. Washington, DC: Office of Food Additive Safety, 2000.
35. U.S. EPA. Harmonization in Interspecies Extrapolation: Use of $BW^{3/4}$ as Default Method in Derivation of the Oral RfD. External Review Draft, EPA/630/R-06/001. U.S. Environmental Protection Agency. Washington, DC: Risk Assessment Forum Technical Panel, 2006.
36. Renwick AG. Data-derived safety factors for the evaluation of food additives and environmental contaminants. Food Addit Contam 1993; 10:275–305.
37. Meek ME, RenwickA, Ohanian E, et al. Guidelines for application of compound specific adjustment factors (CSAF) in dose/concentration response assessment. Comm Toxicol 2001; 7:575–590.
38. IPCS (International Programme on Chemical Safety). Chemical-Specific Adjustment Factors (CSAFs) for Interspecies Differences and Human Variability: Guidance Document for the Use of Data in Dose/Concentration Response Assessment. Harmonization Project Document No. 2. Geneva, Switzerland: World Health Organization, 2005. Available at http://www.whqlibdoc.who.int/publications/2005/9241546786_eng.pdf.
39. CTC (Concurrent Technologies Corporation), TERA (Toxicology Excellence for Risk Assessment). International Toxicity Estimates for Risk Database, Toxicology Excellence for Risk Assessment. 2005. Accessed Feb. 14, 2006 at http://www.tera.org/iter.
40. Hughes K, Meek ME, Walker M, et al. 2-Butoxyethanol: hazard characterization and exposure–response analysis. Environ Carcinogen Ecotoxicol Rev 2001; C19:77–104.
41. Clewell HJ III, Andersen ME. Physiologically based pharmacokinetic modeling and bioactivation of xenobiotics. Toxicol Ind Health 1994; 10:1–24.
42. El-Masri HA, Bell DA, Portier CJ. Effects of glutathione transferase theta polymorphism on the risk estimates of dichloromethane to humans. Toxicol Appl Pharmacol 1999; 158:221–230.
43. Gentry PR, Hack CE, Haber LT, et al. An approach for the quantitative consideration of genetic polymorphism data in chemical risk assessment: examples with warfarin and parathion. Toxicol Sci 2002; 70:120–139.
44. Conolly RB, Kimbell JS, Janszen DB, et al. Human respiratory tract cancer risks of inhaled formaldehyde: dose–response predictions derived from biologically-motivated computational modeling of a combined rodent and human dataset. Toxicol Sci 2004; 82:279–296.
45. Moolgavkar SH, Knudson AG Jr. Mutation and cancer: a model for human carcinogenesis. J Natl Cancer Inst, 1981; 66:1037–1052.
46. Conolly RB, Butterworth BE. Biologically based dose–response model for hepatic toxicity: a mechanistically based replacement for traditional estimates of noncancer risk. Toxicol Lett 1995; 82–83:901–906.
47. Slikker W Jr, Scallet AC, Gaylor DW. Biologically-based dose–response model for neurotoxicity risk assessment. Toxicol Lett 1998; 429:102–103.

48. Lau C, Anderson ME, Crawford-Brown DJ, et al. Evaluation of biologically based dose–response modeling for developmental toxicity: a workshop report. Regul Toxicol Pharmacol 2000; 31:190–199.
49. Environment Canada, Health Canada. Canadian Environmental Protection Act, 1999. Priority Substances List Assessment Report: Formaldehyde. Ottawa, Ontario: Health Canada, 2001. Available at http://www.hc-sc.gc.ca/ewh-semt/pubs/contaminants/psl2-lsp2/formaldehyde/index_e.html.
50. Clewell HJ III, Andersen ME, Barton HA. A consistent approach for the application of phrmacokinetic modeling in cancer and noncancer risk assessment. Environ Health Perspect 2002; 110:85–93.
51. Kimbell JS, Subramaniam RP. Use of computational fluid dynamic models for dosimetry of inhaled gases in the nasal passages. Inhal Toxicol 2001; 13:325–334.
52. Asgharian B, Menache MG, Miller FJ. Modeling age-related particle deposition in humans. J Aerosol Sci 2004; 17:213–224.
53. Ginsberg G, Foos BP, Firestone MP. Review and analysis of inhalation dosimetry methods for application to children's risk assessment. J Toxicol Environ Health A 2005; 68:573–615.
54. Jarabek AM, Asgharian B, Miller FJ. Dosimetric adjustments for interspecies extrapolation of inhaled poorly soluble particles (PSP). Inhal Toxicol 2005; 17:317–334.

2

Guidance for the Development of Chemical-Specific Adjustment Factors: Integration with Mode of Action Frameworks

Bette Meek

Existing Substances Division, Health Canada, Ottawa, Ontario, Canada

Andrew Renwick

School of Medicine, University of Southampton, Southampton, U.K.

INTRODUCTION

Chemical-specific adjustment factors (CSAFs) provide for the incorporation of quantitative data on interspecies differences or human variability in either toxicokinetics or toxicodynamics [mode of action (MOA)] to replace appropriately weighted components of default uncertainty factors commonly adopted in defining health-based guidance values. This requires subdivision of default uncertainty factors for interspecies differences and interindividual variation into toxicokinetic and toxicodynamic components.

CSAFs represent an intermediate step in a continuum of increasingly chemical-specific data-informed approaches to extrapolation of dose–response in animals as a basis to provide guidance regarding acceptable limits of exposure for humans. Their application permits incorporation of partial information in the absence of the considerable data required to support full biologically based dose–response models. They can be compound-related (e.g., based on more generic information on physiological processes such as glomerular filtration rates for substances where renal clearance is limiting) or chemical-specific (e.g., based on experimental data on kinetics or dynamics such as metabolic rate constants specific for the individual chemical of interest).

In this chapter, guidance on CSAFs developed in a project of the International Programme on Chemical Safety (IPCS) initiative on harmonization of approaches to the assessment of risk from exposure to chemicals is considered. The final guidance (1), takes into account experience and comments received since the initial posting of the draft document on the IPCS Web site in 2001.

The guidance for adequacy of data to replace the default values for the toxicokinetic and toxicodynamic components of interspecies differences and human variability is presented in the context of several aspects including:

- determination of the active chemical species,
- choice of the appropriate kinetic measurement or parameter, and
- experimental data, including

  1. relevance of population,
  2. relevance of route,
  3. relevance of dose/concentration, and
  4. adequacy of number of subjects/samples.

Development of a CSAF is necessarily predicated on the basis of understanding of the MOA by which a specific chemical induces its effect(s) (e.g., through action of the parent compound or metabolite). MOA is a description of the hypothesized processes that lead to induction of the relevant endpoint of toxicity for which systematic consideration of the weight of evidence supports plausibility. It is distinguished from "mechanism of action", which implies a more detailed molecular description of causality, for which there is rarely sufficient information. This chapter explores the application of CSAFs in the context of recently developed extensions to MOA frameworks to consider weight of evidence for human relevance through presentation of a relevant case study. These frameworks outline a basis for transparent delineation of expert informed judgment and as such contribute to harmonization of approaches to hazard characterization for cancer and noncancer effects.

The guidance developed in this project and described in part, herein, addresses criteria for the adequacy of quantitative data on toxicokinetics and toxicodynamics; as a result, it is also applicable to other approaches to dose–response analyses (e.g., linear extrapolation from estimates of potency close to the experimental range or development of tumorigenic potencies). Development of CSAFs also results in delineation of appropriate avenues of research to enable more reliable assessments. The approaches described in this chapter are also amenable to presentation in a probabilistic context (rather than development of single measures for dose/concentration–response), where available data are sufficient to meaningfully characterize the distributions of interest.

DEVELOPMENT OF CHEMICAL-SPECIFIC/
COMPOUND-RELATED ADJUSTMENT FACTORS:
THE CONSTRUCT

Traditionally, in assessment of noncancer effects, default safety/uncertainty factors have been applied to develop health-based guidance values based on a measure of dose–response for critical effects. Most commonly, a value of 100 has been applied to no- or lowest-observed-adverse-effect levels (NOAELs or LOAELs) or benchmark doses (BMDs) for critical effects in chronic studies in animals to derive acceptable or tolerable daily intakes (TDI or ADI), or reference doses (RfD) for the general population (2–5).

The NOAEL or BMD/concentration is selected generally to be at or below the threshold in animals; uncertainty factors are then applied to estimate the subthreshold in sensitive human populations, with a 10-fold default factor addressing interspecies differences (i.e., the variation in response between animals and a representative healthy human population) and another 10-fold factor accounting for interindividual variability in humans (the variation in response between a representative healthy human population and sensitive subgroups). While additional factors are sometimes applied to account for deficiencies of the database, the 100-fold default value is common.

For practical purposes, the continuous process between external dose and toxic response (i.e., MOA) can be subdivided into steps related to the fate of the chemical in the body and those related to the actions of the chemical on the body. These different aspects of the overall process represent major sources of interspecies differences and of human variability, the latter defining susceptible subgroups within the population. Measurements of response of humans to external doses in vivo represent both toxicokinetics and toxicodynamics and would be used directly as a basis for the NOAEL, LOAEL, or BMD, obviating the need for consideration of interspecies differences.

Development of CSAFs requires subdivision of each of the usual default uncertainty factors of 10 for interspecies differences and human variability into subfactors which address toxicokinetics and toxicodynamics. For application in this context, the continuum of processes leading to chemical toxicity was split at the level of delivery via the general circulation of the parent compound or of an active metabolite to the target tissue/organ; events up to this point were considered as toxicokinetics and events within the target tissue/organ were considered as toxicodynamics. This split was chosen because the subdivision was derived largely from physiological differences between rodents and humans for interspecies differences and from the clinical pharmacology literature for human variability, based on plasma concentration measurements (toxicokinetics) and data from in vitro studies or from modeling of data from in vivo studies in humans (toxicodynamics). In consequence, the data to replace a default subfactor for toxicokinetics will usually be based on the concentrations of the chemical or active metabolite in the general circulation.

The quantitative split proposed by Renwick (6) and subsequently modified by an international review group (7) is consistent with data on kinetic parameters and pharmacokinetic–pharmacodynamic (PKPD) modeling for a range of pharmacological and therapeutic responses to pharmaceutical agents (8). Based on such analyses, the 10-fold default factor for interspecies differences is subdivided into factors of 4 ($10^{0.6}$) for toxicokinetics and 2.5 ($10^{0.4}$) for toxicodynamics. This is consistent with the approximately four-fold difference on a body surface area basis between rats (the most commonly used test species) and humans in basic physiological parameters such as cardiac output, and renal and liver blood flows, which are major determinants of clearance and elimination of chemicals. The factor for human variability is divided evenly into two subfactors, each of $10^{0.5}$ (3.16 or 3.2) (Fig. 1).

Application of relevant data in the framework is predicated on the basis that measurements of the concentrations of the parent compound or its metabolites in the general circulation reflect major sources of interspecies differences and human variability in tissue/organ delivery. In consequence, the data to replace a default subfactor for toxicokinetics or toxicodynamics should be based on the concentrations of the chemical or active metabolite in the general circulation; if not, then the defaults for the remaining subfactors that were not replaced by a CSAF would need to be reconsidered.

This framework, then, allows the incorporation of quantitative chemical-specific data, relating to either toxicokinetics or toxicodynamics, to replace parts of the usual 100-fold uncertainty factor, but collapses back to the usual 100-fold default in the absence of appropriate data. The framework is based, therefore, on a pragmatic split of defaults, for which the delineation between toxicokinetics and toxicodynamics is necessarily a function of the data from which it was derived. However, framing of the construct in this context does not prevent application, e.g., of the output of physiologically based pharmacokinetic (PBPK) models which incorporate bioactivation and/or detoxification processes within the target tissue/organ. Rather, the fact that the model addresses additionally this aspect needs to be taken into consideration in applying both the chemical-specific values which replace the toxicokinetic default and the remaining toxicodynamic default subfactor, for which information may not be available. For this reason, it must be determined whether quantitative modeling or measures of various parameters or endpoints represent purely toxicokinetics, or toxicokinetics and respective parts or all the toxicodynamic processes as defined in the context of this framework. Their impact to replace the toxicokinetic and potentially a proportion or all of the toxicodynamic defaults needs then to be carefully considered.

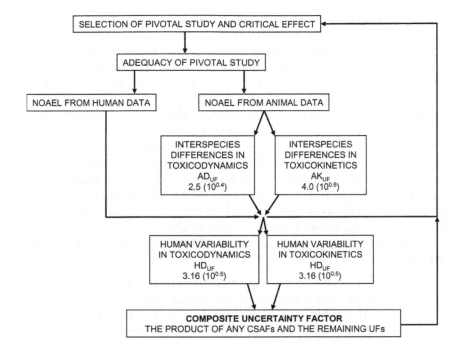

A = the animal to human extrapolation factor (based on quantification of interspecies differences)
H = the human variability factor (based on quantification of interindividual differences)
K = differences in toxicokinetics
D = differences in toxicodynamics
AF = the adjustment factor calculated from chemical-specific data
UF = the uncertainty factor, a default value that is used in the absence of chemical-specific data.

Figure 1 Introducing quantitative toxicokinetic and toxicodynamic data into dose/ concentration–response assessment. *Source*: From Ref. 7.

Similar reconsideration of application in the context of the construct would be necessary if the model related to an effect at the site of contact. While the database on which the values for the subfactors was developed related to systemic effects produced after oral or intravenous dosage, the use of CSAFs and the approach described herein are applicable also to effects at the site of contact, but taking into account that the toxicokinetic component would be direct delivery and not via the general circulation.

Chemical-Specific/Compound-Related Toxicokinetic Adjustment Factors—[AK$_{AF}$, HK$_{AF}$]

The CSAFs for the toxicokinetic components of interspecies differences and interindividual variability are ratios of measurable metrics for internal exposure to the active compound such as area under the plasma or blood concentration–time curve (AUC), the maximum observed concentration (C_{max}), or clearance (CL). For interspecies differences, this may be determined on the basis of comparison of the results of in vivo kinetic studies with the active compound in animals and a representative sample of the healthy human population. Estimates can also be derived from in vitro enzyme studies combined with suitable scaling to determine in vivo activity or by the scaling of in vivo data from animals to predict human equivalent values. Alternatively or necessarily (when metabolism or tissue uptake is nonlinear), they are based on PBPK models in which data on partition coefficients for different tissues are combined with the organ blood flows for animals and for humans to predict delivery to and (often) the concentrations within the target tissue/organ. Partitioning of the chemical between the general circulation and target tissue/organ in PBPK models is usually based on measurements of the partition coefficient in animal tissues or other in vitro models.

In the context of the development of CSAFs, PBPK models may be subdivided into two types:

1. those that estimate the target tissue/organ dose of the parent compound or a circulating active metabolite, and
2. those that additionally incorporate bioactivation and detoxification processes that occur within the target tissue/organ.

Type 1 PBPK models are purely "toxicokinetic" in nature and consistent with the types of data on which the subdivision of the default uncertainty factors of 10 into toxicokinetic and toxicodynamic subfactors was based. Type 2 PBPK models include parts of the overall process (Fig. 2), such as bioactivation or detoxification within the cells of the target tissue, that are not reflected in plasma-based toxicokinetic measurements and therefore reflect processes affecting the tissue "response" and in this context should be considered as part of toxicodynamics.

In addition to these PBPK models for systemic delivery, mathematical models can define delivery when the target tissue/organ is the site of contact, e.g., with the lungs following inhalation, as well as uptake and metabolic processing within the target tissue/organ. Such models differ from the database of systemic effects resulting from oral or intravenous exposure on which the values for the subfactors presented in Figure 1 are based. Although the approach described herein is applicable to effects at the site of contact, the toxicokinetic component would be related to direct delivery versus delivery via the general circulation and consideration would need to be given to the appropriateness of the default values for the particular model on a case-by-case basis.

For interindividual variability, while this adjustment factor could potentially be addressed on the basis of chemical-specific in vivo kinetic studies in a sufficiently broad range of subgroups of healthy and potentially susceptible populations to adequately define the population distribution, this may not be practicable or even possible. The population distribution for the

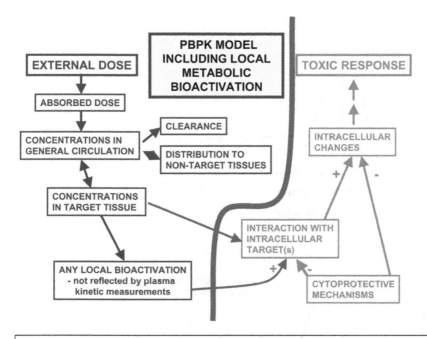

Figure 2 Components of a toxic response. *Abbreviation*: PBPK, physiologically based pharmacokinetic.

relevant metric (e.g., AUC, C_{max}, renal clearance) for the active entity is analyzed and the CSAF (HK_{AF}) calculated as the difference between the central values for the main group and given percentiles (such as 95, 97.5, and 99th) for the whole population (Fig. 3). These differences are analyzed separately for any potentially susceptible subgroup. Often chemical-specific data are not available, but the factors responsible for the clearance mechanisms are identified [renal clearance, cytochrome P450 (CYP)-specific metabolism, etc.]; in such circumstances, a compound-related adjustment factor could be derived based on measured or PBPK-modeled human variability in the relevant physiological and biochemical parameters (10,11).

Chemical-Specific Toxicodynamic Adjustment Factors [AD_{AF}, HD_{AF}]

CSAFs for the toxicodynamic components are, most simply, ratios of the concentrations which induce the critical toxic effect or a measurable related response (based on understanding of MOA) in vitro in relevant tissues of animals and a representative sample of the healthy human population

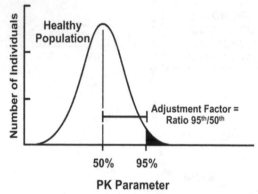

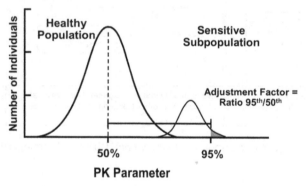

Figure 3 Development of chemical-specific adjustment factors for interindividual variability. *Abbreviation*: PK, pharmacokinetic. *Source*: From Ref. 9.

(interspecies differences) or in average versus sensitive humans (interindividual variability). At its simplest, then, replacement of the dynamic component of the default factor for interspecies differences is the ratio of the effective concentrations in critical tissues of animals versus humans. If the concentration associated with a 10% response was selected (EC_{10}), then the ratio for the CSAF would be ($EC_{10\,animal}/EC_{10\,human}$) for interspecies differences and ($EC_{10\,average}/EC_{10\,sensitive}$) for interindividual variability in healthy human and susceptible subpopulations. Measurements should be derived under experimental conditions where there has been control for variations in toxicokinetics.

Hence, CSAFs for interspecies differences and interindividual variability in toxicodynamics could be derived from in vitro studies, from in vivo studies in which the toxicokinetic component has been delineated (e.g., by kinetic–dynamic link models in which concentrations or amounts in the general circulation or at the site of action are related to the response by an empirical mathematical link-function formula), or from ex vivo experimentation (i.e., studies in which measurements are made in vitro following an in vivo exposure).

Not all in vitro or in vivo biological measurements represent processes that are critical to the development of the in vivo toxic response. There are frequently numerous sequential steps in producing a toxic response, and biomarkers of early changes may not reflect the critical toxicodynamic process. In order to serve as a surrogate marker for toxic effect, the measurements should be representative, both qualitatively and quantitatively, of the critical toxic endpoint, based on an understanding of the MOA by which specific chemicals induce their effects.

Chemical-Specific/Compound-Related CF for Interspecies Differences and Human Variability

The composite factor (CF) is the product of four different factors, each of which could be a chemical-specific or compound-related adjustment factor (AF) or a default uncertainty factor (UF):

$$
\begin{aligned}
CF = [AK_{AF} \text{ or } AK_{UF}] &\times [AD_{AF} \text{ or } AD_{UF}] \\
&\times [HK_{AF} \text{ or } HK_{UF}] \times [HD_{AF} \text{ or } HD_{UF}]
\end{aligned}
\tag{1}
$$

where A is the interspecies, H the human variability, K the kinetics, and D the dynamics.

CFs should be developed for several effects which might be considered critical to ensure that resulting tolerable, acceptable, or reference intakes/concentrations are sufficiently protective.

It is important to recognize that depending on the nature of the data, the CF can be either greater than or less than the usual 100-fold default. If the CF for an effect considered potentially critical based on assessment of the entire database is similar to or exceeds the normal default (i.e., 100), then the resulting tolerable intake or concentration should be protective for most other toxic effects. If, however, the CF for a potentially critical effect is less than the normal default, a different toxic effect with a higher NOAEL/NOAEC combined with a full default uncertainty factor could become critical as a basis for a tolerable intake or concentration (Fig. 4).

In the vast majority of cases, the quantitative toxicokinetic or toxicodynamic data necessary to define a CSAF will not be available, necessitating the usual NOAEL/uncertainty factor approach. The default uncertainty factors (AK_{UF}, AD_{UF}, HK_{UF}, and HD_{UF}) are based on the usual default values (10 for each of interspecies differences and human variability), so this guidance remains compatible with the current default procedures.

Application of such a framework even in the absence of relevant data to replace default values is encouraged in order to focus attention on gaps in

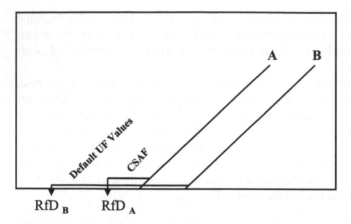

Figure 4 Quantitative relationship between RfD values for effect A based on CSAF values and for effect B based on full default UF values. *Abbreviations*: RfD, reference dose; UF, uncertainty factor; CSAF, chemical-specific adjustment factor.

the available information which, if filled, would permit development of more predictive measures of dose/concentration–response. In this manner, it should contribute to a better common understanding of the appropriate nature of relevant data, thereby facilitating their development and incorporation in dose/concentration–response assessment for regulatory purposes.

GUIDANCE FOR DEVELOPMENT OF CSAF

The development of CSAF requires an understanding of MOA for the critical effect(s) under consideration, in particular, whether or not effects are induced by the parent compound or a metabolite. "MOA," the hypothesized processes that lead to induction of the relevant endpoint of toxicity for which systematic consideration of the weight of evidence supports plausibility, is described in the context of "key events," which are measurable parameters critical to the effect under consideration. Key events may include, e.g., metabolism to the active entity, cytotoxicity and associated regenerative proliferation, deposition of crystals in the target tissue, hormonal perturbations, etc.

Data for application in the four components of the framework for CSAFs must relate to the active form of the chemical, i.e., taking into consideration early metabolic key events. Information that is relevant in this regard includes data on the mode of toxicity of structural analogues, the effects of metabolites administered directly, the influence of induction or inhibition of metabolism of the chemical on the critical effect and variations in patterns of toxicity with metabolic profiles across species, strains, and sexes.

For the components of the framework addressing toxicokinetics [AK_{AF}], [HK_{AF}], choice of the appropriate kinetic metric is an essential first step. Observation of the effect only following administration of an intravenous bolus dose or administration by gavage compared with continuous administration in diet or drinking water may indicate the importance of dose rate (i.e., C_{max} being the appropriate metric). In the absence of such data, a reasonable assumption is that effects resulting from subchronic or chronic exposure are related to the AUC, especially for chemicals with long half-lives, while acute toxicity could be related to either the AUC or the C_{max}. Alternatively, the AUC is a reasonable default because there are likely to be greater species differences or human variability in AUC or $1/CL$ than in C_{max} (12).

Choice of the appropriate endpoint is critical for the components addressing toxicodynamics [AD_{AF}], [HD_{AF}]. The selected measured endpoint must either be the critical effect itself or intimately linked thereto (with similar concentration-response and temporal relationships) based on an understanding of MOA.

As indicated above, the framework for CSAF is based on a pragmatic split of default uncertainty factors, for which the defined delineation between toxicokinetics and toxicodynamics has been determined by the nature of the quantitative data on which the subdivision was based. However, for all datasets, the relevant parameter for toxicokinetics or the measure of effects for toxicodynamics being considered as a basis for quantitative comparison to replace default for either interspecies differences or human variability needs to be carefully considered in relation to the delivery of the active chemical (parent compound or metabolite) to the target organ. Apparently toxicokinetic measurements may represent purely toxicokinetics (e.g., area under the curve) or toxicokinetics and part or all of the toxicodynamic processes (e.g., a PBPK model which incorporates tissue metabolism). This may necessitate consideration, in some cases, of replacement of the toxicokinetic and potentially a proportion or all of the toxicodynamic default for interspecies differences or human variability. For example, development of a factor developed on the basis of a PBPK model which incorporates quantitative data on metabolic activation within the target tissue would replace the toxicokinetic and the toxicodynamic default if metabolic activation was the critical step in determining overall sensitivity of the species or individual.

The relevance of the population studied must also be considered in the development of a CSAF. For the subfactors in kinetics [AK_{AF}] [HK_{AF}], for which the data are normally derived in vivo, the human population investigated must be sufficiently representative of the subpopulation at risk for the adverse effect detected in the animal studies (e.g., males if the critical effects are those on the testes, pregnant females if critical effects are developmental, relevant age group). If not, the impact of any discrepancy on the validity of the calculated ratio needs to be considered. For in vitro studies which inform primarily dynamic components [AD_{AF}] [HD_{AF}], the quality of the samples should be considered and evidence provided that they are representative of the target population, e.g., viability, specific content, or activity of marker enzymes.

The relevance of the route of exposure needs to be considered in relation primarily to in vivo kinetic studies in animals and humans. If the route of exposure for the toxicokinetic study in either animals or humans does not match that of the toxicity study on which the effect level or BMD/concentration is based (which should also be the route by which humans are normally exposed), then the impact of route-to-route extrapolation will need to be critically assessed in relation to the development of a CSAF.

The relevance of dose/concentration needs also to be considered for all components. Ideally, CSAFs for toxicokinetics [AK_{AF}, HD_{AF}] are based on comparison of kinetic parameters determined in animals exposed to doses similar to the critical NOAEL or BMD/concentration to those determined in human kinetic studies, where exposure is similar to the human equivalent concentration. Any discrepancies should be assessed for their potential impact on the dose metric and the validity of the resulting CSAF. For in vitro investigations which inform primarily CSAFs for toxicodynamic aspects [AD_{AF}] [HD_{AF}], relevant studies should include a suitable number of concentrations to adequately characterize the concentration–response relationship. Where the concentration–response curves in animals and humans or in different human subgroups are parallel, selection of the point for comparison can be anywhere between 10% and 90% response on the dose–response curve. Where the curves are not parallel, the point for comparison should be the lowest point on the concentration–response curve that provides reliable information without extrapolation below the experimental data (e.g., the EC_{10}).

The adequacy of numbers of subjects/samples needs also to be considered in relation to all components of the framework. For interspecies comparisons in both kinetics and dynamics [AK_{AF}] [AD_{AF}], the numbers of animals and humans should be sufficient to ensure that the data allow a reliable estimate of the central tendency for each species. As pragmatic advice, it is suggested that for both in vivo and in vitro measurements, the number of subjects within the population, or within the major subgroups (if there are two or more groups on which the estimate of central tendency is based), should be such that the standard error (standard deviation of the sample divided by the square root of the sample size) is less than 20% of the mean. Based on practical experience, this would normally translate to a minimum of five subjects/sample, unless the coefficient of variation is very low. For considerations of factors related to within human variability (HK_{AF}, HD_{AF}), the numbers of humans should be sufficient to ensure that the data allow a reliable estimate of the central tendency and of the population distribution.

INTEGRATING DEVELOPMENT OF CSAF WITH MOA
FRAMEWORKS—AN EXAMPLE

Examples of the development of CSAFs have been presented previously (9,13–15). In the period since the posting of the initial guidance on CSAFs, however, frameworks for systematic and transparent consideration of the weight of evidence for humans of the relevance of hypothesized MOA in animals have been developed. These frameworks set the scene for more fulsome consideration of relevant implications for dose–response, including development of CSAFs (16–18).

An example is presented here which integrates consideration of relevant data in a weight of evidence framework for human relevance of cancer and precursor key events (in hazard characterization) and, subsequently, development of a CSAF for AK_{AF} that addresses the total combined oral, dermal, and inhalation exposures to chloroform. The example includes information from two previous analyses (19,20) where external doses were translated to an internal dose (dose metric) through PBPK modeling as a basis for species extrapolation.

For chloroform, the weight of evidence for genotoxicity, sex, and strain specificity and concordance of sustained cytotoxicity, persistent regenerative proliferation, and tumors (liver and kidney tumors in mice and kidney tumors in rats) is consistent with the hypothesis that marked cytotoxicity concomitant with a period of sustained cell proliferation likely represents a secondary mechanism for tumor induction following exposure. This cytotoxicity is primarily related to oxidation rates of chloroform to reactive intermediates, principally phosgene and hydrochloric acid (19).

A concordance analysis for human relevance for the hypothesized mode of induction of tumors in animals exposed to chloroform is summarized in Table 1 for the liver and kidney. The information summarized in the concordance tables leads to the conclusion that the weight of evidence for the hypothesized mode of induction of tumors (i.e., metabolism by the target cell population, induction of sustained cytotoxicity by metabolites, and subsequent persistent regenerative cell proliferation) is greatest for liver and kidney tumors in mice, followed by kidney tumors in rats. Uncertainty could be reduced by additional information on metabolism, cytotoxicity, and proliferative response in the strain in which tumors were observed (i.e., Osborne–Mendel rats) following long-term exposure. Additional data on metabolism and chronic (e.g., 2 years) cytotoxic/proliferative response in the kidneys of F344 rats would also contribute to greater confidence in the hypothesized MOA. Though data in humans are limited, based on expected similar response in humans and in the absence of data to the contrary, the MOA for chloroform-induced tumors in animals is considered to be qualitatively applicable to humans. Limited available data in humans

Table 1 Concordance Analysis

Key event	Animals' liver	Humans' liver	Weight of evidence
Metabolism by cyp2E1	Incidence/severity of toxicity correlate with covalent binding of metabolites in rats and mice, more prevalent in necrotic lesions	Irreversible binding to macromolecules in human liver microsomes requires prior metabolism; PBPK model based on human physiological parameters and metabolic parameters in vitro in eight human liver samples	Considerable in animals, limited in humans
Sustained cytotoxicity	In all cases where examined, sustained cytotoxicity (as measured by histopathological effects and release of hepatic enzymes) in the liver of mice at doses that induce tumors	Liver also a target organ in humans based on reports of effects associated with occupational exposure	Considerable in animals, limited in humans
Persistent, regenerative proliferation	In all cases where examined, persistent regenerative proliferation (as measured by labeling indices) in the liver of mice at doses that induce tumors	No data	Considerable in animals, none in humans
Liver tumors	Mice	Inadequate epidemiological data	Considerable in animals, inadequate in humans
Metabolism by cyp2E1	In mice, strain- and sex-related differences	Quantitation in PBPK model based on human	Considerable in mice, unconvincing in

(Continued)

Table 1 Concordance Analysis (*Continued*)

Key event	Animals' liver	Humans' liver	Weight of evidence
	correlates with metabolism; necrosis correlates with the degree of covalent binding; few such data for rats and in F344 rats, nephrotoxicity not correlated with bioactivition	physiological parameters and activity in the microsomal fraction of kidneys to that in the microsomal fraction of the liver supported by data on metabolism of two known substrates of CYP2E1 by microsomal fractions of the kidney and liver from 18 humans	rats and limited in humans
Sustained cytotoxicity	In mice, in all cases where examined, sustained cytotoxicity (as measured by histopathological effects and release of hepatic enzymes), at doses that induced tumors; in rats in critical bioassay, cytotoxicity based on histopathological reexamination	Kidney also a target organ in humans based on reports of renal effects resulting from anesthetic use of chloroform	Considerable in mice though less data than for liver, less in rats and limited in humans
Kidney tumors	Mice and rats	Inadequate epidemiological data	Considerable in animals; inadequate in humans

Abbreviations: **PBPK**, physiologically based pharmacokinetic; **CYP**, cytochrome P450.

confirm that target organs in populations exposed to high concentrations are similar to those in experimental animals (i.e., the kidney and liver).

Toxicokinetic comparisons in both rats and humans indicate that critical to the MOA is the requirement for a metabolized dose sufficient to produce cytotoxicity and subsequent cellular regeneration. This analysis, then, sets the stage for identification of critical precursor noncancer key events for which subsequent quantitation of interspecies differences and interindividual variability in dose–response analysis is relevant. This includes application of a physiologically based pharmacokinetic model as a basis to consider interspecies variation in rates of formation of reactive metabolites in the target tissue, which constitutes a reasonable basis for replacement of the default subfactor for interspecies differences in toxicokinetics with a CSAF (AK_{AF}).

Based on the convincing evidence for the hypothesized MOA, the optimum approach to quantitation of exposure–response, might involve analysis of the incidence of essential noncancer precursor events (cytotoxicity and regenerative proliferation) from interim kills in the critical cancer bioassay (i.e., that in which tumors were observed at lowest dose, following administration by the route most relevant to humans, i.e., continuously in drinking water) on the basis of rates or amounts of oxidative metabolites produced per volume of tissue in the critical organ. Unfortunately, data on precursor lesions were not collected in the critical bioassay. Reexamination of a proportion of the slides from several of the dose groups in the critical study, however, confirmed histopathological changes consistent with the hypothesis that sustained tubular cytotoxicity and regenerative hyperplasia led to renal tubular tumor induction, though the data amenable for quantitation of exposure–response in this investigation were limited.

There have been numerous subsequent short-term investigations of the proliferative response in the liver and kidney of various strains of mice and rats exposed to doses and concentrations of chloroform similar to those administered in the cancer bioassays in which tumors have been observed. However, for renal tumors, most of these investigations have been conducted in a strain of rat (F344) which varies from that, in which increases in renal tumors in the critical bioassay were observed (i.e., the Osborne–Mendel rat). Moreover, limited available data indicate that patterns of response (e.g., sex-specific) between the two strains vary. Available data are also inadequate as a basis of characterization of the relative sensitivity of the two strains to cytotoxicity.

Since quantitative data on the incidence of precursor lesions for cancer in the strain of interest are inadequate to meaningfully characterize exposure–response, a tumorigenic concentration was developed based on the incidence of tubular cell adenomas and adenocarcinomas in the critical bioassay in Osborne–Mendel rats (21).

To describe dose–response in the context of rates of formation of active metabolites in the target tissue (renal cortex), a PBPK model was developed and extended to humans. In the animal component of the PBPK model, the liver and kidney were described as individual sites of metabolism including regions of both high and low activity. The maximum rate of metabolism in the kidney was scaled to the maximum rate in the liver based on relative tissue volumes and proportionality constant. In the human component of the PBPK model, the kinetics were described by a single-compartment, based on simulation of an available study in humans in which metabolized and exhaled chloroform were determined for up to 8 hours following administration to male and female volunteers in olive oil or gelatin capsules. Liver tissue subvolumes were assumed to be the same as in the rat, while the kidney was subdivided into a 70:30 cortex:noncortex ratio, as per reference described by the International Commission on Radiological Protection. Blood flow to the kidney was also split between the cortex (90%) and noncortex (10%). Human metabolic parameters were those determined in vitro in eight human liver samples. Kidney metabolic rate constants were based on the relationship of activity observed in the microsomal fraction of kidneys to the activity observed in the microsomal fraction of the liver based on in vitro results but supported by data on metabolism of two known substrates of CYP2E1 by microsomal fractions of the kidney and liver from 18 humans. The combined incidence of renal adenomas and adenocarcinomas and the more limited histological data on cytotoxicity in the critical study were considered in the context of the mean rate of metabolism (i.e., cumulative metabolite formed per gram of tissue in the kidney cortex). BMDs for tumors and histological lesions were converted to internal dose metrics based on the animal component of the PBPK model (amount metabolized per time per unit volume renal cortex tissue, 4 mg/hr/L). The human component was then run to estimate the total external exposure associated with this fixed value for the internal dose metric. Because it was assumed that all chloroform exposure (oral, dermal inhalation) was due to chloroform in drinking water, and because the critical effect was independent of dose route, the combined equivalent human exposure was expressed in units of mg/kg/day. The external exposure of humans associated with the relevant measure of exposure–response, i.e., the estimated mean rate of metabolism in the kidney cortex associated with a 5% increase in tumor risk (TC_{05}) based on the PBPK model, was 92.8 g/kg b.w./day. This was compared with the equivalent external dose for rats of 41.2 mg/kg b.w./day.

The external dose-adjusted ratio of the human dose metric to the animal dose metric was (4 mg/hr L/92.8 mg/kg/day) divided by (4 mg/hr L/41.2 mg/kg/day) and this factor (0.44) constitutes the basis for development of an AK_{AF}, in lieu of the default value of 4. Since the PBPK model incorporates the amount metabolized per time per unit volume tissue (a component of toxicodynamics in the context of the default subfactors for CSAF), retention of the full default factor of 2.5 for species differences in toxicodynamics is considered conservative. The resulting interspecies CSAF would be $0.44 \times 2.5 = 1.1$. With no data to inform the intraspecies extrapolation, the default value of 10 would be retained for human variability, and the composite factor would be $1.1 \times 10 = 11$.

44 Meek and Renwick

For the PBPK model, among those parameters considered in the sensitivity analysis to have most impact on output, uncertainty was greatest for the metabolic parameters, particularly in the kidney for humans. As a result, additional in vitro data on the metabolism of chloroform in the human kidney and liver would be useful not only to reduce uncertainty in derived values, but potentially to address the issue of variability across the human population.

CONCLUSIONS

Consideration of relevant data in the context of a framework that addresses kinetic and dynamic aspects, explicitly, should result in greater understanding of contributing components and transparency in risk assessment. It is also hoped that consideration in this context will lead to clearer delineation and better common understanding of the nature of specific data required which would permit development of more informative measures of dose–response.

Integration with recently developed frameworks for consideration of the weight of evidence for hypothesized MOA for both cancer and noncancer effects additionally contributes to transparency in risk assessment through explicit delineation and consideration of appropriate key events for subsequent dose–response analysis. The example presented here, which includes consideration of the MOA for induction of tumors and subsequent dose–response analysis for noncancer precursor key events, is illustrative of the manner in which these analytical frameworks for hazard characterization and dose–response analyses are contributing to the harmonization of approaches for cancer and noncancer effects.

REFERENCES

1. IPCS (International Programme on Chemical Safety). Chemical-Specific Adjustment Factors (CSAFs) for Interspecies Differences and Human Variability: Guidance Document for the Use of Data in Dose/Concentration Response Assessment. Harmonization Project Document No. 2, World Health Organization, Geneva, Switzerland, 2005. Available at http://www.whqlibdoc.who.int/publications/2005/9241546786_eng.pdf.
2. Chou C-HSJ, Holler J, De Rosa CT. Minimal risk levels (MRLs) for hazardous substances. J Clean Technol Environ Toxicol Occup Med 1998; 7:1–24.
3. U.S. EPA. Reference Dose (RfD): Description and Use in Health Risk Assessments. U.S. Environmental Protection Agency, Washington DC, 1993. Available at http://www.epa.gov/iris/rfd.htm.
4. U.S. EPA. Methods for Derivation of Inhalation Reference Concentrations and Application of Inhalation Dosimetry. EPA/600/8-90/066F, U.S. Environmental Protection Agency, Office of Research and Development, Washington, DC, 1994. Available at http://nepis.epa.gov/pubtitleORD.htm.

5. Meek ME, et al. Approach to assessment of risk to human health for priority substances under the Canadian Environmental Protection Act. Environ Carcinogen Ecotoxicol Rev 1994; 12:105–134.
6. Renwick AG. Data-derived safety factors for the evaluation of food additives and environmental contaminants. Food Addit Contam 1993; 10:275–305.
7. IPCS (International Programme on Chemical Safety). Assessing Human Health Risks of Chemicals: Derivation of Guidance Values for Health-based Exposure Limits. Environmental Health Criteria 170, International Programme on Chemical Safety, World Health Organization, Geneva, Switzerland, 1994. Available at www.inchem.org/documents/ehc/ehc/ehc170.htm.
8. Renwick AG, Lazarus NR. Human variability and noncancer risk assessment—an analysis of the default uncertainty factor. Regul Toxicol Pharmacol 1998; 27:3–20.
9. Meek ME, Renwick A, Ohanian E, et al. Guidelines for application of compound specific adjustment factors (CSAF) in dose/concentration response assessment. Comm Toxicol 2001; 7:575–590.
10. Dorne JL, Walton K, Renwick AG. Human variability in xenobiotic metabolism and pathway-related uncertainty factors for chemical risk assessment: a review. Food Chem Toxicol 2005; 43:203–216.
11. U.S. EPA. Integrated Risk Information System (IRIS), Online Toxicological Review of Boron and Compounds (CASRN 7440–42-8) entered August 2004. U.S. Environmental Protection Agency, Integrated Risk Information System, Washington, DC, 2006. Available at http://www.epa.gov/iris/subst/0410.htm.
12. Renwick AG. The use of safety or uncertainty factors in the setting of acute reference doses. Food Addit Contam 2000; 17:627–635.
13. IPCS (International Programme on Chemical Safety). Guidance Document for the Use of Data in Development of Chemical-specific Adjustment Factors (CSAF) for Interspecies Differences and Human Variability in Dose/Concentration Response Assessment. World Health Organization, Geneva, 2001. Available at www.who.int/entity/ipcs/publications/methods/harmonization/en/csafs_guidance_doc.pdf.
14. Meek B, Renwick A, Ohanian E, et al. Guidance for derivation of chemical-specific adjustment factors (CSAF)-Development and implementation. Human Ecol Risk Assess 2002; 8:769–782.
15. Meek ME, Renwick A, Ohanian E, et al. Guidelines for application of chemical-specific adjustment factors in dose/concentration–response assessment. Toxicology 2002; 27:115–120.
16. Meek ME, Bucher JR, Cohen SM, et al. A framework for human relevance analysis of information on carcinogenic modes of action. Crit Rev Toxicol 2003; 33:591–653.
17. Sonich-Mullin C, Fielder R, Wiltse J, et al. IPCS conceptual framework for evaluating a mode of action for chemical carcinogenesis. Regul Toxicol Pharmacol 2001; 34:146–152.
18. Seed J, Carney EW, Corley RA, et al. Overview: using mode of action and life stage information to evaluate the human relevance of animal toxicity data. Crit Rev Toxicol 2005; 35:663–672.
19. Meek ME, Beauchamp R, Lang G, et al. Chloroform: exposure estimation, hazard characterization, and exposure–response analysis. J Toxicol Environ Health B 2002; 5:283–334.

20. ILSI (International Life Sciences Institute). An Evaluation of EPA's Proposed Guidelines for Carcinogen Risk Assessment Using Chloroform and Dichloroacetate as Case Studies. Report of ILSI HESI Expert Panel. International Life Sciences Institute, Washington, DC, 1997.
21. Jorgenson TA, Melerhenry EF, Rushbrook CT, et al. Carcinogenicity of chloroform in drinking water to male Osborne–Mendel rats and female B6C3F1 mice Fund. Appl Toxicol 1985; 5:760–769.

3

Derivation and Modeling of Mechanistic Data for Use in Risk Assessment

Ronald L. Melnick

*National Institute of Environmental Health Sciences, National Institutes of Health,
Research Triangle Park, North Carolina, U.S.A.*

INTRODUCTION

Toxicology is generally considered to be the study of poisons with emphasis on the detection and identification of toxic agents, determination of the effects of such agents on the environment or public health, and derivation of antidotes to neutralize or counteract their effects. In addition to the above, modern toxicology is aimed at characterizing quantitative relationships between exposure and internal measures of dose and the molecular and cellular perturbations that contribute to adverse health effects. Of critical importance in toxicology research is the need to better understand and account for interindividual differences in susceptibility (1) due to: (i) genetic factors, (ii) life stage or age of exposure, (iii) health status, diet, and lifestyle, or (iv) exposure to multiple agents. Risk assessment is the process of characterizing the nature and probability of an adverse effect in populations exposed to potentially hazardous agents. This chapter focuses on the derivation and use of mechanistic information for the development of tissue dosimetry models and for evaluations of human health risks associated with exposure to toxic or carcinogenic agents.

EVALUATING HUMAN RISK FROM ANIMAL AND MECHANISTIC DATA

General Approaches for Obtaining Information Indicative of Human Risk

In the absence of sufficient human data on an agent, a variety of approaches are available to determine whether or not a chemical may pose a hazard to human health and the likelihood of risk at defined exposure levels. A commonly used first approach is an evaluation of potential structure/activity relationships (2). Concerns are raised if the chemical under consideration or its known or putative metabolites share structural properties with known toxicants. Health concerns also exist for chemicals that are members of a class of compounds that induce toxic effects in humans or animal models. The identification and quantification of metabolites produced in animals or generated in in vitro systems (e.g., liver microsomal fractions) provide essential information on biotransformation pathways and may strengthen concerns related to potential toxic effects of metabolic products.

A second approach for determining whether or not a particular agent poses a potential hazard to human health would be to subject the compound to a variety of in vitro screening assays that provide information on specific activities of various disease precursor events. For example, numerous short-term in vitro assays have been developed to evaluate the genotoxicity of environmental agents and their metabolites (3). The inclusion of metabolic activation systems (e.g., S9, the supernatant fraction from a liver homogenate that contains microsomal and cystolic enzymes) addresses to some extent contributions from metabolic intermediates of the test agent (4). For the most part, genotoxic assays are designed to identify directly and indirectly acting mutagens and agents that induce chromosomal aberrations or other forms of DNA damage. Other screening assays have been developed to evaluate the effects of environmental agents on cell-cycle kinetics, cell death, gene expression, intercellular communication, enzyme activation or inhibition, oxidative stress, etc.

Short-term assays are useful in prioritizing chemicals for selective in vivo studies. Because of uncertainties in the predictability of short-term assay results for specific disease risks, hypotheses developed from structure/activity analyses or from in vitro assays require rigorous testing in animal models that integrate whole animal dosimetry and precursor events with disease outcome. Results from in vivo studies have been reliably used to develop protective public health strategies to reduce human exposures to toxic environmental or occupational agents (5). The establishment of quantitative relationships between early molecular/cellular perturbations and disease outcome may be useful in future assessments of human risk.

In vivo toxicity studies provide a definitive approach to test hypotheses linking exposure and disease causality. These studies typically include observations of overt toxicity (e.g., clinical signs, body weight changes, and organ weights), histopathological evaluations for tissue alterations, and direct measurements of markers of altered tissue response. The identification of an exposure-related effect is dependent on several aspects of the experimental design and evaluation: (i) the size of the dose groups must be sufficiently large to have adequate power to detect a significant difference between controls and exposed animals, (ii) the range of doses and the number of dose groups must be sufficiently large to perform meaningful analyses of dose–response relationships, (iii) the age of exposed animals is a critical factor if susceptibility varies with life stage, (iv) the duration of exposure is important to distinguish acute from chronic effects and to allow sufficient time for manifestation of diseases of long latency, and (v) the pathological evaluations and other endpoint determinations must be complete and reliable. Because animal studies with relatively small numbers per dose group frequently do not have sufficient power to detect significant effects at human exposure levels, substantially higher doses are generally necessary to identify chemical-related effects and characterize dose–response relationships. In selecting doses for a toxicology study, it would not be appropriate to only use doses above those in which absorption and/or metabolic activation or detoxification are saturated. A complete histopathological evaluation of all organs and tissues is necessary to identify all potential target sites of the agent under study.

Several approaches may be taken to determine if an identified toxic effect is due to the parent compound or its metabolites. Studies of suspected contributing biological activities of the parent compound and metabolites in in vitro systems (e.g., binding to a receptor, induced or suppressed transcriptional activation, mutagenesis, enzyme inhibition, etc.) or a determination of the influence of metabolic activation on such responses can reveal a potential critical role of metabolites in the observed in vitro effect. Such studies would indicate the impact of a direct interaction of the parent compound or its metabolites with cellular or molecular components relevant to disease outcome. The use of genetically modified animals in which the primary metabolizing gene is deactivated would also provide insight on the potential role of metabolism in the toxicological response. Dose–response analyses are also informative on this issue; for example, it might be expected that the dose–response curve for the toxic effect or a precursor event would have a similar shape as the exposure-related tissue dosimetry of active metabolites, if metabolism is important in producing the observed toxic response. For these analyses, pharmacokinetic models can provide estimates of the time- and dose-dependent levels of the parent compound and its metabolites in identified target organs.

Biologically Based Approaches for Assessing Human Risk

Physiologically based pharmacokinetic (PBPK) models can provide a biologically based approach to determine tissue dose metrics of parent chemical and metabolites for use in characterizing dose–response relationships and estimating risks at occupational or environmental exposure levels. If adequately developed, such models can provide essential information on tissue dosimetry of parent compound and metabolites in exposed organisms, and can accommodate physiological differences across species as well as genetic and life stage variability in the expression of enzymes that contribute to the biotransformation of the agent being evaluated. PBPK models typically consist of a series of mass-balance differential equations that are formulated to represent in quantitative terms the complex physiological and biochemical processes that affect the behavior of the agent in the intact animal, i.e., the absorption, distribution, metabolism, and elimination. The development of PBPK models is an iterative process that builds on information in the literature as well as the repeated testing of model-based predictions of the in vivo behavior of the agent in animals and/or humans (6). Mass balance information is critical to ensure that contributions from multiple biotransformation pathways are accurately represented.

Pharmacodynamic modeling provides an approach to estimate rates of tissue responses in relation to the delivered dose. Tissue responses that may be evaluated in these models are frequently those provided in in vitro screening assays described above of potential disease precursor events or in vivo biomarker studies. Pharmacodynamic models may also lead to mechanistic hypotheses, which if properly tested, may shed new insights on cellular processes that are altered by the agent under study. Unfortunately, our understanding of the quantitative relationships between potential tissue precursor events and disease outcome is limited. When considering the implementation of biologically based risk assessment methods, caution is needed to ensure that new approaches do not simply involve the replacement of one set of assumptions with different assumptions that are not adequately protective of public health. The general acceptance of alternative approaches will require rigorous testing of assumptions incorporated into biologically based risk assessment models. With greater information on disease pathogenesis and by integrating validated pharmacokinetic and pharmacodynamic models into the risk assessment process, it may be feasible to develop and utilize biologically based approaches to relate human exposures to disease risk even among subpopulations of differing susceptibility.

DERIVATION OF PHARMACOKINETIC AND PHARMACODYNAMIC MEASURES FROM TOXICOLOGY AND MECHANISTIC INFORMATION

As noted above, valuable information about the toxic effects and biological behavior of an agent can be derived from hypothesis-based toxicology studies and in vitro experiments. The development of a dosimetry model for the genotoxic carcinogen, 1,3-butadiene (BD), and pharmacodynamic models for dioxin-induced receptor-mediated processes provide several examples illustrating how toxicity and mechanistic studies can help elucidate critical information for use in risk assessments.

Dosimetry of DNA-Reactive Metabolites of BD

1,3-BD is a high-production-volume industrial chemical used largely in the manufacturing of synthetic rubber and thermoplastic resins. Studies on the metabolism and pharmacokinetics of BD in rats and mice were performed in an attempt to explain apparent species differences in potency of BD-induced carcinogenicity and organ site specificity. BD is metabolized to three different epoxide intermediates: 1,2-epoxy-3-butene (EB), 1,2,3,4-diepoxybutane (DEB), and 3,4-epoxy-1,2-butandediol (EBD) (7). Oxidation of BD to EB or EB to DEB is catalyzed by cytochrome P-450 monooxigenase (CYP450, predominantly CYP2E1), and each of the epoxide intermediates may undergo hydrolysis catalyzed by epoxide hydrolase (EH) or conjugation with glutathione (GSH) catalyzed by glutathione-*S*-transferase (GST). Because BD is mutagenic only in the presence of metabolic activation systems and because EB and DEB were shown to be mutagenic in *Salmonella typhimurium* (8,9) and carcinogenic in rats and mice (10,11), attention focused on the kinetics of formation and elimination of these DNA-reactive epoxides. In addition, exposure to BD causes glutathione depletion, probably as a function of GST-catalyzed elimination of these metabolites.

In vitro kinetics of the oxidation of BD and EB and of EB hydrolysis or conjugation with glutathione were reported in liver and lung microsomal and cytosolic fractions of rats and mice (12). In addition, gas-uptake studies of BD (13,14) and of EB (15) in rats and mice held in desiccators provided time-course data to model the absorption, distribution, and metabolic elimination kinetics of these chemicals. Initial PBPK models of BD in rats and mice (16–18) provided very good fits to data on the uptake and metabolic elimination of BD and EB from closed chambers. The models also predicted measured blood levels of BD, but overpredicted by approximately an order of magnitude subsequently measured blood EB concentrations following nose-only inhalation exposure (19). This discrepancy was likely due to inaccurate characterization of the metabolic elimination of EB produced during in vivo oxidation of BD, i.e., the kinetics of EB metabolism is different when administered directly versus when it is produced in situ by CYP450-mediated oxidation of BD. Experimental evidence indicates that a transient complex exists between the CYP450 that produces EB from BD and the epoxide hydrolase that consumes this intermediate.

To account for the above scenario, the BD PBPK model was modified (20) to allow newly synthesized EB to be channeled to epoxide hydrolase in preference to its dissociation from the endoplasmic reticulum and release into

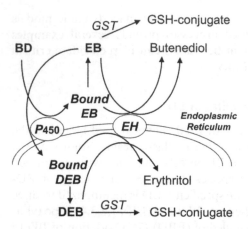

Figure 1 Privileged access of endoplasmic-reticulum-bound epoxide to transiently associated epoxide hydrolase. *Abbreviations*: BD, 1,3-butadiene; EB, 1,2-epoxy-3-butene; GST, glutathione-s-transferase; DEB, 1,2,3,4-diepoxybutane; GSH, glutathione; EH, epoxide hydrolase.

the cytoplasm (Fig. 1). In this scheme, CYP450-mediated oxidation of BD produces endoplasmic-reticulum-bound EB that may undergo hydrolysis via its privileged access to EH or dissociates from the endoplasmic reticulum and undergoes either GST-mediated conjugation with glutathione or EH-mediated hydrolysis. Kinetic parameters for the biotransformation of free EB are available from in vitro data of GST and EH metabolism of EB. In addition, EB may be oxidized to DEB, which would exist in both its bound and free states, with the bound form produced in situ also having privileged access to EH. This model provides a good fit to the BD and EB uptake data as well as to the measured steady-state levels of BD, EB, and DEB in the blood of rats and mice exposed by nose-only inhalation.

    With the publication of a comprehensive mass balance for the disposition of inhaled BD in rats and mice (21), essential data became available to allow expansion of previous PBPK models to include equations for the production and metabolism of other metabolic intermediates, including butenediol and EBD (22). The latter epoxide may be formed by oxidation of butenediol or by partial hydrolysis of DEB. The expanded model predicts higher liver steady-state concentrations of EBD than EB or DEB, in agreement with the observation that most of the DNA adducts in animals exposed to BD arise from this metabolite (23). The replacement of rodent-specific physiological and biochemical parameters with those representing human physiology and biochemical activity should allow the development of a human PBPK model that reflects the range of expected human dosimetry for the mutagenic metabolites formed after exposure to BD.

A Pharmacodynamic Approach to Evaluate the Carcinogenicity of Epoxides and Epoxide-Forming Chemicals

Epoxides and epoxide-forming chemicals, in addition to BD, include several of the highest production chemicals in the United States, e.g., ethylene/ ethylene oxide, propylene/propylene oxide, styrene/styrene oxide, vinyl chloride, acrylonitrile, etc. A proposed pharmacodynamic approach to predict the carcinogenicity of this class of compounds involves the convergence of data on the in vitro mutagenicity of each epoxide (parent compound and metabolites) in putative target cells plus mathematical dosimetry models that predict the time- and dose-dependent in vivo concentrations of the mutagenic epoxides in target tissues of exposed animals. This approach is based on the hypothesis that tumor incidence resulting from exposure to epoxides and epoxide-forming chemicals is correlated with the cumulative number of mutations produced in a tissue at risk. Predictions of site-specific tumor dose–response might be feasible by relating model-based estimates of cumulative mutations in specific tissues to observed tumor incidence for several carcinogenic epoxide chemicals. If additional members of this family of chemicals act by a common mechanism, then the same relationship should hold for these agents as well, i.e., the carcinogenic activity of these agents should be represented by the same tumor incidence versus cumulative mutations curve for each tissue at risk.

An estimate of tissue cumulative mutations could be obtained from a pharmacodynamic model that combines information on tissue dosimetry for each epoxide, data on epoxide reactivity with DNA, tissue-specific information on the efficiency of converting DNA adducts to mutations, and cell division rates. In vitro studies would provide data on epoxide reactivity with DNA and tissue-specific mutational rates for each epoxide; in vivo studies would provide tissue-specific cell division rates. Two related assumptions regarding the utility of an epoxide mutagenesis model for predicting site-specific carcinogenesis are that tumors induced by these chemicals arise from clonal expansion of epoxide-mutated cells and that the mutational rate at a specific locus measured in vitro (e.g., *hprt* gene) reflects the rate of genetic alterations in cancer-related genes.

Receptor-Mediated Mechanisms of Toxicity and Their Application to Risk Assessment

Dose–Response of Dioxin-Induced Enzymes

The interplay between data generation and modeling of cellular and molecular changes induced by 2,3,7,8-tetrachlorodibenzo-*p*-dioxin (TCDD) provides an excellent example of how pharmacokinetic and pharmacodynamic data can be derived from in vivo and in vitro mechanistic studies and applied to risk assessments of environmental toxicants. TCDD induces tumors at multiple organ sites in rats and mice (24). Biological effects of TCDD, as well as its congeners and several other polyhalogenated hydrocarbons, are mediated by binding to and activating the aryl hydrocarbon (Ah) receptor. The activated Ah receptor forms a heterodimer with another transcription factor, Ah receptor nuclear translocator; binding of this complex to regulatory sequences on DNA (dioxin response elements) alters the expression of several proteins, including cytochromes P4501A1 and 1A2 (25). The altered expression of some of these proteins may be involved in the toxicity and carcinogenicity of TCDD. Several PBPK models have been developed to characterize the absorption of TCDD from the gastrointestinal tract, its distribution to tissues, its slow metabolic clearance, and alterations in the concentrations of hepatic proteins that have been suggested as potential biomarkers of TCDD-induced effects (26,27). Biological models that realistically represent physiological and biochemical events involved in TCDD dosimetry and its induced effects can be used to compute dose–response curves and extrapolate TCDD-induced effects.

The physiological models of Kohn et al. (26,28) reproduced measured levels of dioxin in the liver and blood of orally dosed rats, as well as measured changes in hepatic concentrations of CYP1A1 and CYP1A2 (29,30), estrogen receptor, epidermal growth factor receptor (31), and degradation and induction of Ah receptor protein after binding TCDD (32). CYP1A1, which is constitutively expressed in the liver at low levels (30), was induced at levels proportional to TCDD dose, contradicting the opinion that all Ah receptor-mediated events are threshold responses. When data on both mRNA and protein levels became available for genes induced by TCDD (30,33), the model was expanded to include separate steps for gene transcription and translation (28). The observed weaker induction of CYP1B1 by TCDD was predicted to be due to a weaker binding affinity of liganded Ah receptor for CYP1B1 dioxin response elements (34) compared to CYP1A dioxin response elements (35) and a shorter lifetime of CYP1B1 mRNA (28).

The mechanism of tumor induction by TCDD is not known. Based on in vivo and in vitro data showing that TCDD may induce oxidative stress (36), the PBPK model (26) was modified to include hydroxylation of estradiol by CYP1A2 leading to the production of reactive oxygen species that can induce DNA damage (37). This pathway implies that TCDD, which is generally thought of as a pure tumor promoter, may initiate carcinogenesis by indirect mechanisms. Hence, the designation of receptor-mediated carcinogens as nongenotoxic promoters may be an oversimplification of the complex events involved in chemical carcinogenesis.

Other PBPK models for TCDD have been developed that address issues such as regional induction of CYP450 proteins in the liver (38), enzyme induction in multiple tissues (39), and fetal exposures (40). The model of Kohn et al. (28) is being used to estimate target tissue dosimetry in relation to enzyme induction and tumor induction in NTP studies of TCDD and other dioxin-like compounds and to evaluate whether or not cancer risks of complex mixtures of dioxin-like compounds can be predicted from the combined relative potencies and concentrations of the mixture constituents.

Ah Receptor-Dependent Thyroid Hormone Disruption

TCDD also induces uridine diphosphate (UDP)-glucuronosyltransferase (UGT) by an Ah receptor-dependent mechanism (41). This enzyme conjugates thyroxine (3,5,3′,5′-tetraiodothyronine, T4), leading to its clearance. Metabolic depletion of T4 relieves the inhibition of thyroid stimulating hormone (TSH) release from the pituitary by circulating T4, causing a rise in serum TSH concentrations. Increases in UGT mRNA, decreases in serum T4, increases in serum TSH, as well as thyroid follicular cell hypertrophy and hyperplasia have been observed in rats exposed to TCDD (42). Thus, the PBPK model on TCDD disposition was modified (43) to simulate release of thyroid hormones into the blood, binding of thyroid hormones to serum proteins, uptake of thyroid hormones by peripheral tissues, binding of T4 to cytosolic receptors and T3 to nuclear receptors, metabolism of thyroid hormones, and induction of UGT by TCDD via the Ah receptor (Fig. 2). In addition, the model includes regulation of TSH release from the pituitary via effects of T4 on the production and release of the hypothalamic peptides thyrotropin-releasing hormone (TRH) (which stimulates

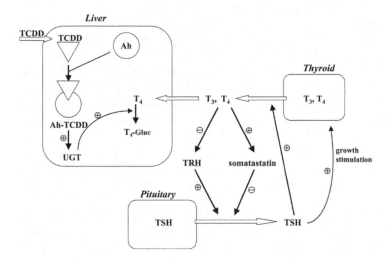

Figure 2 Ah receptor-dependent regulation of thyroid hormones and thyroid growth. *Abbreviations*: Ah, aryl hydrocarbon; TCDD, 2,3,7,8-tetrachlorodibenzo-p-dioxin; UGT, UDP-glucuronosyl transferase; T3, 3,5,3′-triiodothyronine; T4, 3,5,3′,5′-tetraiodothyronine; TRH, thyrotropin-releasing hormone; TSH, thyroid stimulating hormone.

TSH release) and somatostatin (inhibits TSH release). The low-dose linear response for UGT induction and for serum TSH concentrations may reflect the dose–response for thyroid growth and the induction of thyroid tumors by chronic overstimulation of this gland. The linear response is likely due to the fact that these simulations did not require cooperative binding of TCDD to the Ah receptor.

THE NECESSITY TO TEST MECHANISTIC HYPOTHESES

Physiological models can be used to test the consistency between postulated toxicologic mechanisms and observed data and to predict the effects of other agents that act by the same mechanism. Two examples are discussed below in which discrepancies exist between the mechanistic hypothesis and the experimental data: the hypothesis on α2u-globulin (α2u) nephropathy and the hypothesis on liver neoplasms induced by peroxisome proliferators. These examples show how the acceptance and use of untested hypotheses may result in underestimations of human risk.

α2u-Globulin Nephropathy

The hypothesis on α2u-globulin (α2u) nephropathy was developed based on the observation that protein droplets containing α2u accumulate in epithelial cells of proximal tubules of male rats exposed to several hydrocarbons that had been reported to cause kidney cancer in male rats after long-term exposure. α2u is a low-molecular-weight protein (18.7 kDa) synthesized in the liver of mature male rats under androgenic control (44); it is not synthesized by hepatocytes of female rats, mice of either sex, or several other species, including humans. The physiological function of α2u is not known; however, it does bind reversibly to hydrophobic chemicals and may serve as a transport protein for pheromones. α2u is secreted into the blood, filtered through the glomerulus, and partially reabsorbed (~60%) by endocytosis into renal tubule epithelial cells of the P2 segment (45). The reabsorbed protein is hydrolyzed by lysosomal proteinases; however, in vitro studies have shown that binding of xenobiotic ligands to α2u reduces the rate of proteolytic degradation by lysosomal enzymes by about 30% (46). The excessive accumulation of protein droplets containing α2u presumably due to ligand binding has been hypothesized to cause lysosomal dysfunction resulting in cell death (47). The actual mechanism of cell death is unknown. Sloughing of necrotic epithelial cells into the tubule lumen has been observed, and regenerative hyperplasia of epithelial cells in the P2 segment occurs in response to cell loss (48). Although the mechanistic link between cell proliferation and kidney cancer is unknown, it has been suggested that regenerative hyperplasia causes tumor induction in the male rat kidney by increasing the likelihood of fixing presumed spontaneous cancer-initiating

DNA damage into heritable mutations or by promoting the clonal expansion of spontaneously initiated cells (49).

Trimethylpentane (TMP), a nephrotoxic agent found in gasoline, has been used as a model compound to study the mechanism of nephrotoxicity induced by unleaded gasoline in male rats. Disposition studies have shown that 2,4,4-trimethyl-2-pentanol (TMP-2-OH), a metabolite of TMP that binds to α2u, was the major metabolite present in the male rat kidney and that female rats excreted greater concentrations of TMP-2-OH conjugates in the urine than did male rats (50).

Several inconsistencies between experimental data and the above hypothesis have been noted (51). These include (i) binding affinities vary by nearly 1000-fold (52) for compounds that cause similar levels of α2u accumulation in male rat kidneys; (ii) 2,2,4-trimethyl-1-pentanoic acid causes α2u accumulation and renal tubular cell proliferation in male rats (53), but does not bind to α2u (52); (iii) lindane and gabapentin induce α2u nephropathy in male rats at doses that do not induce kidney tumors in chronic exposure studies (54–56); (iv) α2u accumulation is much greater in male rats exposed to unleaded gasoline (57) than those exposed to methyl *t*-butyl ether (MTBE) (58), however, increases in protein droplets and renal epithelial cell proliferation were similar for these two chemicals; (v) proteins other than α2u also accumulate in hyaline droplets in the kidney of rats exposed to unleaded gasoline (59); and (vi) in vivo studies indicate that lysosomal proteolytic enzymes are increased by about 65% after exposure of male rats to agents that induce α2u accumulation (60). Inconsistencies in quantitative relationships among the intermediary events in the α2u nephropathy hypothesis for several hydrocarbons indicate that factors other than simple binding of ligand binding to α2u contribute to the renal accumulation of this protein; consequently, an alternative hypothesis was advanced (51) with α2u acting as a vector to increase the delivery of toxicant or protoxicant to proximal tubule epithelial cells (Fig 3).

With the availability of blood and kidney time-course data on TMP-2-OH and α2u in male rats (61), a PBPK model was created to determine what factors may contribute to the accumulation of α2u in the male rat kidney (62). The model included metabolic elimination of TMP-2-OH by oxidation and glucuronidation, hepatic α2u production, partial resorption of the glomerular filtered protein into kidney tissue by pinocytosis, reversible binding of TMP-2-OH to α2u, degradation of α2u by renal lysosomal cathepsins, with a 30% reduction in proteolytic efficiency for the liganded form (increased K_m) compared to the free form, and induction of cathepsin activity with increased uptake of α2u by proximal tubule epithelial cells. Although the model included the purported mechanistic features of the α2u accumulation hypothesis, simulations of the observed disposition of TMP-2-OH and the α2u data required a transient increase in hepatic synthesis and secretion of α2u, stimulated by the presence of liganded α2u in

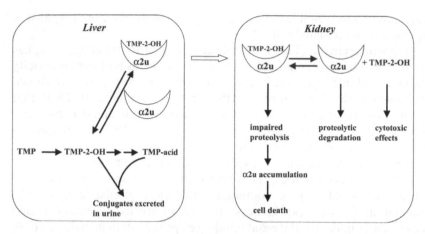

Figure 3 Hypotheses on the role of α2u in trimethylpentane-induced nephropathy in male rats. *Abbreviations*: TMP, trimethylpentane; α2u, α2u globulin; TMP-2-OH, 2,4,4,-trimethyl-2 pentanol.

the blood. Thus, renal accumulation of α2u may not be due only to reduced proteolytic degradation of the liganded protein. The model also predicts that α2u increased the delivery of TMP-2-OH to the kidney. Increased production of toxic TMP-2-OH metabolites and induced lysosomal activity may contribute to the nephrotoxicity observed in male rats exposed to α2u ligands or their precursors.

The physiological model was also used to examine the dependence of renal α2u accumulation on ligand-binding affinity and ligand dosimetry. For high-binding affinity ligands, the model predicts that accumulation of α2u in the kidney saturates even at relatively low cumulative concentrations of unbound ligand in the blood. For chemicals with low-binding affinities such as MTBE (63), little renal accumulation of α2u is predicted even at high doses. This prediction is consistent with the experimental data on renal α2u accumulation in male rats exposed to MTBE (58). Thus, the extent of proximal tubule necrosis and increased renal cell proliferation in rats exposed to MTBE (58) is likely not due simply to α2u accumulation and probably involves toxicity from MTBE metabolites (64). Human risk assessments of agents that are associated with α2u accumulation should also consider the possible contribution of toxic metabolites to the nephrotoxic response.

Peroxisome Proliferation and Hepatocarcinogenicity

Reddy et al. (65) suggested that hypolipidemic peroxisome proliferators might represent a novel class of chemical carcinogens, because of an apparent correlation between peroxisome proliferation and hepatocarcinogenesis in laboratory animals. Based on reports that these agents induce peroxisome proliferation in rat or mouse hepatocyte cultures, but not in cultured human hepatocytes, it has been claimed that humans are refractory to the adverse effects of peroxisome proliferators and that liver tumors induced by these agents in rodents are not relevant to assessments of human cancer risk (66).

The discovery of the peroxisome proliferator-activated receptor (PPARα), a ligand-activated transcription factor (67), provides a mechanistic basis for understanding how peroxisome proliferators modulate gene expression. Humans express a functionally active PPARα (68,69) and hypolipidemic fibrates modulate lipid homeostasis in humans through the activation of this receptor (70). Ligand-activated PPARα forms a heterodimer with the retinoid X receptor (RXR) (71) that can bind to peroxisome proliferator response elements (PPREs) in the promoter region of genes induced by these ligands. Factors that may contribute to tissue and species differences in response include tissue dosimetry of the activating ligand, the relative levels of the receptor, the nature of PPREs in targeted genes, crosstalk among nuclear transcription factors for RXR, and levels of coactivators and corepressors of the liganded receptor (72). Species comparisons of the latter two factors have not been reported; however, levels of PPARα mRNA levels were found to be lower in the liver of humans compared to rats or mice (73). The hypothesis that differences in PPARα levels account for species differences in response to PPARα agonists has not been tested (74).

Increased replicative DNA synthesis and cell division rates have been proposed as the mechanism of peroxisome proliferator-mediated liver carcinogenesis (75); however, persistent increases in hepatocyte proliferation do not occur with most carcinogenic peroxisome proliferators. This suggests that other factors must be involved (76), including suppression of apoptosis, inhibition of gap junction intercellular communication, and activation of Kupffer cells. The mode-of-action proposed for rodent liver tumors induced by peroxisome proliferators is ligand activation of PPARα resulting in altered transcription rates of genes that regulate cell proliferation and apoptosis (Fig. 4). Stimulation of cell proliferation and suppression of apoptosis favor the proliferation and persistence of DNA-damaged cells that develop into preneoplastic lesions and eventually progress to tumors (74).

Proposed mechanisms of liver tumor induction by peroxisome proliferators have not been adequately tested nor have biologically based dose–response models been created that account for the multiple steps that may be involved. This is in part due to limited information on the nature and time dependence of critical interactive events. Thus, there is a need to quantify hepatocyte proliferation and suppression of apoptosis as a function of exposure to several peroxisome proliferators and compare dose–response relationships for these events with dose–responses for liver tumor incidence.

A biologically based dose–response model for peroxisome proliferator-induced cell proliferation and suppression of apoptosis is needed. Such a model should include the contribution of growth factors produced in

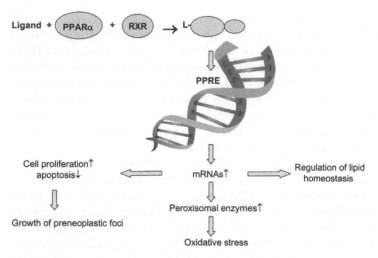

Figure 4 Peroxisome proliferation hypothesis: role of peroxisome proliferator-activated receptor. *Abbreviations*: PPARα, peroxisome proliferator-activated receptor; RXR, retenoid X receptor.

Kupffer cells by PPARα-independent events. Parzefall et al. (77) and Hasmall et al. (78) showed that peroxisome proliferators had no effect on DNA synthesis but still induced peroxisomal acyl CoA oxidase activity in rat or mouse hepatocytes that had been purified of contaminating Kupffer cells. Kupffer cells, which are resident macrophages in the liver, are a major source of growth factors that induce DNA synthesis or suppress apoptosis in hepatocytes. Kupffer cells do not express PPARα (79), but are activated by peroxisome proliferators (80). Activation of Kupffer cells by peroxisome proliferators requires functionally active nicotinamide adenine dinucleotide phosphate (NADPH) oxidase (81,82); this leads to oxidant-dependent activation of the transcription factor NF-κB and subsequent production of mitogens that can induce hepatocyte proliferation (Fig. 5). Increased superoxide production by treatment with Wy-14,643 or DEHP occurs similarly in wild-type mice and in PPARα-null mice (79,83). Inhibition of Kupffer cell NADPH oxidase activity prevents peroxisome-proliferator-induced cell proliferation, but does not affect peroxisomal acyl CoA oxidase induction (81,82). Thus, hepatocyte proliferation and peroxisome proliferation occur by different mechanisms and PPARα-independent stimulation of hepatocyte growth factor production in Kupffer cells is required for increased hepatocyte DNA synthesis and suppression of apoptosis in response to peroxisome proliferators. Because peroxisome proliferators do not induce cell proliferation in PPARα-null mice, it is essential that the interplay between PPARα activation and Kupffer cell activation be properly addressed in future biological models of peroxisome proliferator-induced hepatocarcinogenesis. This information is

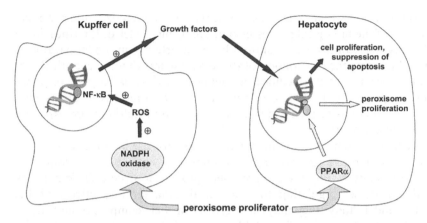

Figure 5 Role of Kupffer cells in signaling hepatocyte proliferation by peroxisome proliferators. *Abbreviations*: NADPH, nicotinamide adenine dinucleotide phosphate; PPARα, peroxisome proliferator-activated receptor; RXR, retenoid x receptor.

critical for risk assessments of peroxisome proliferators because humans express a functionally active PPARα.

CONCLUSION: PHARMACOKINETIC AND PHARMACODYNAMIC MODELING IS AN ITERATIVE PROCESS

For physiological models to be useful and reliable for risk assessment, they must accurately represent the biochemical and physiological processes that produce the observed pharmacokinetic behavior and tissue responses of the agent in the intact animal. Thus, if an agent generates signals leading to cellular proliferation or suppression of apoptosis (e.g., peroxisome proliferators, TCDD), then the model must include explicit mathematical representations of the processes that determine tissue dosimetry and the consequent signals affecting rates of cell replication and cell death. For chemicals that cause DNA damage leading to the formation of mutant cells, the corresponding model must include the kinetics of induction of such damage and related repair processes. Models of gene transcription and translation of message into protein should include degradation of mRNA by ribonucleases and of protein products by proteolytic enzymes. In addition, discrepancies in published parameter values of the system being modeled (due to differences in methods of measurement or experimental conditions) must be reconciled (6).

Not only must the model reproduce the experimental data on which it is based, it must also reproduce data sets that were not used to estimate parameters. Confirmation of a model's predictions by subsequent experiments is essential for hypothesis testing and model validation. If several models fit experimental data nearly equally, then model predictions may be useful in identifying new experiments to characterize other critical attributes of the biological processes that are being modeled. Interindividual variability may contribute substantially to uncertainty in model parameter values. Including the distribution of those parameters that affect critical biological processes would facilitate modeling of the variability of response among populations. Linking dosimetry models with models of disease can provide predictions of the incidence of that disease resulting from exposure to particular toxicants.

The development of mathematical models for use in risk assessment is typically an iterative process requiring important contributions from multiple data sources. Physiological models provide a structural framework for the processes regulating the biological behavior and effects of the agent(s) under study; they may vary greatly in their complexity and utility depending on the extent of available data and the complexity of the processes being modeled. The iterative process of model development involves hypothesis formulation based on toxicology and pharmacokinetic observations, designing experiments to test these hypotheses, confirming that model-based predictions reproduce experimental data, and evaluating the consistency of the model with alternative mechanistic hypotheses (Fig. 6). Model refinement may become necessary as new data become available. Because mechanistic modeling involves drawing inferences from experimental data, incorrect interpretations can lead to model misspecifications. To avoid this situation, alternative model structures must be evaluated rigorously for consistency with all available experimental data.

The model of BD disposition in rats and mice provides an example of the iterative process of model development. The BD PBPK model was created to characterize organ-specific dosimetry of the three DNA-reactive

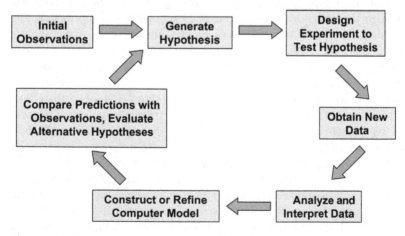

Figure 6 The iterative process of model development.

epoxide intermediates that are formed after inhalation exposure to this gas. These intermediates can produce genetic alterations that are causal to the precursor mutagenic events in BD-induced carcinogenicity. Initial disposition models of BD (16–18) were based on rat and mouse physiological parameters and partition coefficients that control the absorption, internal distribution, and elimination of this gas in rats and mice, and in vitro biochemical parameters for the metabolic activation of BD to mutagenic epoxides and the detoxification (hydrolysis and glutathione conjugation) of those intermediates. Although these models provided good fits to timecourse data on the uptake of BD and EB from closed chambers as well as data on blood concentrations of BD in rats and mice, they overpredicted subsequently measured blood concentrations of EB (19). Evidently, the in vitro kinetic data did not accurately represent processes regulating the in vivo production of EB from BD and release of this epoxide into the blood. After several iterations, models of BD disposition are in good agreement with data on the metabolic elimination of BD in rats and mice, blood levels of BD and its epoxide intermediates, changes in tissue glutathione levels, and DNA adduct levels (22).

To be considered as reasonable representations of the in vivo behavior of an agent, physiological models of chemical disposition must fit all available pharmacokinetic data, and model-based predictions must be verified with new experimental data that had not been used in the development of the model. More complete enzymological data and time courses of BD and its three epoxide intermediates in blood and tissues of different aged animals in gas-uptake experiments and more reliable values of measurable parameters to which the model is most sensitive would permit specification of a more definitive model. By replacing rodent physiological parameters with those for humans and applying biochemical parameters that capture the range of their distribution in human populations, a verified rodent model of BD disposition could be useful for characterizing the dosimetry of BD and its mutagenic metabolites in humans.

ACKNOWLEDGMENT

This chapter is dedicated to the memory of my deceased colleague, Dr. Michael Kohn, who developed most of the kinetic models presented in this manuscript.

REFERENCES

1. Perera FP. Molecular epidemiology: on the path to prevention? J Natl Cancer Inst 2000; 92:602–612.
2. Richard AM. Structure-based methods for predicting mutagenicity and carcinogenicity: are we there yet? Mutat Res 1998; 400:493–507.

3. Tennant RW, Margolin BH, Shelby MD, et al. Prediction of chemical carcinogenicity in rodents from in vitro genetic toxicity assays. Science 1987; 236: 933–941.
4. Ames BN, Lee FD, Durston WE. An improved bacterial test system for the detection and classification of mutagens and carcinogens. Proc Natl Acad Sci 1973; 70:782–786.
5. U.S. EPA. An examination of EPA risk assessment principles and practices. EPA/100/B-04/001. U.S. Environmental Protection Agency, Office of the Science Advisor, Washington, DC, 2004.
6. Kohn MC. Achieving credibility in risk assessment model. Toxicol Lett 1995; 79:107–114.
7. Malvoisin E, Roberfroid M. Hepatic microsomal metabolism of 1,3-butadiene. Xenobiotica 1982; 12:137–144.
8. de Meester C, Poncelet F, Roberfroid M, et al. The mutagenicity of butadiene towards *Salmonella typhimurium*. Toxicol Lett 1980; 6:125–130.
9. Wade MJ, Moyer JW, Hine CH. Mutagenic action of a series of epoxides. Mutat Res 1979; 66:367–371.
10. Van Duuren BL, Nelson L, Orris L, et al. Carcinogenicity of epoxides, lactones, and peroxy compounds. J Natl Cancer Inst 1963; 31:41–55.
11. Van Duuren BL, Langseth L, Orris L, et al. Carcinogenicity of epoxides, lactones, and peroxy compounds. IV. Tumor response in epithelial and connective tissue in mice and rats. J Natl Cancer Inst 1966; 37:825–838.
12. Csanady GA, Guengerich FP, Bond JA. Comparison of the biotransformation of 1,3-butadiene and its metabolite, butadiene monoepoxide, by hepatic and pulmonary tissues from humans, rats and mice. Carcinogenesis 1992; 13:1143–1153.
13. Kreiling R, Laib RJ, Filser JG, et al. Species differences in butadiene metabolism between mice and rats evaluated by inhalation pharmacokinetics. Arch Toxicol 1986; 58:235–238.
14. Bolt HM, Filser JG, Störmer F. Inhalation pharmacokinetics based on gas uptake studies. V. Comparative pharmacokinetics of ethylene and 1,3-butadiene in rats. Arch Toxicol 1984; 55:213–218.
15. Kreiling R, Laib RJ, Filser JG, et al. Inhalation pharmacokinetics of 1,2-epoxybutene-3 reveal species differences between rats and mice sensitive to butadiene-induced carcinogenesis. Arch Toxicol 1987; 61:7–11.
16. Kohn MC, Melnick RL. Species differences in the production and clearance of butadiene metabolites: a mechanistic model indicates predominantly physiological, not biochemical, control. Carcinogenesis 1993; 14:619–628.
17. Johanson G, Filser JG. A physiologically based pharmacokinetic model for butadiene and its metabolite butadiene monoxide in rat and mouse and its significance for risk extrapolation. Arch Toxicol 1993; 67:151–163.
18. Medinsky MA, Leavens TL, Csanady GA, et al. In vivo metabolism of butadiene by mice and rats: a comparison of physiological model predictions and experimental data. Carcinogenesis 1994; 15:1329–1340.
19. Himmelstein MW, Turner MJ, Asgharian B, et al. Comparison of blood concentrations of 1,3-butadiene and butadiene epoxides in mice and rats exposed to 1,3-butadiene by inhalation. Carcinogenesis 1994; 15:1479–1486.
20. Kohn MC, Melnick RL. The privileged access model of 1,3-butadiene disposition. Environ Health Perspect 2000; 108:911–917.

21. Richardson KA, Peters MM, Wong BA, et al. Quantitative and qualitative differences in the metabolism of ^{14}C-1,3-butadiene in rats and mice: relevance to cancer susceptibility. Toxicol Sci 1999; 49:186–201.
22. Kohn MC, Melnick RL. Physiological modeling of butadiene disposition in mice and rats. Chem Biol Interact 2001; 135:285–301.
23. Koc H, Tretyakova NY, Walker VE, et al. Molecular dosimetry of N-7 guanine adduct formation in mice and rats exposed to 1,3-butadiene. Chem Res Toxicol 1999; 12:566–574.
24. Huff JE, Salmon AG, Hooper NK, et al. Long-term carcinogenesis studies on 2,3,7,8-tetrachlorodibenzo-*p*-dioxin and hexachlorodibenzo-*p*-dioxins. Cell Biol Toxicol 1991; 7:67–94.
25. Whitlock JP Jr. Induction of cytochrome P4501A1. Ann Rev Pharmacol Toxicol 1999; 39:103–125.
26. Kohn MC, Lucier GW, Clark GC, et al. A mechanistic model of effects of dioxin on gene expression in the rat liver. Toxicol Appl Pharmacol 1993; 120: 138–154.
27. Andersen ME, Mills JJ, Gargas ML, et al. Modeling receptor-mediated processes with dioxin: implications for pharmacokinetics and risk assessment. Risk Anal 1993; 13:25.
28. Kohn MC, Walker NJ, Kim AH, et al. Physiological modeling of a proposed mechanism of enzyme induction by TCDD. Toxicology 2001; 162:193–208.
29. Tritscher AM, Goldstein JA, Portier CJ, et al. Dose–response relationships for chronic exposure to 2,3,7,8-tetrachlorodibenzo-*p*-dioxin in a rat tumor promotion model: quantification and immunolocalization of CYP1A1 and CYP1A2 in the liver. Cancer Res 1992; 52:3436–3442.
30. Walker NJ, Portier CJ, Lax SF, et al. Characterization of the dose–response of CYP1B1, CYP1A1, and CYP1A2 in the liver of female Sprague–Dawley rats following chronic exposure to 2,3,7,8-tetrachlorodibenzo-*p*-dioxin. Toxicol Appl Pharmacol 1999; 154:279–286.
31. Sewall CH, Lucier GW, Tritscher AM, et al. TCDD-mediated changes in hepatic epidermal growth factor receptor may be a critical event in the hepatocarcinogenic action of TCDD. Carcinogenesis 1993; 14:1885–1893.
32. Sloop TC, Lucier GW. Dose-dependent elevation of Ah receptor binding by TCDD in rat liver. Toxicol Appl Pharmacol 1987; 88:329–337.
33. Vanden Heuvel JP, Clark GC, Kohn MC, et al. Dioxin-responsive genes: examination of dose–response relationships using quantitative reverse transcriptase–polymerase chain reaction. Cancer Res 1994; 54:62–68.
34. Zhang L, Savas U, Alexander DL, et al. Characterization of the mouse Cyp1B1 gene: identification of an enhancer region that directs aryl hydrocarbon receptor-mediated constitutive and induced expression. J Biol Chem 1998; 273:5174–5183.
35. Denison MS, Yao EF. Characterization of the interaction of transformed rat hepatic cytosolic Ah receptor with a dioxin responsive transcriptional enhancer. Arch Biochem Biophys 1991; 284:158–166.
36. Stohs SJ. Oxidative stress induced by 2,3,7,8-tetrachlorodibenzo-*p*-dioxin (TCDD). Free Radic Biol Med 1990; 9:79–90.
37. Kohn MC, Portier CJ. A model of effects of TCDD on expression of rat liver proteins. Prog Clin Biol Res 1994; 387:211–222.

38. Andersen ME, Birnbaum LS, Barton HA, et al. Regional hepatic CYP1A1 and CYP1A2 induction with 2,3,7,8-tetrachlorodibenzo-*p*-dioxin evaluated with a multicompartment geometric model of hepatic zonation. Toxicol Appl Pharmacol 1997; 144:145–155.
39. Santostefano MJ, Wang X, Richardson VM, et al. A pharmacodynamic analysis of TCDD-induced cytochrome P450 gene expression in multiple tissues: dose- and time-dependent effects. Toxicol Appl Pharmacol 1998; 151:294–310.
40. Emond C, Birnbaum LS, DeVito MJ. Physiologically based pharmacokinetic model for developmental exposures to TCDD in the rat. Toxicol Sci 2004; 80: 115–133.
41. Bock KW. Roles of UDP-glucuronosyltransferases in chemical carcinogenesis. Crit Rev Biochem Mol Biol 1991; 26:129–150.
42. Sewall CH, Flagler N, Vanden Heuvel JP, et al. Alterations in thyroid function in female Sprague–Dawley rats following chronic treatment with 2,3,7,8-tetra-chlorodibenzo-*p*-dioxin. Toxicol Appl Pharmacol 1995; 132:237–244.
43. Kohn MC, Sewall CH, Lucier GW, et al. A mechanistic model of effects of dioxin on thyroid hormones in the rat. Toxicol Appl Pharmacol 1996; 136:29–48.
44. Roy AK, Neuhaus OW. Androgenic control of a sex-dependent protein in the rat. Nature 1967; 214:618–620.
45. Neuhaus OW, Flory W, Biswas N, et al. Urinary excretion of α2μ-globulin and albumin by adult male rats following treatment with nephrotoxic agents. Nephron 1981; 28:133–140.
46. Lehman-McKeeman LD, Rivera-Torres MI, Caudill D. Lysosomal degradation of alpha 2μ-globulin and alpha 2μ-globulin-xenobiotic conjugates. Toxicol Appl Pharmacol 1990; 103:539–548.
47. Swenberg JA, Short B, Borghoff S, et al. The comparative pathobiology of alpha$_{2\mu}$-globulin nephropathy. Toxicol Appl Pharmacol 1989; 97:35–46.
48. Short BG, Burnett VL, Cox MG, et al. Site-specific renal cytotoxicity and cell proliferation in male rats exposed to petroleum hydrocarbons. Lab Invest 1987; 57:564–577.
49. Short BG, Steinhagen WH, Swenberg JA. Promoting effects of unleaded gaso-line and 2,2,4-trimethylpentane on the development of atypical cell foci and renal tubular cell tumors in rats exposed to *N*-ethyl-*N*-hydroxyethylnitrosamine. Cancer Res 1989; 49:6369–6378.
50. Charbonneau M, Lock EA, Strasser J, et al. 2,2,4-Trimethylpentane-induced nephrotoxicity. I. Metabolic disposition of TMP in male and female Fischer 344 rats. Toxicol Appl Pharmacol 1987; 91:171–181.
51. Melnick RL. An alternative hypothesis on the role of chemically induced protein droplet (α2μ-globulin) nephropathy in renal carcinogenesis. Regul Toxicol Pharmacol 1992; 16:111–125.
52. Borghoff SJ, Miller AB, Bowen JP, et al. Characteristics of chemical binding to alpha 2μ-globulin in vitro—evaluating structure–activity relationships. Toxicol Appl Pharmacol 1991; 107:228–238.
53. Charbonneau M, Lock EA, Strasser J, et al. Nephrotoxicity of 2,2,4-trimethyl-pentane (TMP) metabolites in male Fischer 344 rats. Toxicologist 1987; 7:89.
54. Dietrich DR, Swenberg JA. Lindane induces nephropathy and renal accumu-lation of alpha 2μ-globulin in male but not in female Fischer 344 rats or male NBR rats. Toxicol Lett 1990; 53:179–181.

55. NCI (National Cancer Institute). Bioassay of lindane for possible carcinogenicity (CAS No. 58–89-9). NCI-Technical Report No. 14. National Institute of Health, Bethesda, MD, 1977.
56. Dominick MA, Robertson DG, Bleavins MR, et al. Alpha 2μ-globulin nephropathy without nephrocarcinogenesis in male Wistar rats administered 1-(aminomethyl)cyclohexaneacetic acid. Toxicol Appl Pharmacol 1991; 111:375–387.
57. Borghoff SJ, Youtsey NL, Swenberg JA. A comparison of European High Test gasoline and PS-6 unleaded gasoline in their abilities to induce alpha 2μ-globulin nephropathy and renal cell proliferation. Toxicol Lett 1992; 63:21–33.
58. Prescott-Mathews JS, Wolf DC, Wong BA, et al. Methyl *tert*-butyl ether causes alpha2μ-globulin nephropathy and enhanced renal cell proliferation in male Fischer-344 rats. Toxicol Appl Pharmacol 1997; 143:301–314.
59. Olson MJ, Garg BD, Murty CV, et al. Accumulation of alpha 2μ-globulin in the renal proximal tubules of male rats exposed to unleaded gasoline. Toxicol Appl Pharmacol 1987; 90:43–51.
60. Murty CV, Olson MJ, Garg BD, et al. Hydrocarbon-induced hyaline droplet nephropathy in male rats during senescence. Toxicol Appl Pharmacol 1988; 96:380–392.
61. Borghoff S, Gargas ML, Andersen ME, et al. Development of a mechanism-based dosimetry model for 2,4,4-trimethyl-2-pentanol-induced alpha 2μ-globulin nephropathy in male Fischer 344 rats. Fundam Appl Toxicol 1995; 25:124–137.
62. Kohn MC, Melnick RL. A physiological model for ligand-induced accumulation of α2μ-globulin in male rat kidney: roles of protein synthesis and lysosomal degradation in the renal dosimetry of 2,4,4-trimethyl-2-pentanol. Toxicology 1999; 136:89–105.
63. Prescott-Mathews JS, Poet TS, Borghoff SJ. Evaluation of the in vivo interaction of methyl *tert*-butyl ether with alpha2μ-globulin in male F-344 rat. Toxicol Appl Pharmacol 1999; 157:60–67.
64. Melnick RL, Kohn MC. Possible mechanisms of induction of renal tubular cell neoplasms in rats associated with α2μ-globulin: role of protein accumulation versus ligand delivery to the kidney. In: Capen CC, Dybing E, Rice JM, Wilbourn JD, eds. Species Differences in Thyroid, Kidney and Urinary Bladder Carcinogenesis. Vol. 147. Lyon: IARC Scientific Publication, 1999:119–137.
65. Reddy JK, Azarnoff DL, Hignite CE. Hypolipidaemic hepatic peroxisome proliferators form a novel class of chemical carcinogens. Nature 1980; 283: 397–398.
66. Doull J, Cattley R, Elcombe C, et al. A cancer risk assessment of di(2-ethylhexyl)phthalate: application of the new U.S. EPA Risk Assessment Guidelines. Regul Toxicol Pharmacol 1999; 29:327–357.
67. Issemann I, Green S. Activation of a member of the steroid hormone receptor superfamily by peroxisome proliferators. Nature 1990; 347:645–650.
68. Sher T, Yi HF, McBride OW, et al. cDNA cloning, chromosomal mapping, and functional characterization of the human peroxisome proliferator activated receptor. Biochemistry 1993; 32:5598–5604.
69. Mukherjee R, Jow L, Noonan DJ, et al. Human and rat peroxisome proliferator activated receptors (PPARs) demonstrate similar tissue distribution but different responsiveness to PPAR activators. J Steroid Biochem Mol Biol 1994; 51: 157–166.

70. Schoonjans K, Staels B, Auwerx J. The peroxisome proliferator activated receptors (PPARS) and their effects on lipid metabolism and adipocyte differentiation. Biochim Biophys Acta 1996; 1302:93–109.
71. Kliewer SA, Umesono K, Noonan DJ, et al. Convergence of 9-*cis* retinoic acid and peroxisome proliferator signalling pathways through heterodimer formation of their receptors. Nature 1992; 358:771–774.
72. Qi C, Zhu Y, Reddy JK. Peroxisome proliferator-activated receptors, coactivators, and downstream targets. Cell Biochem Biophys 2000; 32:187–204.
73. Tugwood JD, Aldridge TC, Lambe KG, et al. Peroxisome proliferator-activated receptors: structures and function. Ann NY Acad Sci 1996; 804:252–265.
74. Klaunig JE, Babich MA, Baetcke KP, et al. PPAR agonist-induced rodent tumors: modes of action and human relevance. Crit Rev Toxicol 2003; 33: 655–780.
75. Bentley P, Calder I, Elcombe C, et al. Hepatic peroxisome proliferation in rodents and its significance for humans. Food Chem Toxicol 1993; 31:857–907.
76. Melnick RL. Is peroxisome proliferation an obligatory precursor step in the carcinogenicity of di(2-ethylhexyl)phthalate (DEHP)? Environ Health Perspect 2001; 109:437–442.
77. Parzefall W, Berger W, Kainzbauer E, et al. Peroxisome proliferators do not increase DNA synthesis in purified rat hepatocytes. Carcinogenesis 2001; 22: 519–523.
78. Hasmall S, James N, Hedley K, et al. Mouse hepatocyte response to peroxisome proliferators: dependency on hepatic nonparenchymal cells and peroxisome proliferator activated receptor alpha (PPARalpha). Arch Toxicol 2001; 75:357–361.
79. Peters JM, Rusyn I, Rose ML, et al. Peroxisome proliferator-activated receptor alpha is restricted to hepatic parenchymal cells, not Kupffer cells: implications for the mechanism of action of peroxisome proliferators in hepatocarcinogenesis. Carcinogenesis 2000; 21:823–826.
80. Rose ML, Rivera CA, Bradford BU, et al. Kupffer cell oxidant production is central to the mechanism of peroxisome proliferators. Carcinogenesis 1999; 20:27–33.
81. Rose ML, Germolec D, Arteel GE, et al. Dietary glycine prevents increases in hepatocyte proliferation caused by the peroxisome proliferator WY-14,643. Chem Res Toxicol 1997; 10:1198–1204.
82. Rusyn I, Yamashina S, Segal BH, et al. Oxidants from nicotinamide adenine dinucleotide phosphate oxidase are involved in triggering cell proliferation in the liver due to peroxisome proliferators. Cancer Res 2000; 60:4798–4803.
83. Rusyn I, Kadiiska MB, Dikalova A, et al. Phthalates rapidly increase production of reactive oxygen species in vivo: role of Kupffer cells. Mol Pharmacol 2001; 59:744–750.

4

Empirically Observed Distributions of Pharmacokinetic and Pharmacodynamic Variability in Humans—Implications for the Derivation of Single-Point Component Uncertainty Factors Providing Equivalent Protection as Existing Reference Doses

Dale Hattis

George Perkins Marsh Institute, Clark University, Worcester, Massachusetts, U.S.A.

Meghan Keaney Lynch

Boston University School of Public Health, Boston, Massachusetts, U.S.A.

INTRODUCTION

The International Programme on Chemical Safety (IPCS) framework for data-derived uncertainty factors (1) that is described and used extensively in other chapters of this volume is a welcome effort to open up the traditional system of risk assessment for noncancer effects. It provides guidance for modification of the traditional adjustment factors where more chemical-specific information is present than usual on either human interindividual variability or interspecies dose projections for the parent chemical or metabolite responsible toxicity.

Welcome as this goal is, the derivation of the IPCS guidance does not draw on as extensive a database of pharmacokinetic (PK) and pharmaco-dynamic (PD) variability observations as is now available [2,3; also see our Web site (4)]. Nor does it derive its specific guidance for modifying the existing uncertainty factors based on a quantitative analysis of the "baseline" uncertainties in human PK and PD variabilities for individual chemicals. (By the "baseline" uncertainty, we mean the uncertainty one should have in the absence of chemical-specific information, based on prior observations with putatively analogous chemicals.)

This chapter provides a start to such an analysis. Ultimately, we derive preliminary conclusions for allocation of total interindividual variability (and uncertainty in this variability) among PK and PD components. This analysis is done with and without new information that hypothetically removes all uncertainty in the amount of human PK or PD variability for responses to particular toxicants (this is the value of "perfect information").

Briefly, we found in our previous analyses of human interindividual variability data (2,3,5) that PD variability is generally larger than PK variability. The analysis in this chapter extends this with new estimates of the uncertainty in PK versus PD variability. From this we find that the information benefit from perfect information on PK variability is less than what would be expected from simple application of the formulae in the most recent IPCS guidance. A final section provides numerical guidance on changes to the default factor for human interindividual variability that would lead to greater consistency in the health protection afforded by reference doses (RfDs) based on quantal versus continuous toxicity endpoints, and larger versus smaller overall traditional uncertainty factors [including factors introduced for subchronic versus chronic studies; interspecies projections, use of lowest-observed-adverse-effect level (LOAELs) versus no observable adverse effects level (NOAELs), etc.]. Use of these point estimates can provide a bridge between the pure distributional analysis methods in our earlier "straw man" paper (2) and the single-point-estimate uncertainty factors now generally used for noncancer risk assessments.

Below we first briefly recapitulate the IPCS guidance (reviewed in more detail in Chapter 2). We then review the framework developed in our previous work (2) to replace each of the traditional "uncertainty" factors with empirical distributions, and use those distributions in Monte Carlo simulations of the likely risk performance of 18 randomly selected RfDs from the IRIS database[a]). (All these calculations assume that the human interindividual

---

[a] A more extended version of this chapter, containing tables describing the uncertainty factors making up the 18 randomly selected RfDs, the empirical observations of variability in AUC and C_{max} in our expanded database, unweighted observations of variability in various types of parameters, and detailed tables of results is contained in an EPA report available from the first author via email at dhattis@aol.com.

variability in thresholds for effect are lognormally distributed, i.e., although the diverse population of potentially exposed individuals differs widely in their individual vulnerability/sensitivity because of a wide variety of physiological factors and ongoing pathologies, there is not, in general, any finite "population threshold" that can be assumed to produce a true zero incidence of harm for the whole group). The results of these simulations are then compared with our hypothesized "straw man" risk management criterion:

> that there should be 95% confidence that the projected risks for the general population from exposure at the RfD should be less than an incidence of 1/100,000 for mild adverse effects.

The aim is to make a preliminary assessment of what values for the traditional single-point uncertainty factors, when included in a revised framework for inferring RfDs from basic toxicological data, would provide approximately equivalent risk performance for 18 sampled RfD chemicals. This analysis is conducted with and without additional measurements that would hypothetically provide perfect information about the extent of either PK or PD variability.

To do this (Fig. 1), we first summarize the observed distributions of interindividual variability measurements of area under the curve (AUC) (36 drugs) and C_{max} (29 drugs) parameters.[b] This, combined with analogous data from the existing PD database, leads to basic inferences for the split of the assumed lognormal variance in overall human susceptibility into PK and PD components. Then we analyze the comparative uncertainty in these two components. In each case, we correct as best we can for the spread in chemical-specific estimates for both PK and PD parameters that is attributable to the limited sample sizes in the studies in our database. After this subtraction, the remaining differences among chemicals in estimated PK and PD variability represents our estimates of the uncertainty in the real PK and PD variabilities for a randomly drawn RfD chemical or drug. It is these uncertainties that could be reduced by experimental measurements of PK or PD variability for the random RfD chemical. Perfect information is modeled by reducing to zero the uncertainty in one or both of these variability estimates for the sample RfD chemicals within the straw man risk simulation system.

---

[b] The summary measure for each chemical is a log(GSD)—the standard deviation of the base 10 logarithms of the measured AUC or C_{max} parameter for each person studied. Because the log(GSD) values for individual chemicals appear to be themselves approximately lognormally distributed, we express the spread of variability measurements for different chemicals as a log[log(GSD)-the standard deviation of the log(GSD) measures for the various chemicals/drugs.

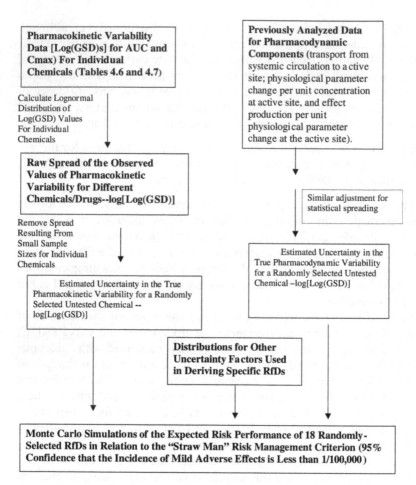

Figure 1   Flow chart for the analysis in this paper.

IPCS ASSUMPTIONS OF SINGLE-POINT FACTORS REPRESENTING OVERALL INTERINDIVIDUAL VARIABILITY AND PK/PD COMPONENTS

In the IPCS framework (1), chemical-specific adjustment factors (CSAFs) are intended to be used in place of the traditional 10-fold factors when there is a judgment that enough suitable quantitative toxicokinetic (PK) and/or toxicodynamic (PD) data are available. An important innovation in the IPCS framework is a split of each of the traditional interspecies and inter-individual variability uncertainty factors between PK and PD components. These splits allow more readily available PK data to be practically used to replace a portion of the total uncertainty factor(s) even when more difficult to obtain PD information is not available. For human interindividual variability an even split of 3.2 (approximately the square root of 10) for both the kinetic and dynamic factors is proposed. For practical applications using real PK data [for the AUC of the concentration × time plot for the active chemical species for a chronic toxic agent, and the maximum systemic concentration (C_{max}) for an acute toxic agent] the framework proposes the use of multiplicative distance between some distant percentile, i.e., the 95th percentile and the 50th percentile of the human population as the "data derived" replacement for the 3.2-fold kinetic factor.[c]

DESCRIPTION AND UPDATE OF PREVIOUS WORK TO CONSTRUCT A STRAW MAN PROPOSAL FOR A QUANTITATIVE DISTRIBUTIONAL GOAL FOR DEFINING REFERENCE DOSE PROTECTION OBJECTIVES

In previous work (2), we developed a straw man probabilistic system for deriving RfDs. The foundation of this system is that quantitative insights into particular sources of uncertainty for untested chemicals can be developed by assembling empirical data for putatively analogous chemicals for which toxicity information is available. The untested chemical is then regarded as a random draw from the set of analogous chemicals. This essentially reduces an uncertainty to a directly observable variability (among the analogous chemicals) for the uncertainty component in question.

Where possible, structural, mechanistic, or other relevant categorization information may be used for construction of the analogous sets for a particular toxicant under consideration for an RfD. For example, in the current version of the system, this is most developed for the uncertainty factor for interspecies projection. In this case, distributions are adjusted for the composition and number of species with data that were used in choosing the "most sensitive" species for assessment of the LOAEL or NOAEL or benchmark dose. Where the "most sensitive" species has been selected based on data from four or five species, this clearly represents a more "conservative" projection, and the central estimate and distribution of likely human relative potencies are adjusted accordingly based on analogous cases where the most sensitive species has been selected from a similarly sized database for the chemicals in the analogous reference set. Other sources of information for forming analogous sets could include types of elimination mechanisms for PK data (e.g., specific CYP enzymes), or anatomical site and putative mode of action for PD data.

---

[c] The multiplicative distance between the 95th and 50th percentiles is erroneously referred to in the guidance as 2 geometric standard deviations of a lognormal distribution, whereas it actually represents only about 1.64 standard deviations. A point that is 2 geometric standard deviations above the median corresponds to approximately the 97.5th percentile of a lognormal distribution.

Briefly, after making preliminary estimates of the animal ED_{50} for minimally adverse effects from LOAELs (where available),[d] the straw man system applies the following distributions for various uncertainty factors:

- Distributions for the subchronic/chronic data projection (where needed) derived from an analysis by Baird et al. (6) of subchronic to chronic NOAELs from Weil and McCollister (7) and Nessel et al. (8). These data included 61 sets of comparative experiments involving 51 different toxicants.
- Distributions for the effects of incomplete data (missing either a reproductive study or a chronic toxicity study) in deriving an NOAEL, based on distributions inferred from a set of 35 pesticides with complete databases analyzed by Evans and Baird (9).
- Lognormal distributions for interspecies projection based on observations of human maximum tolerated doses for 61 anticancer agents in relation to animal predictor data. [Price PS, Swartout JC, Kennan RE. Characterizing inter-species uncertainty using data from studies of anti-neoplastic agents in animals and humans. Human Ecol Risk Assess. In preparation].
- Distributions of the extent of human interindividual variability based on our collection of both PK and PD interindividual variability data.

These latter data come primarily from studies of pharmaceuticals in clinical settings, although there are also some epidemiological observations of PD parameters based on biomarkers of internal exposure for methylmercury and cadmium. Overall, we have interindividual variability observations for 471 "data groups" [where a "data group" is a measurement of a particular parameter for a particular chemical, sometimes including a combination of the variability observed in several studies as described previously (5)]. Eleven of these data groups were for uptake parameters, 363 for PK parameters, and 97 parameters that include some PD variability information. The current database reflects a modest expansion from the database

---

[d] There is an important distinction in this initial derivation of animal ED_{50} between quantal and continuous endpoints. For continuous endpoints, such as changes in fetal growth, LOAELs are essentially determined at the point where the group average value differs from control values by an amount (often as much as 5% for fetal growth) that is deemed statistically and toxicologically significant. Therefore, for these continuous parameters where the average member of a studied group shows a significant effect, we do not further adjust the LOAEL to estimate an animal ED_{50}. On the other hand, for quantal parameters (e.g., the incidence of some manifestly adverse yes-no effect), such as cardiomyopathy or kidney lesions, we do adjust the LOAEL dose upward to account for the difference in dose between that causing a barely statistically detectable incidence (about 5–20%) and an ED_{50}. See Ref. 2 for details of this adjustment, and the larger and more uncertain adjustment needed for the few cases where the data did not contain a LOAEL and we needed to use a NOAEL as the starting point.

Table 1 A Scale for Understanding Lognormal Variability—Fold Differences Between Particular Percentiles of Lognormal Distributions

log(GSD)	Probit slope [1/log(GSD)]	Geometric standard deviation	95/50th percentile ratio (1.645 standard deviations) (fold)	99/50th percentile ratio (2.326 standard deviations) (fold)
0.1	10	1.26	1.46	1.71
0.2	5	1.58	2.13	2.92
0.3	3.33	2.0	3.11	4.99
0.4	2.5	2.5	4.55	8.52
0.5	2	3.2	6.64	14.6
0.6	1.67	4.0	9.70	24.9
0.7	1.43	5.0	14.1	42.5
0.8	1.25	6.3	20.7	72.6
0.9	1.11	7.9	30.2	124
1	1.0	10.0	44.1	212

Abbreviation: GSD, geometric standard deviation.

used in the latest previous report (2), where there were a total of 447 data groups, including 93 with PD variability information.

The primary unit we use to summarize interindividual variability in specific parameters is the log(GSD)—the standard deviation of the common (base 10) logarithms of the parameter values, as most of the data are well described by lognormal distributions (3). Table 1 provides a handy key to translation of these values into other terms that may be more familiar to the reader, and which relate to the benchmark 95% percentile/median ratios highlighted in the IPCS guidance.

We focus our analysis of PK variability on AUC and C_{max} observations because these parameters are reasonable summary estimators of the full pathway from exposure to systemic availability of relevant chemical species for chronic and acute systemic toxicity, respectively.[e] Using the

---

[e] The AUC is the integrated area under the concentration × time curve of the toxicant or drug in the blood, in all cases for our data set normalized to the mg/kg of oral dose. This is the most frequently used internal dose metric for effects that occur as the result of the gradual buildup of relatively slowly reversible effects from chronic or repeated exposures. Similarly the C_{max} is the maximum blood concentration achieved after a single acute exposure—most often used for effects that occur as the result of a pharmacodynamic process that quickly proceeds to clinically detectable impairment or is quickly reversed following individual exposures. These parameters are also the ones specifically recommended in the IPCS guidance for assessment of data-derived uncertainty factors for the pharmacokinetic component of interindividual variability in susceptibility. We have not, however, evaluated whether all of our data sets meet the IPCS guidance that the standard error of the mean should be less than 20% of the mean. To apply this arbitrary cutoff, we would systematically exclude relatively small data sets that happen to indicate particularly large amounts of individual variability, thus biasing the overall estimates. We deal with the issue of statistical sampling error in our estimates of human variability more explicitly and quantitatively, as described below.

log(GSD) summarization measure, the geometric mean of the log(GSD) for AUC interindividual variability observations for the 37 drugs in our sample is 0.167 [Table 1, the expected ratio of AUCs for 95th percentile people to median people for this drug is about $10^{(0.167 \times 1.645)} = 1.88$]. Some drugs have more or less variability than this geometric mean, of course. The spread between the 10th percentile drug and the 90th percentile drug in our sample of 37 drug AUCs corresponds to log(GSD) of 0.084 to 0.334, respectively— corresponding to ratios of 95th percentile to median people of 1.37 to 3.55. In general, as has been noted in prior work (3,5), the log(GSD) observations themselves appear to be approximately lognormally distributed. The C_{max} interindividual observations in the 29 drugs in our sample with this kind of PK data are similar in both geometric mean and spread.[f] A minority of the data sets includes children under 12, and many have quite limited sample sizes (the minimum size for inclusion in the database is five).

Among the more prevalent types of observations of interindividual variability in parameters that include PD processes, there are 21 observations of differences among people in external doses needed to produce specific local contact-site responses [geom. mean log(GSD)= 0.655; 10–90th percentile spread of 0.37–1.16], 23 observations of variability in the internal systemic doses associated with a defined physiological parameter changes [geom. mean log(GSD) of 0.235; 10–90th percentile spread of 0.098–0.566], and 18 observations of the variability in internal dose associated with a specific clinical response [geom. mean log(GSD) = 0.25, 10–90th percentile spread of 0.097–0.644].

Given the raw observations, the straw man system utilized a regression-like procedure to construct a weighted allocation of interindividual variability among various steps in the process from toxicant exposure through to the production of biological responses of various degrees of severity, and for broad categories of organ systems. In general, the data indicated that local tissue responses, responses rated relatively mild on a severity scale, and responses of the immune system showed greater PD variability than other responses. Table 2 shows the allocation of variability among the PD steps for non-immune systemic responses using the most recent updated data. For the current analysis, estimates of PK variability for comparison will be drawn only from the observations of AUC and C_{max} parameters.

A final step in the straw man procedure is the combination of information about the uncertainty in all of the factors used for particular RfDs (including uncertainty in the interindividual variability for a randomly selected chemical, discussed in more detail below) in Monte Carlo simulations of the uncertainty in risk as a function of dose, assuming that simple unimodal

---

[f] A full report, containing tables of the raw log(GSD) observations, as well as the details of the population size and types of people studied, will be made available at the web site listed in Ref. 4.

Table 2 Statistically Weighted Optimum Allocation of Human Interindividual Variability Among Steps in the Causal Process for Different Subsets of the Database

	Subset of the database			
	All 447 data groups, incl. direct contact PD observations	All 365 PK + 54 systemic PD data sets—excluding direct contact PD[a]	All 365 PK + 27 systemic neurological PD variability—excluding direct contact PD	All 365 PK + 23 systemic cardiovascular PD variability—excluding direct contact PD
Active site availability/general systemic availability	0.093	0.099	0.092	0.105
Non-immune physiological parameter change/active site availability	0.205	0.199	0.229	0.184
Immune physiological parameter change/active site availability	0.551	0.548	No data	No data
Reversible non-immune mild functional reserve capacity—change in baseline physiological parameter needed to pass a criterion of abnormal function	0.338	0.238	0.220	No data
Non-immune moderate reversible or irreversible functional reserve capacity	0.226	0.230	0.102	0.253
Non-immune severe and irreversible functional reserve capacity	0.000	0.000	0.000	No data
Reversible immune functional reserve capacity	0.636	No data	No data	No data

Entries are optimized central estimates of log(GSD) for individual steps; the columns represent different subsets of the overall interindividual variability data groups.

[a]The data set analyzed also included the 11 media intake parameters in each case, although the casual steps covered by these parameters are not relevant to the determination of the uncertainty factors included in the determination of RfDs.

Abbreviations: PD, pharmacodynamic; PK, pharmacokinetic.

unbounded lognormal distributions describe the overall population variability in thresholds for responses for mild adverse effects (this is essentially a log-probit projection of risk). The Monte Carlo simulation runs are done in Excel spreadsheets, and results summarized as the average of three parallel runs of 5000 trials each (each trial consisting of a selection of random values from the distributions defined for each uncertainty factor contributing to an RfD, and calculation of expected risks as a function of dose). These results are then compared with our hypothesized straw man risk criterion that there should be 95% confidence that the projected risks for the general population from exposure at the RfD should be less than an incidence of 1/100,000 for mild adverse effects. The same system can be used to produce estimates predicated on a number of other potential risk management criteria, including the arithmetic mean "expected value" of the incidence of mild adverse responses above and below current RfD levels.

It should be understood that these risk projections carry great uncertainty. It is likely that in many cases the population has subgroups (both genetic, developmental, and from other origins) that may contribute variability that is not fully captured in the observations of small relatively healthy subgroups that form the basis of our data. This could cause the actual variability to be larger and of somewhat different statistical form than our simple unimodal lognormal estimates, leading to some tendency to underestimate population risks. On the other hand, there has been no effort in the current system to subtract variance due to experimental measurement errors from the observed spread of the basic PK and PD variability observations. Other things being equal, this would tend to cause our interindividual variability estimates to be larger than the real variation among people for these parameters for the generally restricted populations that were studied. This leads to a potential for overestimation of real variability and therefore an overestimation of expected incidences of adverse effects. It is not clear which of these partially offsetting sources of uncertainty is greater, and thus it is not clear whether the net result is likely to be an over- or an underprediction of real interindividual variability and low dose risks.

Finally, some have observed that the sets of drugs that gave rise to our estimates of variability may not be perfectly analogous to sets of environmental chemicals that are the subjects of risk assessment efforts by environmental and food regulatory authorities. The drugs may tend to be more water-soluble than the environmental chemicals, and drug testing, particularly for the responses that give rise to our PD variability estimates, undoubtedly was carried out at larger doses in relation to effect levels than would generally be the case for environmental exposures. These considerations produce some additional uncertainty that is not quantified in our system. However, we would suggest that the set of interindividual variability observations we have compiled and analyzed provides a much closer analogy to the desired information for untested environmental chemicals than

Table 3 Statistically Weighted Regression Equations for the Effect of the Presence of Children Under 12 in Groups Studied for Interindividual Variability in AUC and C_{max} PK Parameters

	log[log(GSD)]		log(GSD)	
	AUC[a]	C_{max}[b]	AUC	C_{max}
Without children under 12	−0.805	−0.796	0.157	0.160
With children under 12	−0.694	−0.704	0.202	0.198

[a]AUC log[log(GSD)] = −0.805 + 0.111 (1 if group contains children under 12; 0 if no children) (± 0.079 SE of the 0.111 estimate).
[b]C_{max} log[log(GSD)] = −0.796 + 0.0925 (1 if group contains children under 12; 0 if no children) (± 0.123 SE of the 0.0925 estimate).

the relatively arbitrary bases for the traditional uncertainty factor for human variability [dating back to the original proposal of a 100-fold safety factor for both interspecies projection and human variability by Lehman and Fitzhugh (11)], and the PK and PD components that have more recently been postulated (3,5).

EXPANSION OF THE ESTIMATED PHARMACOKINETICS VARIABILITY FOR SOME AGE-RELATED DIFFERENCES

In the AUC and C_{max} data used here, as in observations analyzed earlier for elimination half-lives,[g] there is a tendency for data groups that include children under 12 to show somewhat more variability than data groups that do not include children. However, with the limited number of chemicals where these parameters have been measured in groups including children, this suggested increment fails to exert a large or consistent enough effect to reach conventional criteria for statistical significance (Table 3). In other work (12), we have seen consistent statistically significant reductions in clearance and increases in elimination half-lives only for age groups less than 6 months. Nevertheless, in order to avoid understating the PK portion of the variability (and therefore its significance for risk, relative to PD variability) we choose to include the somewhat expanded estimates of variability derived from the groups with children. We did this by adding the central estimate of the increment in variability associated with the inclusion of children in the study groups in our central population estimates of the log(GSD) for the AUC and C_{max} parameters.

---

[g] The elimination half-life is the time it takes for half of the toxicant to be removed from the blood in the terminal phase of exponential elimination from the systemic circulation, usually determined from the final slope of a fitted straight line in a plot of the log of blood concentration versus time.

BASIC INFERENCES FOR THE SPLIT OF LOGNORMAL VARIANCE IN INTERINDIVIDUAL VARIABILITY LOG(GSD) BETWEEN PK AND PD PORTIONS OF THE PATHWAY FROM ORAL ADMINISTRATION TO BIOLOGICAL RESPONSES

The various PK and PD components of interindividual variability in susceptibility act multiplicatively to determine individual susceptibility. Mathematically, the multiplicative interaction of the components is equivalent to adding their logarithms. Then, because the variance[h] of a sum is the sum of the corresponding variances, the central estimate for the total interindividual variability in susceptibility for a median chronic toxicant can be found as the square root of the sum of the lognormal variances from the PK component from those listed in Table 3 and the three PD components whose central estimates are given in the second column of numbers in Table 2 .

log(GSD) for overall interindividual variability in susceptibility for a non-immune systemic adverse effect at the lowest adverse severity level is

$$\{[\log(GSD)]^2 \text{for PKs(AUC)} + [\log(GSD)]^2 \text{for active site}$$

availability/general systemicavailability $+ [\log(GSD)]^2$ for an

nonimmune physiologicalparameter change/active site

availability $+ [\log(GSD)]^2$ for differences in the amount of

physiological parameter change needed for a mild-severity

$$\text{non-immune systemic effect}\}^5 = (PK = 0.202^2 + PD = 0.099^2$$
$$+ 0.199^2 + 0.238^2)^{0.5} = 0.383 \qquad (1)$$

For this case, therefore, from the same calculations as are shown in Table 1, the ratio of the 95/50th percentile individual sensitivity would be $10^{(1.645 \times 0.383)} = 4.3$.

For the PD steps alone, the central estimate for the interindividual log(GSD) is

$$(0.099^2 + 0.199^2 + 0.238^2)^{0.5} = 0.326 \qquad (2)$$

The percentage of the overall variance attributable to the PK portion of the pathway is therefore

$$100 \times 0.202^2/(0.202^2 + 0.099^2 + 0.199^2 + 0.238^2) = 27.8 \qquad (3)$$

Thus, if one wished to retain an overall factor for interindividual variability factor of 10-fold, and make a split between two multiplicative

---

[h] The variance of a distribution is simply the square of the standard deviation. The lognormal variance is therefore just the square of the log(GSD).

components according to the central estimates of PK and PD variance derived here, then the factor allocated to PKs would be

$$10^{0.278} = 1.9\text{--fold} \tag{4}$$

and the complementary factor allocated to PDs would be

$$10^{0.722} = 5.3\text{--fold} \tag{5}$$

Thus, even without considering in detail the uncertainties in our estimates of the various components of overall variability in susceptibility, some adjustment away from the even split between PK and PD components of interindividual variability in the IPCS guidance is indicated—to approximately two-fold for PKs and five-fold for PDs. This analysis could require modification in cases where there are appreciable dose-dependent changes in absorption, transport, or elimination of a specific chemical in the range of concentrations covered in the underlying PK data.

UNCERTAINTIES IN CURRENT ESTIMATES OF PK AND PD VARIABILITY FOR AN UNTESTED CHEMICAL

In decision theory, the "value" of a particular piece of new information (such as data on the actual human PK variability for a particular compound) depends on the "baseline" or "prior" uncertainty in the measured parameter in the absence of the added information. Therefore, in order to understand the improvement for RfD decision making that is potentially available from improved information about PK variability for specific compounds, we need to have a quantitative baseline assessment of how much uncertainty is likely to be present for an untested chemical in the absence of any compound-specific PK variability measurements.

As described previously, our straw man simulations of risk at the RfD are based on random draws from the database of PK and PD variability components. The lognormal spread of each of these distributions of log(GSD) represents our estimate of the likelihood that the true value of the PK or PD variability will differ from the median estimate by various amounts.

Earlier we described the empirical spread of 10 to 90th percentile estimates of AUC PK interindividual variability among drugs. It would be possible to take these observations as indicating our uncertainty in the AUC or C_{max} variability in a randomly selected untested drug (or by extension, environmental chemical). However, those observations are far from ideal representations of the underlying truth about the PK variability for those specific chemicals, or the real chemical-to-chemical differences in that variability for chemicals in general. The measurements are based on limited sample sizes, the groups of people studied differ in their representativeness of the overall human population, and the procedures for estimating C_{max} and AUC variability from individual blood concentration × time profile data undoubtedly include some imperfections. All these sources of random (and possibly some systematic) error for the PK variability estimates for individual chemicals tend to spread the individual chemical values apart from one another to a greater extent than would be observed if perfect information for the same chemicals was available (even if they would not be expected to affect the central tendency). This artifactual spreading, however, does not affect the real underlying differences among chemicals in individual variability, and therefore, the real differences among chemicals in risks at doses that are very much lower than the ED_{50} for specific effects. To put it simply, if chemicals are really not very different in their actual human PK or PD interindividual variability, then specific information about the variability for a particular chemical cannot reduce very much the spread of our errors in estimates of variability for individual compounds.

We can, at least, make a preliminary analysis to allow us to offset the likely effects of small sample sizes on the spread of PK and PD variability observations for individual chemicals. The PD portion of this analysis is taken from our earlier straw man effort (2). Briefly, we first calculated the uncertainty in the estimated variability for each PD observation by a cross-validation technique:

- We reestimated the parameters in our overall variability allocation model excluding the data point whose uncertainty was being assessed.
- We then calculated the square of the deviation between the log[log(GSD)] predicted for that data point and the actual observation.
- Finally, we did a linear regression analysis of this "observed" variance versus the variance that would be predicted based on simple sampling error,[i] resulting in

$$\text{Observed variance} = 0.0259(\pm 0.0092 \text{ standard error}) + 3.05 \\ \times \text{ statistically predicted variance} \tag{6}$$

---

[i] For continuous parameters, such as directly measured doses causing a particular degree of change in a physiological parameter, or continuous pharmacokinetic parameters such as AUC or C_{max}, we derived an approximate relationship for the statistically predicted variance in the log[log(GSD)] that was simply dependent on sample size (3): Predicted sampling-error variance in log[log(GSD)] $= 1/(10.6N{-}10.33)$ where N is the number of people for whom the continuous parameter was measured. For quantal parameters (such as the percentages of people showing a particular response as a function of dose, e.g., nausea or vomiting in response to consumption of a copper solution) the statistical sampling error was derived from a determination of 5–95% confidence limits on the probit slope, as derived using a binomial likelihood-based spreadsheet dose response modeling system adapted from the one developed by Haas (13).

where 0.0259 and 3.05 are the intercept and slope, respectively, of the regression line found in the analysis described in the final point above. The intercept of this relationship (at zero "statistically predicted variance") is essentially an estimate of the variance at infinite sample size. In our modeling, we therefore represent the standard deviation of our lognormal distribution of overall PD variation as having a median log(GSD) of 0.326 (from Equation 2) and an uncertainty corresponding to a standard deviation of the log[log(GSD)] of the square root of $0.0259 = 0.161$.

Unfortunately, when we applied this same technique to the AUC and C_{max} PK data we did not obtain a central estimate for the intercept that was greater than zero. We did obtain an upper 95% confidence limit for the intercept of 0.163. This is comparable to the central estimate of our uncertainty for the PD variability parameters. We use this upper confidence limit value in our sensitivity analysis for our final results.

For our central estimate of real chemical-to-chemical variability in PK parameters, we reverted to an older technique in which we plotted the data in groups, and took the variance observed in the group with the strongest data (the 12 observations with the largest sample size) as representing our best judgment of the variation in PK variability estimates that would remain even if the variability data were based on infinite-sized samples.[j] Without giving the detailed statistical analysis here, we can simply summarize that the average variance seen for the 12 PK observations with the largest samples sizes led us to a central estimate of the uncertainty in PK variability corresponding to a standard deviation of 0.092 in the log[log(GSD)] for AUC or C_{max}.[k] The first two columns of Table 4 show the consequences of the central and upper confidence limit estimates of our uncertainty in true PK variability [i.e., the spread of PK log(GSD) that would be seen among chemicals if they were very large sample-size observations without appreciable measurement error]. The estimated uncertainty distribution for PD interindividual variability is shown for comparison in the final column of Table 4 .

Clarification of the uncertainties in real chemical-to-chemical PK variability apparent in this analysis will need to await compilation of a more extensive database of AUC and C_{max} variability measurements based on larger and perhaps more diverse samples of people. For the analysis shown below, however, we utilize both a central estimate of 0.092 and an upper confidence limit of 0.164 for the standard deviation in the log[log(GSD)] for the uncertainty in PK variability to show the sensitivity of the results to this parameter.

---

[j] A full report, containing tables of the raw log(GSD) observations, as well as the details of the population size and types of people studied will be made available at the web site listed in Ref. 4.

[k] This is reduced from the spread of the raw interindividual variability observations of 0.255.

Table 4 Uncertainty Distributions for PK and PD Interindividual Variability

| Percentile of confidence/ uncertainty distribution | Log(GSD) for PK variability[a] (median = 0.202) | | Log(GSD) for PD variability (baseline PD uncertainty log[log(GSD)] = 0.161) |
	Baseline PK uncertainty log[log(GSD)] = 0.092	95% upper confidence limit for PK uncertainty log[log(GSD)] = 0.164	
1	0.123	0.084	0.137
5	0.142	0.109	0.177
10	0.154	0.125	0.203
25	0.175	0.157	0.254
50	0.202	0.202	0.326
75	0.233	0.261	0.418
90	0.265	0.328	0.524
95	0.287	0.376	0.599
99	0.331	0.486	0.772

[a]The numbers in the second and third columns of this table are log(GSD) for PK variability under the central estimate versus upper confidence limit assumptions for the true spread of chemical-to-chemical differences in PK variability. For example, under the central estimate log[log(GSD)] = 0.092 uncertainty assumption, the 95th percentile chemical would have a log(GSD) for PK variability of 0.287. Under the high-uncertainty-estimate assumption, log[log(GSD)] = 0.092, the 95th percentile chemical would have a log(GSD) for PK variability of 0.376.
Abbreviations: PK, pharmacokinetics; PD, pharmacodynamics.

EFFECTS OF PERFECT KNOWLEDGE OF OVERALL AND COMPONENT INTERINDIVIDUAL VARIABILITY ON DOSES NEEDED TO ACHIEVE THE STRAW MAN PROTECTION GOAL

There is no easy formula for evaluating the desirability or appropriateness of specific recipes for data-derived uncertainty factors in isolation from the operation of the rest of the system for setting RfDs. Until recently, the official definition of the RfD has been "an estimate (with uncertainty spanning perhaps an order of magnitude) of a daily exposure to the human population (including sensitive subgroups) that is likely to be without an appreciable risk of deleterious effects during a lifetime."

Because of the quantitative vagueness of this definition it is simply impossible to evaluate the extent to which the current system of uncertainty factors achieves any specific interpretation of its risk goal, or how the IPCS proposals for data-derived uncertainty factors for PK variability would modify the system's performance in protecting against specific degrees of risk for the ensemble of RfD chemicals with different kinds and quality of available toxicological data.

Therefore, this section first develops a context for evaluating the performance of the interindividual variability factor by developing updated estimates of the performance of the system as a whole. As one plausible social policy benchmark for this process we use the straw man protection goal that we articulated in earlier work (2)—for lifetime continuous daily exposure at the RfD there should be 95% confidence that there will be no more than a 1/100,000 incidence of mild adverse effects. For testing purposes, we evaluate the same set of 18 randomly chosen sample RfDs as was used in our earlier work. As before, we use Monte Carlo simulations of the effects of uncertainties in all of the factors that appeared as numerical uncertainty factors in the corresponding analyses published in the EPA's IRIS database, including database and modifying factors where they were used.

In all cases, the data presented in the summary tables (Tables 5 and 6) represent averages from three sets of Monte Carlo simulation runs of 5000 trials each. On each trial, random values are chosen from all of the distributions representing the various uncertainty factors that were included in the derivation of each RfD. The distribution of results on each set of 5000 trials represents the confidence distribution for the risk results. For example, the 95th percentile of the estimated risk at the RfD is defined as the 250th highest among the 5000 values for the predicted population incidence of moderate-severity adverse effects (assuming, on each trial, a lognormal distribution of individual thresholds for response defined by the combined effects of PK and PD variability).

One important difference from the procedure used in the earlier work (2) is that for the present simulations we separated the estimated uncertainties in the human interindividual variability components representing PKs and PDs. This has the effect, e.g., of allowing a randomly chosen high value for the PD variability component to be associated with a randomly chosen low realization for PK variability. The procedure used in earlier work (2)— deriving a global estimate of human interindividual variability and applying to it a global estimate of uncertainty derived from an analysis of uncertainty in the dynamic parameters—inadvertently treated the extent of kinetic and dynamic variability as highly correlated. Because we know of no reason why people with greater susceptibility due to dynamic factors should also have systematically greater AUCs or C_{max} values per unit dose, we think it is more appropriate to model the kinetic and dynamic variability components as uncorrelated. Treating the two variability components as uncorrelated has the effect of greatly reducing the estimated frequency of extremely high values of overall human interindividual variability in susceptibility and corresponding risks at low RfD exposure levels.

Table 5 shows the results of these simulations for the base-case estimates of variability and uncertainties in each source of variability (second and fourth columns of Table 4). In addition to the findings relative to the

Table 5 Distributional Results for the Base Case

Chemical	Estimated median (50% confidence) risk at RfD	Estimated arithmetic mean risk at RfD	Upper 95% confidence limit on risk at the RfD	Fold increase in RfD needed to cause 1/100,000 risk with 50% confidence	Fold reduction in RfD needed to achieve 95% confidence that risk is under 1/100,000
4-(2,4-Dichloro-phenoxy) butyric acid	Very small	2.2×10^{-4}	2.9×10^{-3}	18	0.54
Tridiphane	5.2×10^{-7}	1.5×10^{-3}	2.5×10^{-2}	2.0	3.4
Sodium azide	1.8×10^{-8}	7.0×10^{-4}	1.1×10^{-2}	1.9	4.6
S-Ethyl dipropylthio-carbamate	8.8×10^{-7}	2.1×10^{-3}	3.2×10^{-2}	1.8	4.2
1,2,4,5-Tetrachloro-benzene	Very small	4.6×10^{-5}	9.5×10^{-4}	51	0.42
Ethephon	1.5×10^{-5}	2.1×10^{-3}	3.6×10^{-2}	0.9	5.2
Tetraethyl-dithiopyro-phosphate	8.2×10^{-9}	5.7×10^{-4}	8.8×10^{-3}	2.2	4.3
2,4,6-Trinitro-toluene	Very small	2.5×10^{-4}	3.6×10^{-3}	4.8	2.0
Butyl benzyl phthalate	3.2×10^{-9}	4.8×10^{-4}	7.4×10^{-3}	2.6	3.4
Octabromo-diphenyl ether	4.3×10^{-7}	2.6×10^{-3}	4.0×10^{-2}	2.1	4.8
Metolachlor	4.6×10^{-6}	2.9×10^{-3}	4.7×10^{-2}	1.2	5.8
Dichloromethane	9.4×10^{-8}	9.7×10^{-4}	1.6×10^{-2}	2.8	2.4
Acetophenone	Very small	1.0×10^{-3}	1.3×10^{-2}	8.3	1.52
Nickel, soluble salts	1.3×10^{-9}	1.0×10^{-3}	1.1×10^{-2}	6.3	1.58
Methoxychlor	Very small	7.3×10^{-4}	1.1×10^{-2}	7.7	1.35
Acetochlor	8.1×10^{-9}	5.5×10^{-4}	8.5×10^{-3}	4.6	1.52
Dacthal	5.2×10^{-9}	6.0×10^{-4}	9.0×10^{-3}	4.9	1.59
Methyl metha-crylate	5.1×10^{-6}	6.8×10^{-3}	1.0×10^{-1}	1.1	16.1
Geometric mean reduction to RfDs needed to meet the straw man criterion of < 1/100,000 risk with 95% confidence					2.54
Fold reduction to RfDs for needed for 17/18 RfDs to meet the straw man criterion of < 1/100,000 risk with 95% confidence					5.8

Central estimate of PK variability = log(GSD) 0.202 with log[log(GSD)] uncertainty = 0.092; central estimate PD variability = 0.325 with log[log(GSD)] uncertainty = 0.161. Data represent means of three Monte Carlo simulation runs of 5000 trials each.
Abbreviation: RfD, reference dose.

Table 6 Fold Reductions in RfD Needed to Meet the Straw Man Risk Management Criterion, Given Different Changes in the Information Status on PK and PD Variability

	Base case	95% Upper confidence limit for PK uncertainty	Perfect information for PK variability	Perfect information for PD variability	Perfect information for both PK and PD variability
Central log(GSD) for PK variability	*0.202*	*0.202*	*0.202*	*0.202*	*0.202*
log[log(GSD)] uncert. in PK variability	*0.092*	*0.164*	*0*	*0.092*	*0*
Central log(GSD) for PD variability	*0.325*	*0.325*	*0.325*	*0.325*	*0.325*
log[log(GSD)] uncert. in PD variability	*0.161*	*0.161*	*0.161*	*0*	*0*
4-(2,4-Dichlorophenoxy) butyric acid	0.54	0.61	0.49	0.18	0.17
Tridiphane	3.4	4.3	3.4	1.00	0.92
Sodium azide	4.6	5.7	4.4	1.52	1.41
S-Ethyl dipropylthiocarbamate	4.2	5.3	4.0	1.28	1.12
1,2,4,5-Tetrachlorobenzene	0.42	0.49	0.39	0.14	0.13
Ethephon	5.2	6.7	5.1	0.73	0.45
Tetraethyldithiopyrophosphate	4.3	5.2	4.3	1.41	1.30
2,4,6-Trinitrotoluene	2.0	2.3	1.9	0.67	0.60
Butyl benzyl phthalate	3.4	4.3	3.3	1.21	1.04
Octabromodiphenyl ether	4.8	6.0	4.3	1.67	1.59
Metolachlor	5.8	7.1	5.9	1.70	1.52
Dichloromethane	2.4	3.1	2.4	0.65	0.60
Acetophenone	1.52	1.78	1.46	0.54	0.51
Nickel, soluble salts	1.58	1.85	1.48	0.47	0.44
Methoxychlor	1.35	1.59	1.37	0.51	0.46
Acetochlor	1.52	1.71	1.47	0.42	0.38
Dacthal	1.59	1.78	1.47	0.44	0.40
Methyl methacrylate	16.1	17.3	14.4	4.7	4.2
Geometric mean RfD reduction needed to meet the straw man criterion	*2.54[a]*	*3.03*	*2.43*	*0.767*	*0.685*
Fold reduction needed for 17/18 RfDs to meet the straw man criterion	*5.8*	*7.1*	*5.9*	*1.7*	*1.6*

Data represent means of three Monte Carlo simulation runs of 5000 trials each.
[a]Three significant figures are shown in this row in order to facilitate reproduction of the comparative calculations across columns that are reported and discussed in the text.
Abbreviations: PK, Pharmacokinetics; PD, Pharmacodynamics..

straw man risk management criterion in the final column of Table 5, other columns show a few other expectations for the risk distributions at RfD exposure levels. The differences between median (50th percentile), arithmetic mean (the average of all the 5000 projections risks at the RfD in each simulation), and the 95th percentile risks emphasize the great uncertainty in these projections, which derives in part from projections over several standard deviations of hypothesized perfectly lognormal distributions of sensitivity from estimated human ED_{50} levels to RfD exposure levels.

Despite the great uncertainties in individual risk projections for particular agents, these calculations provide a consistent framework for judging the likely relative protection performance of the current RfD system as a whole. The results in the final column of Table 5 indicate that current RfDs may generally come within several fold of meeting our straw man risk criterion. The final column of results has nine values under 3 and nine values over 3. Therefore, an approximate three-fold reduction is projected to be needed to make the median RfD meet the straw man criterion. Similarly, a six-fold reduction is projected to bring all but one of the 18 RfDs within the same standard.

With this background, Table 6 shows the changes in RfDs relative to the straw man criterion that would be expected to result from various changes to the "baseline" estimates of uncertainty in PK and PD variability (refer to columns 2 and 4 of Table 4 for the quantitative definition of the "baseline"). The first column of numbers recapitulates the baseline findings from Table 5. Then, the second column of numbers shows the effects of increasing the baseline uncertainty in PK variability to our estimated 95% upper confidence limit as derived in the previous section (third column of Table 4). It can be seen that this makes a modest difference—increasing the adjustment needed to reach the straw man criterion by about 20% for the two cases highlighted at the bottom of the table. Similarly, the third column of numbers shows the expected effects of perfect information on PK variability—holding the central estimate of PK variability constant, but assuming that a wonderful study has reduced the uncertainty in PK variability to zero from the 0.092 baseline estimate. In this case, the changes in the expected changes to meet the straw man criterion are small enough, relative to the base case, that one can see the effects of stochastic uncertainties in the simulations for the different RfDs. Overall, averaging the multiplicative changes for all the chemicals relative to the baseline, the geometric mean difference is about 5% [100 (1−2.43/2.54)]. Relative to the expanded baseline corresponding to the upper confidence limit of PK uncertainty, the geometric mean for the perfect PK information column indicates a change in the distance between RfDs and the straw man criterion comes to about 20% [100 (1− 2.43/3.03)]. In contrast, the fourth column of numbers shows the expected effects of perfect information (0 uncertainty) for PD variability amount to a little more than a three-fold reduction in the amount of adjustment calculated to be needed to meet the straw man risk management criterion. The last column, showing the effects of perfect information for both PK and PD variability, indicates approximately a 10% further change in the dose expected to reach the comparative benchmark risk management goal.

All the results in Table 6 presume that the research that yields improved estimates of PK variability reduces uncertainty, but converges on the same value derived for the central estimate of PK variability for all compounds. Other model calculations have been made based on alternative possibilities in which the hypothetical new research is assumed to have produced estimates of PK variability that correspond to either a 5th percentile or a 90th percentile chemical, respectively. This range of plausible outcomes from PK measurements leads to about a two-fold overall difference in the RfD adjustments needed to meet the straw man risk management criterion (data not shown—detailed results available on request).

These results allow us to make comparisons with the changes in allowable dosage that would be indicated under the IPCS guidance. From calculations similar to those in Table 1, our central estimate log(GSD) for PK variability of 0.202 would correspond to a ratio of the 95 to 50th percentile values of 2.15. If this factor were substituted for the 3.16-fold default factor for PK interindividual variability, it would result in an increase in allowable RfD/ADI doses of about 47% (3.16/2.15 = 1.47). In contrast, the results in Table 6 indicate that, depending on the detailed specification of the risk management goal, the enhanced information on PK variability should lead to central estimate (geometric mean) adjustments in allowable dosage of about +5% [i.e., 100(2.54/2.43 − 1)], relative to the case where there is no compound-specific information about PK interindividual variability. [As much as a +54% adjustment would be indicated if the new "perfect information" leads to a revised estimate of human interindividual variability as low as a log(GSD) of 0.1, rather than the central estimate of 0.202 for the median chemical].

LESSONS FOR THE DERIVATION OF SINGLE-POINT FACTORS TO REPRESENT OVERALL INTERINDIVIDUAL VARIABILITY AND PK/PD COMPONENTS

It is desirable to take one further step to make the kinds of assessments that we illustrate here more accessible to working risk assessors in the field. We do not imagine that, at least in the near term, many risk assessment agencies will want to make the investment necessary to become comfortable in doing the combined analyses of variability and uncertainty that we have pursued here—although we have taken care to develop our system in such a way that no more software is required than simple Excel spreadsheets. In this section, we use the results of our probabilistic modeling for 18 RfDs to derive a series of simple formulae that allows approximate derivation of analogous results for other RfDs. In the course of this, we derive approximate guidance for interim alterations of the traditional single-point uncertainty factors.

Figure 2 shows a plot of our prime dependent variable—the dose reduction needed to change current RfDs into values that are expected to meet the straw man 1×10^5 risk with 95% confidence criterion—versus the overall uncertainty factor assigned by the EPA in light of the information available for each RfD chemical. The data are plotted separately for RfDs based on observations of quantal endpoints (where we made dose adjustments for the animal data corresponding to our estimates of the differences between LOAEL/NOAEL and ED_{50} levels) versus RfDs based on observations of continuous endpoints (where no such adjustments were made, because the levels were judged to already reflect toxicologically significant changes in average individuals in the exposed groups of animals).

As was observed previously (2), it can be seen in Figure 2 that there is some tendency for the RfDs that incorporate larger overall uncertainty factors to require less multiplicative reduction to meet the straw man criterion than RfDs where smaller uncertainty factors were deemed necessary in the IRIS analysis. This is apparently because the uncertainty factors that are added above the traditional 10-fold factors for interspecies projection and interindividual variability carry some added "conservatism" beyond what is assessed to be strictly needed to offset the added uncertainty of the projections.

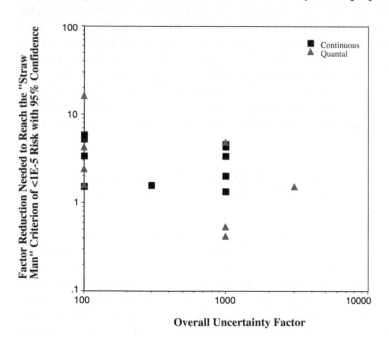

Figure 2 Multiplicative reduction in the RfD that would be needed to reach the "straw man" risk criterion—continuous versus quantal endpoints.

Table 7 Results of a Multiple Regression Analysis for the Effects of Various Independent Variables on the Log of the Multiplicative Reduction Needed for Current RfDs to Meet the Straw Man Risk Management Criterion

Term	Regression estimate	SE	t Ratio	Prob. $> \lvert t \rvert$
Intercept	1.348	0.337	−4	0.0013
Quantal? 1 if yes 0 if no	−0.414	0.164	2.52	0.0243
log(Unc. fact)	−0.356	0.130	2.73	0.0163
No LOAEL available	0.760	0.219	−3.48	0.0037

A multiple regression analysis of these same data (Table 7) found three statistically significant determinants of the differences among RfDs in the multiplicative distance from the straw man criterion:

- The quantal versus continuous nature of the endpoint used to determine the RfD. RfDs based on continuous endpoints required $10^{0.414}$ (about 2.6-fold) greater reductions, other things being equal, to meet the straw man criterion.
- The magnitude of the overall uncertainty factor (i.e., the combination of all uncertainty factors used in the derivation of the RfD). Where this was 100, e.g., a $10^{0.356}$ which is about 2.27-fold greater reduction would be needed to meet the same risk standard, compared to the cases where the combined uncertainty factors were 1000.
- Where there was no LOAEL available, and our risk inferences needed to be based on more uncertain NOAEL information, an additional factor of $10^{0.76}$ which is about 5.75-fold was required on average for adjustment to the risk standard.

If we apply the regression estimates in Table 7 to the case of an RfD based on a quantal endpoint, to which only the two traditional interindividual and interspecies uncertainty factors were applied, and for which an LOAEL was available, we obtain

$\log(\text{RfD multiple to meet the straw man risk criterion})$
$= 1.348 - (1 \text{ for quantal})0.414 - [2 = \log(100)\text{for UF of } 100]$
$0.356 = 1.348 - 0.414 - 0.712 = 0.222, \text{implying}$
a multiplicative distance from the straw man criterion of
$10^{0.222} = 1.67$ \hfill (7)

For convenience, and with the caveats previously discussed, this factor could be incorporated into current risk assessment practice for analogous chemicals by changing the traditional 10-fold factor for interindividual variability to 16.7. Similarly, for an RfD based on a continuous end point with an overall uncertainty factor of 100, meeting the straw man criterion could be accomplished by changing the traditional uncertainty factor for interindividual variability from 10 to $16.7 \times 10^{0.414} = 43$. For analogous chemicals where an overall uncertainty factor of 1000 would normally be assigned, the replacements for the usual 10-fold interindividual variability uncertainty factor would become 7.4 and 19, for RfDs based on quantal and continuous endpoints, respectively.

It should be understood that no official body has, as yet, adopted the straw man risk management standard as its management goal. However, the articulation of numerical guidance in terms of this standard shows how quantitative specification of risk goals in terms of variability and uncertainty can lead to practical and implementable changes toward more consistency in defining the noncancer risks that are considered "acceptable" by risk management agencies. It also allows more natural and quantitatively defensible rules for incorporation of improved quantitative information about risks. Finally, the same tools allow some, albeit highly uncertain, estimates of the potential health benefits of reductions in population exposures. These latter estimates are potentially helpful for analyses juxtaposing the benefits and costs of public policy measures to reduce population exposures.

REFERENCES

1. Meek ME, Renwick AG, Chanian E, et al. Guidelines for application of chemical-specific adjustment factors in dose/concentration-response assessment. Toxicology 2002; 27:115–120.
2. Hattis D, Baird S, Goble R. A straw man proposal for a quantitative definition of the RfD. Drug Chem Toxicol 2002; 25:403–436.
3. Hattis D, Banati P, Goble R. Distributions of individual susceptibility among humans for toxic effects. How much protection does the traditional tenfold factor provide for what fraction of which kinds of chemicals and effects? Ann NY Acad Sci 1999; 895:286–316.
4. http://www2.clarku.edu/faculty/dhattis.
5. Hattis D, Banati P, Coble R, et al. Human interindividual variability in parameters related to health risks. Risk Anal 1999; 19:711–720.
6. Baird SJS, Cohen JT, Graham JD, et al. Noncancer risk assessment: a probabilistic alternative to current practice. Human Ecol Risk Assess 1996; 2:79–102.
7. Weil CS, McCollister DD. Relationship between short and long-term feeding studies in designing an effective toxicity test. Agric Food Chem 1963; 11:486–491.
8. Nessel CS, Lewis SC, Stauber KL, et al. Subchronic to chronic exposure extrapolation: toxicologic evidence for a reduced uncertainty factor. Human Ecol Risk Assess 1995; 1:516–524.

9. Evans JS, Baird SJS. Accounting for missing data in noncancer risk assessment. Human Ecol Risk Assess 1998; 4:291–317. [b12].
10. Lehman AJ, Fitzhugh OG. 100-fold margin of safety. Assoc Food Drug Off 1954; 18:33–35.
11. Hattis D, Ginsberg G, Sonawane B, et al. Differences in pharmacokinetics between children and adults. II. Children's variability in drug elimination half-lives and in some parameters needed for physiologically-based pharmacokinetic modeling. Risk Anal 2003; 23:117–142.
12. Haas CN. Dose–response analysis using spreadsheets. Risk Anal 1994; 14:1097–1100.

5

Use of Classical Pharmacokinetic Evaluations in Drug Development and Safety Assessment

Rakesh Dixit

Toxicology Department, Johnson & Johnson, PRD, San Diego, California, U.S.A.

Peter Ward

Drug Metabolism Department, Johnson & Johnson, PRD, San Diego, California, U.S.A.

INTRODUCTION

The extrapolation of experimental data to predict efficacy (e.g., pharmacodynamic activity) and toxicity from preclinical species to humans is one of the major steps in drug development. Given the differences in physiological, pharmacological, and pharmacokinetic processes between laboratory animals and humans, the extrapolation based on doses given in mg/kg or mg/m^2 (based on species-specific surface area) is mostly inaccurate (1). It is now well-recognized that interspecies extrapolation of toxicity and efficacy is greatly improved when adequate information on toxicokinetics of absorption, distribution, metabolism, and excretion processes are known across laboratory species and humans (1). As a result, toxicokinetic investigations have become a critical and integral part of drug safety evaluation, including an assessment of acute, subchronic and chronic safety, genotoxicity, carcinogenicity, and reproductive and developmental toxicity evaluations. The toxicokinetic data, when juxtaposed with target organ toxicity (TOT) or clinical adverse effects, can be very valuable in predicting what blood concentrations may be expected to result in adverse effects and provide appropriate safety margins to administer a new drug safely into humans. Overall, the critical goal is not to harm patients or healthy volunteers (2).

CLASSICAL PHARMACOKINETICS AND TOXICOKINETICS

The combined process of the absorption (A) of a drug into the systemic circulation, its distribution (D) to organs and tissues, metabolism (M) to other active or inactive chemical species, and its elimination (E) out of the body is collectively known as ADME (3–5).

The term "pharmacokinetics" refers to the kinetics of ADME processes employed at relevant low pharmacological doses where pharmacokinetics generally follow first-order kinetics and kinetic processes are expected to be linear (3–5). Toxicokinetics represent the study of kinetic processes of ADME under the conditions of preclinical toxicity testing where both first- and zero-order kinetics are expected, and kinetic process can substantially change over a wide range of doses (2,6). The application of toxicokinetics has evolved in the pharmaceutical industry and toxicokinetics often refer to exposure assessment in drug safety assessment (SA) studies. Pharmacodynamics refers to effects elicited by the drug and active metabolites at low pharmacological doses, whereas toxicodynamics refers to toxic effects related to doses (systemic exposures) used in toxicity testing (3–5).

An understanding of dose–response, including dose administered and circulating drug levels (systemic exposure), and their relationship to toxicity are critical to interpretation of data in toxicity studies where doses can vary by a log order of magnitude. Therefore, in contrast to pharmacokinetics studies where ADME processes are assumed to be first-order and linear, toxicokinetics must also consider zero-order and nonlinear processes in interpretation. Nonlinearity in toxicokinetics may result from drug-specific biopharmaceutical factors and toxicodynamics. Biopharmaceutical factors include alterations in drug absorption at different doses (e.g., from low solubility of drug), differences in blood or tissue distribution (e.g., due to saturable protein binding, changes in tissue pathology), differences in metabolism (e.g., saturable metabolic enzymes kinetics), and in drug elimination (e.g., urinary and fecal excretion) (2,6). In relating blood levels to toxicodynamics or pharmacodynamics, the most practical surrogate or measure of dosimetry is through the determination of the time course of the drug and its major metabolite(s) in plasma. The time course of plasma drug concentration provides information on the kinetics of absorption, distribution, metabolism, and elimination of a given drug. When the concentration of a drug in blood or plasma is in equilibrium with concentrations in tissues, one can assume plasma as a reasonable practical surrogate for tissue(s) exposure to drug. Under these circumstances, changes in plasma drug concentrations reflect changes in tissue drug concentrations over time, and relatively simple pharmacokinetic calculations or models can be extremely useful to describe the behavior of that drug in the body.

Dose Selection for General Toxicity Studies

The major goal of a successful nonclinical toxicology program is to identify potential TOT and no observed adverse effect levels (NOAEL) for humans. The following discussion provides a perspective on the dose selection concepts.

Dose-Limiting Toxicity

The dose-limiting toxicity (DLT) is the dose that may produce the maximally tolerable adverse effects and further dose escalation beyond the DLT may produce serious adverse effects. Adverse effects could be based on physical signs, body weight, clinical pathology, or histopathology-based endpoints. Doses that cause lethality should be avoided because adverse effects associated with morbidity and mortality in animals will not be relevant for humans in clinical settings.

Maximum Tolerated Dose

The maximum tolerated dose (MTD) is the dose at which toxicity is relatively moderate; however, further dose escalation may not be compatible with long-term health or survival. Typically, the MTD is one notch lower than the DLT in the dose versus toxicity response. In principle, MTD can be defined as the dose that causes minimal reversible toxicities (e.g., 10% decrease in body weight gain) and/or overt pharmacological adverse effects; however, these effects must be compatible with long-term conduct of a study and must not decrease the natural lifespan of rodents.

Target Organ Toxicity

The TOT dose is typically the dose that produces histopathology-based overt structural and functional adverse effects. The supportive evidence for the TOT could be derived from physical signs, body weight changes, and clinical pathology changes.

Limit Dose

The limit dose (LD) is typically a predefined maximal dose that can be used in the absence of a clear evidence for toxicity. Although there is no consensus on the LD, a dose of 2000 mg/kg has been accepted as the LD for the general toxicity study.

Maximum Absorbed Dose (Saturable Absorption Dose)

The maximal absorbable dose is the dose that results in saturation of absorption which limits systemic exposure to the drug and its major metabolites.

Saturable absorption is expected to be attained when there is a clear plateau between two/three doses (one above and one below the indicated saturable absorption dose).

Maximal Feasible Dose

The maximal feasible dose (MFD) is the maximal dose that can be administered on the basis of practical considerations, which include but are not limited to physico-chemical considerations (e.g., homogeneity of suspension, viscosity, and stability) and the ability to dose animals on a chronic basis. Typically, it is the MFD concentration multiplied by specific dosing volume for a given formulation.

Preclinical Early Dose Range-Finding Toxicity Studies (DLT or Tolerability Studies)

Typically, the exploratory range-finding studies (DLT or tolerability) are conducted to assist in dose selection for the good laboratory practice (GLP) toxicity studies needed to support phase I clinical trials. For a large number of drugs, dose selection is difficult for the main toxicity studies, therefore, the dose range-finding studies are of great value. These early studies can be from a minimum of 5 to 15 days in duration, and depending upon the availability of the drug, typically three to five doses may be used. Selected doses may range from low doses to high doses ranging up to the LD of 2000 mg/kg/day. Typically, the low dose should be selected to provide two to fivefold multiples of the projected clinical efficacious exposure and the high dose should be high enough to produce some toxicity or provide substantial (e.g., 10–100-fold) or be at the LD of 2000 mg/kg/day. Early pharmacokinetic/toxicokinetic data can be very helpful in selecting dose levels, especially the low and middle doses. To conserve the drug, these studies can be conducted in a very small of number of animals (five per group for rats) of one sex provided that there are no sex-related differences in metabolism. In nonrodents, an ascending 5 day dosing may be conducted at 1× (e.g., low dose corresponding to a two to fivefold multiples of expected clinically efficacious exposures adjusted for differences in in vitro pharmacodynamic activity), 2.5×, 5×, 10×, 20×, etc., or until dose-limiting adverse effects are observed. Toxicity end points may involve histopathology of only major organs of toxicity (typically liver, kidneys, heart, pancreas, lung, brain, gastrointestinal tissues, sex organs, and limited primary and secondary immune organs), limited clinical pathology, and toxicokinetic (TK) investigations.

Repeated Dose Toxicity Studies

High Dose

Prior to exposing humans to a new drug entity, especially belonging to a new pharmacological mechanism and/or structure, it is critical to know what adverse effects may occur and whether they are reversible. In principle, the high dose should be selected principally based on toxicological considerations. Ideally, the high dose should produce frank toxicity, identify TOT, and establish a positive correlation between clinical pathology-based biomarkers and TOT. Identification of toxicity helps to formulate a better safety monitoring program in phase I and II clinical trials. This also provides assurance that if humans (healthy human volunteers and targeted sick patients) are more sensitive than animals to certain toxicities, appropriate safety monitoring is included in the clinical plans.

Given the dissimilarity in quantitative metabolism, pharmacological mechanisms, and sensitivity to certain toxicity, it is expected that preclinical safety species may not respond to certain pharmacological mechanisms. When toxic effects are not observed with increasing doses, toxicokinetics can be very useful in setting the high dose. If a drug is not getting absorbed with increasing doses, then the lowest dose that provides maximal exposure to the drug and its metabolites can be considered as the high dose.

Saturable absorption or dissolution-limited drug absorption can be demonstrated by establishing that systemic exposure to parent drug and major metabolites has plateaued over the range of doses. In practice, a plateau may be considered when the plasma area under the curve (AUC) for the drug and each of its major metabolites does not increase by more than 20% with a two-fold increase in dose. When there is a large interanimal variability in exposures with increasing insoluble doses, appropriate statistical tests may be necessary to establish a plateau in exposures. If saturable absorption cannot be demonstrated, then the high dose may be based on MFD or the LD (generally not to exceed 1500–2000 mg/kg/day). Figure 1 illustrates a step-wise process of selecting the high dose and some important considerations in selection of the middle and low doses for general toxicity studies.

Middle Dose

In designing the toxicity study, it is also important to fully understand the dose–response in toxicity, including the threshold dose for inducing the toxicity. The middle dose selection should consider multiples of projected human maximally efficacious dose, threshold in toxicity response, pharmacodynamic response, and mechanisms of toxicity.

Low Dose

Given the importance of NOAEL in toxicity studies, the low dose should be selected to provide an NOAEL. Ideally, the low dose should be set to provide a two to fivefold safety margin over the estimated/projected maximal clinical efficacious dose/exposure.

Panel A : DLT Observed

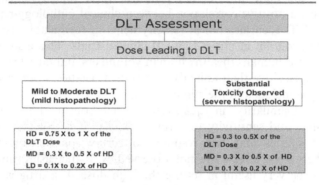

Panel B : No DLT Observed

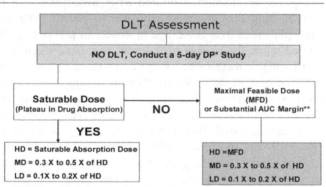

• DLT = Dose limiting toxicity; DP = Dose proportionality
• ** AUC margins ranging from 5 to 100-fold

Figure 1 Selection of doses for general toxicity studies. Dose selection for the repeated dose toxicity is described. Prior to dose selection for the GLP toxicity studies, typically a short-term (up to 2 weeks) DLT or tolerability studies are conducted to make rapid go/no-go decisions. Based on the results of the DLT study, a dose proportionality study to understand the dose versus exposure assessment may be desirable. Additionally, if saturable absorption is not attained with increasing doses, it may be important to assess the MFD; however, if the selected MFD is excessive, the high dose may be based on substantial multiples (10–100-fold) of projected human efficacious exposures. In principle, the high dose for the repeated dose toxicity study could be a fraction of the dose producing DLT, dose providing saturable absorption, or MFD or multiples of projected human exposure. (**A**) A recommended strategy for dose selection when DLT is observed in the preliminary range finding study. (**B**) A strategy for dose selection when DLT is not observed in the preliminary range finding study. For additional details, see text. *Abbreviations*: DLT, dose-limiting toxicity; MFD, maximal feasible dose; LD, limit dose; AUC, area under curve.

Carcinogenicity Bioassay Studies in Rodents

The conduct of the two-year carcinogenicity bioassays remains a major challenge. The U.S. Food and Drug Administration (FDA) has placed a high importance on the process of dose selection and protocol approvals for the carcinogenicity studies. The FDA document has provided details of the information that is needed to seek approval of dose selection and protocol approval prior to initiating carcinogenicity studies (Pharmacology/ Toxicology, Procedural, Carcinogenicity Study Protocol Submissions) (7). Additionally, the International Conference on Harmonization (ICH) has provided guidance on dose selection for the carcinogenicity studies. Figure 2 describes recommended toxicokinetics-based approaches to high dose selection and a brief discussion follows.

Use of Systemic Exposure Margins in Selecting the High Dose for Carcinogenicity Bioassays

For drugs that are nongenotoxic and well-tolerated, an MTD may not be attained until excessive unrealistic doses are used. To alleviate the problem of selecting excessive doses, U.S. FDA scientists analyzed carcinogenicity databases available from the U.S. FDA, National Toxicology Program and International Agency for Research on Cancer (IARC), and scientists. Based on the evaluation of carcinogenicity studies of 35 compounds belonging to various pharmacological mechanisms, it was concluded that in the absence of an MTD, an exposure margin of at least 25-fold between systemic exposure in rodents and humans at the maximally recommended clinically efficacious dose will allow an adequate margin of safety. The ratio of 25 will likely allow the detection of a positive neoplastic response in the vast majority of two-year bioassays for all IARC class I definitive human carcinogen and 2A possible human carcinogens.

As outlined in Figure 2, the dose selection based on the 25-fold AUC margin should consider the following criteria: (i) the drug should have low degree of toxicity, (ii) the drug must not be nongenotoxic, (iii) the drug must be metabolized similarly in both rodents and humans, (iv) if the protein binding is greater than 80%, the AUC should be corrected for protein binding in both humans and in animals, (v) exposure margins should be justified on the basis of the parent drug, parent drug plus major metabolite(s), or solely on metabolites, and (iv) exposure data in humans must be attained at the maximum recommended human daily dose.

Use of Uncertainty Factor in Exposure Margin-Based Carcinogenicity Dose Selection

In practice, maximum recommended human dose (MRHD) is often not established at the time of selecting dose levels for the carcinogenicity studies. Also, maximal efficacious exposures are not known until the drug has been in

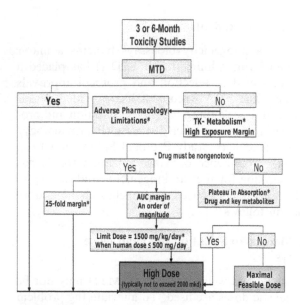

Figure 2 Step-wise approach to high dose selection in carcinogenicity studies. A step-wise approach to dose selection for two-year carcinogenicity studies is described. For compounds that produce toxicity in rodents, the first choice often is maximum tolerated. For drugs that are generally well tolerated and non-genotoxic, TK-based approaches such as 25× multiple of AUC or saturable absorption can be used for dose selection. LD and MFDs (e.g., 2000 mg/kg/day) are additional possibilities. In certain cases, dose-limiting adverse pharmacodynamic effects may be used as a criteria for dose selection. For details, see text. *Abbreviations*: AUC, area under the curve; LD, limit dose; MFD, maximal feasible dose; MTD, maximum tolerated dose.

phase III testing. Under the conditions of the uncertainty in prediction of MRHD and associated exposures, it is advisable to select an uncertainty factor of 2–5 when using the 25-fold margin. If extrapolations to human suggest that the bioavailability will be good and the drug will be metabolically stable, it will be practical to consider an AUC margin of 50–125-fold (e.g., 2–5-fold greater than the targeted 25-fold margin) over the projected exposures at the MRHD. The factor of 2–5 may depend on the following factors: (i) medical need for the targeted therapy, (ii) chronic exposure versus intermittent exposure, (iii) chronic safety of drug in preclinical species, (iv) previous experience with the class of drugs, (v) structure–activity relationship, and (vi) interhuman variability in drug exposures.

TISSUE DRUG TOXICOKINETICS

Plasma drug levels have been routinely used as surrogates for tissue(s) exposure and this has served well in many cases; however, drugs that are distributed in peripheral compartments, have long terminal half-life, or have high affinity for tissue macromolecules, may show preferential drug uptake in certain tissues. Drugs with a large volume of distribution or reactivity towards cellular macromolecules may get preferentially sequestered in certain tissues leading to delayed toxicities. It is likely that plasma exposures may misrepresent tissue drug exposures for compounds that have a potential to accumulate in tissues. Limited tissue drug toxicokinetic studies can be of great value in interpreting toxicity. As outlined in ICH Guideline S3B, the following conditions may require the conduct of tissue distribution studies as described below:

1. When the projected target tissue half-life exceeds the dosing interval by at least two-fold (e.g., target tissue half-life of 48 hours for once-a-day drug) from a single dose, in tissue distribution studies or projected by modeling.
2. When the plasma half-life after repeated dosing is significantly longer than after a single dose, indicating the possibility of tissue sequestration.
3. When TOT (histopathological lesions) poorly relates to plasma drug concentration and the toxicity is suspected to result from the tissue sequestration of the drug in target organs of toxicity.
4. When drugs are intended for tissue-specific delivery and tissue distribution.

Because most drugs attain steady state by 4 weeks of repeated dosing, typically 1 to 3 weeks of tissue distribution study in a relevant rodent species should be sufficient to monitor steady-state tissue drug concentrations at critical time points (C_{max} and trough levels). In some cases, a full tissue drug AUC may be necessary to interpret TOT. Typically, liver and target organs of toxicity are sufficient for tissue distribution studies.

QUALITATIVE AND QUANTITATIVE ASSESSMENT OF METABOLITES

The qualitative similarity in metabolic profiles in nonclinical safety species and humans provides assurance regarding the validity of selected species for human risk assessment. The qualitative similarity in metabolites can be demonstrated as follows:

1. Comparison of in vitro metabolic profile of the radio-labeled drug in liver slices, hepatocytes, and hepatic microsomes from toxicology species and humans.
2. Comparison of plasma or urinary metabolites profile in preclinical species and humans.

Metabolites can be responsible for the desired efficacy as well as the adverse effects. While qualitatively both phase I and II metabolic reactions are fairly conserved across laboratory safety species and humans, there are often important quantitative differences in the rate and the extent of metabolism between safety species and humans. If a metabolite is responsible for toxicity, the quantitative differences in exposure to a major toxic metabolite between humans and laboratory animals can lead to misrepresentation of safety margins based on the parent drug exposure. Nonclinical SA studies are critical to identifying potential toxicity signals for humans and for providing a reasonable assurance related to the test drug's safety at proposed clinical doses. Given the use of high doses (large multiples of the projected human doses) in safety studies and the fact that the rates of most phase I and II reactions are relatively greater in nonclinical safety species than in humans, it is expected that in the vast majority of cases there will be adequate coverage of metabolites, both qualitatively and quantitatively, in safety studies. In a small number of cases, when the major versus minor pathways of metabolism are dissimilar across species, there may be human-specific "unique" or "dominant metabolites" which may not have been adequately tested in nonclinical safety studies.

The U.S. FDA has recently issued "Guidance for Industry: Safety Testing of Drug Metabolites." The document makes a serious regulatory attempt to define the major human metabolite(s) and provides scientific guidance on the safety testing of the major human metabolites in nonclinical studies. However, the guidance document has also raised ambiguity and concerns regarding the definition of the major metabolites as well as the extent of safety studies with human dominant or unique metabolites. In this document, the FDA recommends that efforts should be made to identify differences in drug metabolism in animals (used in nonclinical SAs) and humans as early as possible during the drug development process. Furthermore, attempts should be made to identify metabolites that may be unique to humans. The lack of understanding of the safety of human unique metabolites could have possible implications for marketing approval. The FDA recommends that metabolites identified in human plasma that account for greater than 10% of drug-related material (administered dose or systemic exposure, whichever is less) be considered for SA. The rationale for setting the level at greater than 10% for characterization of metabolites reflects consistency with other FDA and EPA regulatory guidelines. From the human health risk assessment perspective, major human metabolites at levels greater than those measured in the laboratory animals test species used for toxicological assessment will be of important concern.

Although plasma is mostly the choice of matrix for measuring drug or metabolite concentrations, other biological matrices, such as urine, feces, or bile can be used to demonstrate exposure.

The following additional points should be considered when deciding about the quantitation of metabolites to support the safety of drug-related substances:

1. When a prodrug is converted (nonenzymatically or enzymatically) to bioactive metabolite(s), both prodrug and active metabolite should be quantified, as there may be species differences in conversion of prodrug to active metabolite between animals and humans.
2. When metabolites constitute the predominant circulating drug-related moieties, systemic exposure to these metabolites should be assessed.

Conclusions

Classical toxicokinetics and drug metabolism studies when conducted to support nonclinical SA studies have greatly reduced the uncertainty in interpretation of preclinical toxicity data and its utility in human health risk assessment. The classical noncompartmental toxicokinetics and drug metabolism-based data have been particularly useful in selecting the dose levels for short- to long-term toxicity and carcinogenicity studies. Furthermore, these assessments have allowed great improvements in selection of relevant species, formulation, and doses for safety studies. The greatest utility of preclinical toxicokinetics data has been in the interspecies comparison of product toxicity. It is now widely accepted that toxic effects can be better extrapolated from animals to humans when these comparisons are based on toxicokinetics and disposition (absorption, distribution, metabolism, and excretion) data in preclinical species and humans. In this context, the safety margin that is based on the ratio of animal AUC at NOAEL to human AUC at an efficacious dose is the key predictor of human toxicity risk. It is generally accepted that when the AUC ratio is large, the expected risk of toxicity in humans is low. Although model-independent or compartment-based plasma/blood toxicokinetics have served well as a practical means of assessing systemic exposure, they provide no information on time course of exposure of target organs to drug or metabolites. Overall, toxicokinetics have greatly enhanced our understanding of interspecies differences in toxicity and significance of safety margins.

DRUG–DRUG INTERACTIONS

Introduction

Pharmacokinetic-based drug–drug interactions comprise a majority of the cases leading to the loss of efficacy or an unexpected toxicity from the medication (8). Any alteration in the absorption, distribution, or metabolism of the drug by another drug can lead to a drug–drug interaction. Distribution- and metabolism-mediated drug–drug interactions may also result in altered elimination of the drug.

Absorption-mediated drug–drug interactions may occur by: (i) changes in gastrointestinal motility (e.g., the absorption rate of acetaminophen is increased by metoclopramide and delayed by propantheline), (ii) chelation (e.g., cholestyramine can form complexes with warfarin, nonsteroidal anti-inflammatory drugs, or sulfonamides), or (iii) changes in pH (e.g., increased pH due to H_2-blockers and antacids can reduce the absorption of ketoconazole) (9).

Distribution-mediated drug–drug interactions are possible if: (i) an extensively protein-bound drug of low volume of distribution is coadministered with a high-affinity displacing drug that is present at a concentration approaching or exceeding the molar concentration of the protein-binding sites (e.g., displacement of warfarin by trichloroacetic acid, the major metabolite of chloral hydrate) (8) or (ii) the distribution of a drug (and possibly elimination) mediated by a transporter is displaced (e.g., the *P*-glycoprotein inhibitor, quinidine, increases plasma levels of digoxin) (9,10).

Metabolic-mediated drug–drug interactions are possible if: (i) induction or increased synthesis of one or more drug-metabolizing enzymes that leads to enhanced metabolism of a drug (11) or (ii) inhibition or decreased metabolism for a specific drug-metabolizing enzyme that results from competition between drugs for the enzyme's binding sites, covalent modification of the enzyme by a reactive metabolite, or formation of a catalytically inactive reversible complex between the enzyme and a drug or its metabolite (11).

Of these type of drug–drug interactions, the clinical implications of metabolic-based drug–drug interactions are sometimes the most numerous and severe, especially for anticoagulant, antidepressant, or cardiovascular drugs that have a narrow therapeutic index (11). For example, coadministration of terfenadine, an antihistamine agent, with ketoconzole led to fatal ventricular arrhythmias in some patients (12). The metabolic basis of this interaction was that ketoconazole inhibited cytochrome P450 (CYP) 3A4-mediated metabolism of terfenadine which resulted in excessive plasma levels of terfenadine and the resultant toxicity (10,13,14). Therefore, the pharmaceutical industry has focused on the evaluation and prediction of clinically relevant metabolic-based drug–drug interactions. In the following section, methods and systems for assessment of metabolic-based drug–drug interactions will be briefly reviewed, and the predictability of these systems will be discussed.

Assessment of Metabolic-Based Drug–Drug Interactions for CYP Induction and Inhibition

Prediction of in vivo drug–drug interactions has moved from the drug development stage to the discovery stage with focused attention on the CYP enzymes (15). Human liver microsomes and human hepatocytes are the current standard systems to study the potential of a drug candidate to inhibit and induce CYP enzymes, respectively. The predictability of these models is still under investigation, but, in general, the predictability of these systems is suitable to highlight drug candidates with drug–drug interaction potential. The development of recombinant systems (e.g., cDNA CYP isoforms with fluorescent substrates and reporter gene assays for determination of inhibition and induction of CYP isoforms, respectively) has enabled increased screening of large libraries of compounds in the drug discovery stage. However, the predictive ability of these systems has not been proven.

CYP Inhibition

For assessment of potential inhibition of CYP isoform activity by the drug candidate, the focus test model has been human liver microsomes. Initially, the effect of drug candidates on the activity of three major CYP isoforms (e.g., CYP3A4, CYP2D6, and CYP2C9) should be investigated, and an IC_{50} (concentration that inhibits activity of the CYP isozyme by 50%) of the drug candidate should be determined. Assessment of the select activities of each CYP isoform is determined by the measurement of the formation of a specific metabolite by mainly liquid chromatography–mass spectrometry (LC–MS). Some recommendations for experimental conditions are: (i) incubation times should just be long enough to detect the product formation in the linear range, e.g., incubations of midazolam should rarely exceed a few minutes due to the loss of velocity after long incubation time (16); (ii) minimal amount of microsomal protein to detect metabolite should be used (preferably $< 0.25 \, mg/mL$) to minimize the chance of nonspecific protein binding of the substrate or inhibitor; (iii) the K_m (concentration of substrate that is at $\frac{1}{2} \, V_{max}$ of product formation) of the selected substrate should be used in the experiment—ensure that the K_m is determined in your system and compared to a reasonable range of values in the literature to verify that your system is correctly established; (iv) determine the IC_{50} of a known selective inhibitor of the CYP isoform (e.g., ketoconazole for CYP3A4 and quinidine for CYP2D6) and compare to a reasonable range of literature values; (v) add NADPH as the final step in the incubation for initiation of the reaction to avoid premature NADPH-mediated metabolism of the CYP-selective substrate or the drug candidate (except to investigate time- or mechanism-based inhibition); (vi) if the drug candidate has low aqueous solubility, dilute the drug candidate in the liver microsomes to increase the solubility, and (vii) if the drug candidate is subject to NADPH-independent metabolism (e.g., esterases), ensure that the drug candidate is not exposed to the liver microsomes for a prolonged period of time before the initiation of the assay with NADPH. Once confidence in this system has been established, verification of steps (iii) and (iv) may only be occasionally determined; for example, the inclusion of the known selective inhibitor of the CYP isoform may be reduced to one concentration at the predicted IC_{50}.

To bypass resources required for LC–MS analysis, evaluation of the metabolism of a blocked fluorescent substrate to fluorescent product (e.g., resorufin or coumarin) in CYP recombinant systems may be possible. As this system is a lot more artificial than the microsomal system, results from this system should always be spot checked with microsomal results to ensure that false-positive/negatives are not concluded. For example, limitations that may generate spurious results include: (i) fluorescence quenching, (ii) interference by NADPH, (iii) lack of probe specificity, and (iv) interference by test compounds or metabolites (17). However, with adequate quality controls, the fluorescent CYP inhibition assays are efficient and powerful tools for generation of rank order comparison of compounds within a series without impacting analytical resources.

Extrapolation of IC_{50} values obtained for drug candidates in the human liver microsomal system for potential of a significant clinical drug–drug interaction is sometimes difficult. In drug discovery, the mode of inhibition (e.g., competitive or noncompetitive) is not known and the clinical pharmacokinetics of the drug candidate (i.e., plasma protein binding, metabolism, distribution into hepatocyte) may be very difficult to predict. However, some practical rules may be applied: (i) inhibition potential for drug candidates with IC_{50} values above $10\,\mu M$ is very low; (ii) drug candidates with IC_{50} values between 1 and $10\,\mu M$ have possible inhibition potential; however, this result alone should not "kill" a potential drug candidate; and (iii) drug candidates with IC_{50} values below $1\,\mu M$ may have significant inhibition potential and, especially for IC_{50} values below $0.5\,\mu M$, the validity of these drug candidates proceeding into drug development must be questioned.

In drug development, the reversible inhibition constant, K_i, should be determined. K_i is independent of the concentration of substrate, which will affect the apparent inhibition of the CYP isoform (i.e., IC_{50}) by the drug candidate if interaction with the substrate at the active site is competitive. Experimental design for determination of K_i is an expanded form of the IC_{50} determination where not only the effect of different concentrations of the drug candidates on the specific metabolism of a CYP selective substrate at a single K_m concentration is determined, but this effect is expanded to different concentrations of the substrate (i.e., multiple IC_{50} experiments). At least four different substrate concentrations that span a 3–5-fold range below and above the K_m of the substrate should be investigated. From this data, a Dixon plot, which is plotted as the reciprocal of the velocity of formation of CYP-specific metabolite on the y-axis versus the concentration of the drug candidate on the x-axis for each concentration of substrate, can be generated; the absolute value of the concentration of drug candidate where the lines on the Dixon plot intersect is the approximate K_i value. This determination assumes that the inhibition is competitive, if the inhibition is more noncompetitive or uncompetitive (rare occurrence) in nature, then determination of the intersection of the line from the Dixon plot may not be correct (Fig. 3).

Determination of the mode of inhibition (i.e., competitive, noncompetitive, or mixed) can be accomplished with the Eadie-Hofstee plot, which is plotted as the velocity on the y-axis versus the velocity divided by the substrate concentration on the x-axis for each concentration of the drug candidate. If the mode of inhibition is competitive or mixed, the lines on the Eadie-Hofstee plot will intersect on the y-axis or away from the y-axis, respectively. If the mode of inhibition is noncompetitive, the lines on the

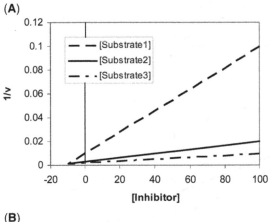

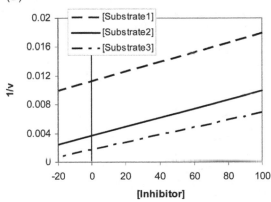

Figure 3 Representative Dixon plots of **(A)** competitive and **(B)** uncompetitive inhibition. In **(A)**, the K_i would be the absolute value of the inhibitor concentration at the intersection point of the lines (i.e., approximately $10\,\mu M$).

Eadie-Hofstee plot will not intersect (Fig. 4). Verification of the mode of inhibition can also be obtained by Michaelis–Menton plots and determining the K_m and V_{max} at each concentration of the drug candidate. Changes only in K_m and V_{max} as a function of drug candidate concentration suggest that the mode of inhibition is competitive and noncompetitive, respectively. Changes in both these parameters suggest mixed inhibition.

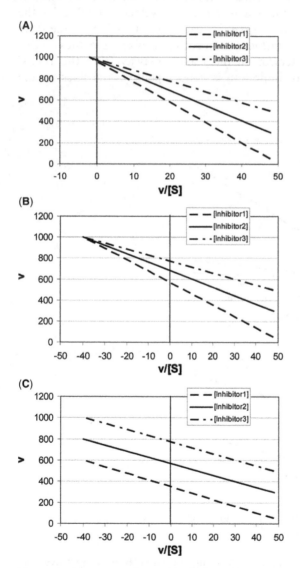

Figure 4 Representative Eadie-Hofstee plot of (**A**) competitive, (**B**) mixed, and (**C**) noncompetitive inhibition.

To predict the theoretical fold increase in exposure (i.e., AUC) of a CYP-specific substrate in vivo from the in vitro K_i determination for both competitive and noncompetitive inhibition, the following equation may be used:

$$\frac{AUC_i}{AUC} = 1 + \frac{[I]}{K_i} \tag{1}$$

in which AUC_i and AUC refer to the exposure of substrate with and without the CYP inhibitor, respectively, and $[I]$ refers to the concentration of the CYP inhibitor in the hepatocyte. If the mode of inhibition is competitive in nature, the assumption of this formula is the $[S] \lll < K_m$ for its specific CYP isoform; otherwise, as the $[S]$ increases the fold increases in exposure will decrease.

Another potential evaluation in drug discovery is to determine whether the compound is a time- or mechanism-based inhibitor (15). This early screen can be accomplished with preincubation of the drug candidate in NADPH for approximately 30 minutes, then addition of the CYP-selective substrate to initiate the assay as described above. Any reduction in IC_{50} values with preincubation suggests that the inhibition is time- or mechanism-based and the potential of that drug candidate to cause a clinically significant drug–drug interaction is likely (18). For example, when irreversible CYP3A4 inhibitors such as erythromycin and clarithromycin are coadministered with other drugs such as terfenadine, astemizole, or pimozide, patients may experience torsades de pointes (19–21). Rhabdomyolysis has occurred when simvastatin was combined with erythromycin or ritonavir (22). Symptomatic hypotension may occur when mechanism-based CYP3A4 inhibitors are combined with some dihydropyridine calcium channel antagonists (23), as well as with the phosphodiesterase inhibitor sildenafil (24). In addition, ataxia can occur when carbamazepine is coadministered with mechanism-based CYP3A4 inhibitors such as macrolide antibacterials, isoniazid, verapamil, and diltiazem (25,26).

CYP Induction

For assessment of potential induction of CYP isoform activity by the drug candidate, hepatocytes and ex vivo studies are utilized. Ex vivo studies usually involve dosing a rat with the drug candidate for approximately 5 days, removing the liver, isolating the liver microsomes, and then assessing the expression and activity of select CYP family (e.g., CYP3A, CYP1A, CYP2E, and CYP2B). Induction of metabolic systems is apparent if the drug candidates induce their own metabolizing systems (i.e., autoinducer); however, a drug candidate may be an inducer of other enzymes not related to its metabolism. The success of development of drug candidates that are shown to be autoinducers in the rat may be hindered because sometimes high exposures are needed to show sufficient safety margins and/or to reach an MTD in preclinical SA studies. Knowledge of the specific CYP family induced from ex vivo studies can be used to establish a rat hepatocyte assay to screen backup compounds.

Prediction of induction potential of drug candidates from ex vivo induction studies in rat and primary rat hepatocytes may not be directly applied to humans because important interspecies differences exist in the response of other inducible subfamilies of CYP (11). For example, omeprazole is a CYP1A2 enzyme inducer in humans, but has no such inductive effect in mice or rabbits (27,28). Phenobarbital induces predominantly members of the CYP2B subfamily in rats, whereas in humans it appears that the major form induced belongs to the CYP3A subfamily (29). Thus, drugs that induce CYP enzymes in animals should not be assumed necessarily to have enzyme-inducing capacity in humans and vice versa. Therefore, for assessment of potential induction of CYP isoform activity by the drug candidate in humans, the "gold standard" has been primary human hepatocytes.

Culture conditions for primary human hepatocytes that (i) maintain the optimal results, (ii) are the most forgiving to inexperienced scientists, and maintain cuboidal shape and cord-like arrays (instead of forming distinct clusters or aggregates) throughout the culture period are the sandwich-culture condition where hepatocytes are plated on a layer of gelled collagen and then another layer of gelled collagen is placed on top of the plated hepatocytes (30,31). However, the overall benefit of this culture method versus a single gelled collagen layer or even no gelled collagen layer may not be marked to a more experienced scientist; for example, the presence of a gelled collagen layer has little effect on the CYP induction response; furthermore, the contribution of the matrix constituents to the overall protein content must be accounted for if enzyme activities are measured by adding the substrates directly to the cell monolayers (31). This is in contrast to rat hepatocytes, where layering of gelled collagen has been shown to be necessary for xenobiotic-mediated induction response (32–34). Similarly, the type of culture medium used for human hepatocytes seems not to be crucial and thus general types of culture medium like William's E or Chee's medium are acceptable (31). The optimal plating density is probably the most important parameter for human hepatocytes to form intact junctional complexes between neighboring cells (30,35,36). An optimal cell density has been reported to be 125,000–150,000 cells/cm^2 (31).

The use of cryopreserved hepatocytes to evaluate the induction potential of drug candidates has been discouraged by the pharmaceutical community (37). However, recently a study evaluated the utility of cryopreserved hepatocytes as a tool for evaluation of induction potential of drug candidates (38). The advantage of utilization of cryopreserved versus primary hepatocytes is the greater availability of cryopreserved hepatocytes and the ability to standardize the hepatocyte population across each assay (38). Interestingly, the authors showed that induction of CYP1A2, CYP2B6, CYP2C9, CYP2E1, and CYP3A4 was possible in cryopreserved hepatocytes and this system, in some cases, has a more sensitive readout of induction potential than the primary hepatocytes due to the lower basal activity of CYP isoforms in cryopreserved hepatocytes (38). Possibly, the use of cryopreserved hepatocytes with advances in cryopreservation technology may have utility as an increased throughput screen for drug discovery in the near future with the limitations of this system understood (39–41).

Conclusions

The mechanism(s) of pharmacokinetic-based drug–drug interactions can vary from absorption-based to metabolic-based. The most documented pharmacokinetic-based drug–drug interactions have been metabolic-based. In vitro metabolic systems including liver microsomes and hepatocytes are useful in understanding and predicting these interactions. Distribution-based drug–drug interactions are increasing in the literature.

IMPORTANCE OF UNDERSTANDING DRUG METABOLISM IN DISCOVERY AND EARLY DEVELOPMENT

Introduction

The need for understanding drug metabolism and identifying metabolites of drug candidates in the discovery and early development phases is growing (42). The advantages of early metabolite identification include the ability to (i) identify metabolic spots responsible for shortened half-life or low oral bioavailability, (ii) gain insight into reactive or toxic metabolites, and (iii) find metabolites that are active at the target site. Metabolite identification now plays a pivotal role in generating better molecules and ultimately in selecting potential clinical candidates. Especially now since regulatory agencies expect the metabolism of a drug candidate is understood not only in the preclinical or SA species, but in humans before the start of the clinical studies. This requirement demands that predictive metabolism is complete and correct and/or initial human ADME studies be performed in preclinical development. In this section, the benefit of metabolite identification and prediction in each segment of drug discovery/development will be discussed; furthermore, strategies, experiments, and analysis will also be suggested to ensure successful prediction of metabolism in animals and humans (Fig. 5).

Understanding Drug Metabolism in Early Discovery

During the hit to early lead optimization, early prediction of phase I metabolites for a select group of drug candidates will be useful (i) to examine to

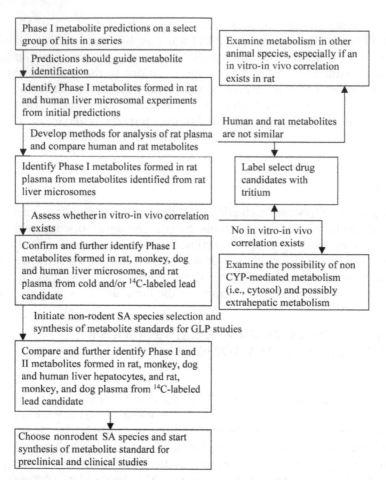

Figure 5 Strategy for the assessment of metabolism of the drug candidate in discovery and early development. Testing the validity of metabolic predictions should start early in drug discovery. The use of radio-labeled drug candidates should be implemented early if there are complications in understanding drug metabolism in preclinical species and/or human. For additional details, see text.

general predictability of the metabolism of the series, (ii) aid in the establishment of early in vitro–in vivo correlations in high-throughput screen (HTS) for rat metabolic stability and pharmacokinetic (PK) prediction, and (iii) develop early human metabolism prediction from rat metabolism and thus investigate the relevance of rat in lead selection of drug candidates. Predictions should focus on common phase I metabolism (i.e., oxidation and demethylation) and only a fraction of the compounds in this series with potential promise for lead candidate selection should be examined.

Suggested matrices for these experiments are rat plasma from initial oral dosing PK studies, and rat and human liver microsomal samples that initially define the metabolic stability of the compounds. Bioanalysis should be initially performed in the microsomal samples since this matrix is simpler compared to plasma (i.e., less endogenous compounds to provide background). Detection of simple mass changes (e.g., $+16$ m/z for oxidation and/or -14 m/z for demethylation) in single quad mode and at least initially, the a priori prediction should guide this analysis. If structure of metabolites from these predictions can be easily elucidated (i.e., a distinct N-demethylation or O-demethylation or oxidation), structural elucidation should proceed. Once metabolites are defined in the microsomal incubation, bioanalysis of the rat plasma may occur and should focus on metabolites identified in the microsomal samples. If possible, MRM methods can be developed from the microsomal bioanalysis and used to analyze the rat plasma to ensure selectivity. Select limited plasma samples (i.e., only two samples may be needed—one close to the C_{max}/T_{max} of the parent and one half-life from this C_{max}/T_{max}) for analysis.

In the author's experience, if a priori predictions of metabolism are true in all of the above matrices, chances of predicting from in vitro to in vivo systems and human metabolism will be strong. If no oxidation products can be detected in the rat plasma, the existence of conjugation products may be determined in the plasma and correlated with rat and human hepatocytes; however, unless the conjugation product is predicted to be a potentially reactive or pharmacologically active metabolite (e.g., acyl glucuronide or sulfonate), the concern about these conjugates in early development of the compound may be minimal.

If metabolite(s) in rat liver microsomes and plasma are similar, but are not similar to metabolite(s) in human liver microsomes, this may be a warning that the metabolites in rat for SA studies may not adequately cover human metabolites, and selection of nonrodent species for this coverage may be critical. Furthermore, compound selection from rat PK may not be a relevant choice and other species may be more relevant for compound selection for human PK. For example, the metabolism of Org 30659, a progestagen, showed considerable difference in rat compared to human (43). In rat, opening of the A-ring resulted in a 2-hydroxy, 4-carboxylic acid, 5alpha-H metabolite (43). In contrast, metabolic route of Org 30659 in human was (i) reduction of the 3-keto moiety to a 3alpha- or 3beta-hydroxy; (ii) reduction of the 4-double bond to a 5alpha-H or 5beta-H; and (iii) glucuronidation of the 17beta-hydroxy group (43). The majority of the metabolites isolated from the monkey were similar to the human metabolism, except in monkeys Org 30659 was glucuronidated at both the C3- and C17-hydroxy groups (43). Therefore, lead candidates with suitable human PK in this structurally related series may be better selected with primate versus rat models. Currently, commercial liver microsomes are available for this further evaluation, which include but are not limited to mouse (ICR/CD-1), guinea pig (Dunkin-Hartley), dog (Beagle), rabbit (New Zealand White), minipig (Yucatan), and monkey (Cynomolgus and Rhesus).

Incongruities in metabolite(s) from rat liver microsomes compared to plasma combined with other factors (comparison of in vitro metabolic stability and PK parameters like bioavailability and/or clearance/half-life) may indicate that rat liver microsomes are not useful systems for early in vitro–in vivo correlations in HTS rat PK prediction. A possible explanation may be that metabolism in the rat is cytosolic and/or not mediated by CYPs. Enzymes in liver cytosol are abundant, e.g., sulfotransferases. If the metabolic stability and/or intrinsic clearance of a drug candidate in rat liver microsomes does not correlate with its apparent elimination (e.g., i.v. clearance) from rat plasma, the drug candidate may not be metabolized by CYPs in vivo. These differences also exist in human drug metabolism. For example, a retrospective study showed that the metabolism of XK469 in human liver microsomes produced one oxidation product which did not correspond to the major oxidation product in human urine; however, metabolism of XK469 in liver cytosol produced the distinct N-oxide metabolite, the major oxidation product in human urine (44). Furthermore, hepatocytes, which contain the complete enzymatic machinery of the liver, metabolized XK469 to both oxidation products (44). To possibly obtain better quantitative estimates from in vitro to in vivo, investigate different concentrations of substrate to ensure that lowering the drug concentration in the liver microsomal system does not shift the metabolizing systems that are low affinity and high capacity to systems that are high affinity and low capacity. Another potential difference for lack of in vitro–in vivo correlation is that the liver may not be the primary site of metabolism of the drug candidate; however, examples of these differences are not prominent in the literature. Many extrahepatic enzymatic systems metabolize drugs, most notably, but not exclusively, intestine, lung, and kidney.

The inability to predict metabolism of a compound series may indicate the need for further investigation into other mechanism(s) of metabolism at an early stage of development for the series. One recommendation may be to investigate metabolism of a radio-labeled drug candidate at an earlier stage of development to assure the accountability of all metabolites in the system. Displacement reaction with tritium for a drug candidate may be useful for obtaining a qualitative analysis of the metabolites in vitro with rat and human microsomes, and rat plasma. As tritium label can be displaced with water, this radio-label is not recommended for quantitative analysis; however, the low expense of this procedure combined with the relative ease of the reaction favors the use of the radiolabel in early lead optimization to obtain a better perspective on metabolism of a series. For example, Hegstad et al. (45) characterized metabolites of benz(j) aceanthrylene in feces, urine, and bile from rat after administration of [^{3}H]-labeled benz(j) aceanthrylene. This method enabled detection of previously unidentified metabolites with HPLC, UV, and electrospray/atmospheric pressure chemical ionization mass spectrometry (45).

In vitro the use of tritium has been found to assess the quantitative significance of metabolic pathways that lead to the formation of reactive metabolites relative to the overall metabolic turnover of the compound (46). Detection of reactive metabolites has been implicated in various drug-induced toxicities and other undesirable drug properties (e.g., mechanism-based enzyme inactivation). Approaches that use LC–MS^n experiments on ion trap mass spectrometer to rapidly identify the sites of bioactivation within the lead structures during routine metabolite identification studies in liver microsomes allow early synthetic chemistry intervention to these sites within the molecule and minimize the potential for the formation of reactive metabolites (46). Reactivity of a drug candidate may be considered negligible if reversible binding is below 5–10 pmol-equiv./mg protein. This parameter may be useful for compound selection in early-to-late lead optimization; however, reactivity may not translate into toxicity of the metabolite, so other parameters of drug candidates should also be considered during this process.

Understanding Drug Metabolism in Late Discovery

During early-to-late lead optimization, detection and characterization of phase I metabolites for the lead candidates will be useful in further establishment of in vitro–in vivo correlations for the understanding of metabolism in SA species and prediction to human. During this stage, the knowledge obtained from these more advanced investigations will allow medicinal chemistry and/or PR&D to start research into synthesis of metabolites as standards for analysis in preclinical and clinical studies. Species selection for SA studies may be aided where knowledge that metabolism of some lead candidate(s) in SA species may not be relevant for human metabolism. Furthermore, early knowledge may be gained for the need of additional resources or alternative plans in early development (i.e., phase I) if the human metabolism cannot be predicted.

Quantitative analysis with ^{14}C-labeled compounds in liver microsomal systems and animal plasma is probably the most useful experiment in understanding metabolism of a drug candidate. The incorporation of a ^{14}C label should be placed on a part of a molecule that has little potential to be metabolized. This label should also ideally be a part of a common intermediate in a series of candidates, so it could be incorporated into many structures. If a metabolism has been predicted well in earlier studies, the incorporation of ^{14}C label into drug candidate(s) may be delayed until preclinical development; however, incorporation and utilization of this label can also be important in further defining metabolites and reactivity of drug candidates at an early stage. Unlike the use of a tritium label to assess metabolite, the proper incorporation ^{14}C label into a drug candidate allows quantitative analysis of the contribution of the metabolites to the overall metabolism of the drug candidate and also allows calculation of mass balance to ensure accountability of metabolism and elimination of the drug candidate.

Understanding Drug Metabolism in Early Development

During late lead optimization to preclinical candidate approval, detection of phase I and II metabolites in SA species facilitates appropriate selection of nonrodent species in SA studies and consolidates prediction to human. This assessment is usually limited to the lead candidate chosen for preclinical and clinical evaluation and ensures that the metabolites that have been chosen to synthesize for analysis in preclinical and clinical studies are complete and correct.

For this evaluation, metabolites of ^{14}C-labeled compounds should be assessed in human and animal (i.e., SA species) hepatocytes or liver slices, and compared with animal plasma. All metabolites (i.e., phases I and II) should be quantified in all models. For example, phase II conjugates, such as acyl glucuronides and sulfonates can be detected, and synthesized for monitoring in SA studies. Possible analysis of spectra from rat, dog, or monkey excreta (i.e., bile and urine), and target organs may also be useful for this evaluation. Determination of exposure of metabolites and parent in plasma is the best way to measure the importance of the metabolite in terms of systemic exposure. When relative exposure of metabolite is compared in all in vitro and in vivo systems, the predictability of the relative exposure and toxicity of metabolite(s) in human is greatly improved. For example, assessment of the pharmacokinetics of [^{14}C]2-[3-[3-[(5-ethyl-4′-fluoro-2-hydroxy[1,1′-biphenyl]-4-yl)oxy]propoxy]-2-propylphenoxy-]benzoic acid ([^{14}C]LY293111), an experimental anticancer agent, suggested long-lived circulating metabolites in rats (47). In vivo metabolites of LY293111 were examined in plasma, bile, urine, and feces of Fischer 344 (F344) rats after oral administration of [^{14}C]LY293111 (47). Metabolites were profiled by high-performance liquid chromatography–radiochromatography, and identified by LC–MS and LC/NMR (47). The major in vivo metabolites of LY293111 identified in rats were phenolic (ether), acyl, and bisglucuronides of LY293111 (47). Measurement of radioactivity in rat plasma confirmed that a fraction of LY293111-derived material was irreversibly bound to plasma protein and that this bound fraction increased over time (47). This was consistent with the observed disparity in half-lives between LY293111 and total radioactivity in rats and monkeys, and is likely due to covalent modification of proteins by the acyl glucuronide (47). In vitro metabolism of [^{14}C]LY293111 in liver slices from CD-1 mice, F344 rats, rhesus and cynomolgus monkeys, and humans indicates that glucuronidation was the primary metabolic pathway in all species (47). The acyl glucuronide was the most prevalent radioactive peak (16% of total ^{14}C) produced by F344 rat slices, whereas the ether glucuronide was the major metabolite in all other species (26–36% of total

^{14}C) (47). The data presented suggest that covalent modification of proteins by LY293111 acyl glucuronide is possible in multiple species, although the relative reactivity of this metabolite appears to be low compared with those known to cause adverse drug reactions (47).

Mass balance studies with ^{14}C-labeled compounds, where radioactivity in vitro with hepatocytes and in vivo with excreta and plasma can be monitored, determine the accountability of metabolites in the experiments. Significant lack of mass balance in the experiment may need further investigation because an elimination and/or metabolic pathway may be missed. Other possible explanations for a lack of mass balance may be the following: (i) ^{14}C label may be eliminated via a gas (e.g., CO_2), (ii) ^{14}C-labeled compound may be covalently bound in the cellular pellet (in vitro) or tissue protein (in vivo), (iii) ^{14}C-labeled compound may stick to plastic (in vitro) or be sequestered in a tissue compartment (in vivo), and (iv) a possible analytical issue (i.e., elusion off column) might exist.

Complete separation of metabolites (i.e., good chromatographic resolution) ensures the complete identification of metabolites. For example, the metabolism of droloxifene, an analog of the antibreast cancer drug tamoxifen, by human liver microsomes is an example of the application of LC–MS in an in vitro metabolism study (48). The metabolites identified were N-desmethyldroloxifene [(M1H) 1, m/z 374], 4-hydroxydroloxifene [(M1H) 1, m/z 404], droloxifene N-oxide [(M1H) 1, m/z 404], and a-hydroxydroloxifene [(M1H) 1, m/z 404] (48). The example demonstrates the importance of chromatographic resolution in LC–MS analysis (48). There were three metabolites with the (M1H) 1 ion at m/z 404 and the unequivocal identification of these metabolites is difficult without complete separation (48).

Conclusions

With the new focus on metabolites and safety of drug candidates, the need for understanding drug metabolism of drug candidates earlier in the discovery/development process is essential. Hypothesis-driven detection of metabolites in early discovery will facilitate identification of potential problems with metabolism early, which may avoid delays in drug development later in the program. Consideration of non-CYP-mediated and/or extrahepatic metabolism may also be needed if in vitro–in vivo correlations are difficult to obtain. Establishment of these correlations is essential for a selection of preclinical species and early prediction of human metabolism for the drug candidate.

REFERENCES

1. Voisin E, Ruthsatz M, Collins JM, et al. Extrapolation of animal toxicity to humans: interspecies comparisons in drug development. Regul Toxicol Pharmacol 1990; 12:107–116.

2. Medinsky MA, Valentine JL. Toxicokinetics. In: Klaassen CD, ed. Casarett and Doull's Toxicology. New York: McGraw-Hill, 2001.
3. Gibaldi M, Perrier D. Pharmacokinetics. 3rd. New York: Marcel Dekker, 1982.
4. Rowland M, Tozer TN. Clinical Pharmacokinetics. 3rd ed. Philadelphia: Lippincott Williams & Williams, 1995.
5. Shargel L, Yu ABC. Applied Biopharmaceutics and Pharmacokinetics. 3rd ed. Norwalk: Appleton & Lange, 1993.
6. Dixit R, Riviere J, Krishnan K, et al. Toxicokinetics and physiologically based toxicokinetics in toxicology and risk assessment. J Toxicol Environ Health B 2003; 6:1–40.
7. U.S. FDA, Guidance documents. U.S. Food and Drug Administration, 2005. http://www.fda.gov/cder/guidance/index.htm.
8. Kedderis GL. Pharmacokinetics of drug interactions. Adv Pharmacol 1997; 43:189–203.
9. Fleisher D, Li C, Zhou Y, et al. Drug, meal and formulation interactions influencing drug absorption after oral administration: clinical implications. Clin Pharmacokinet 1999; 36:233–254.
10. Yu DK. The contribution of *P*-glycoprotein to pharmacokinetic drug–drug interactions. J Clin Pharmacol 1999; 39:1203–1211.
11. Lin JH, Lu AY. Inhibition and induction of cytochrome P450 and the clinical implications. Clin Pharmacokinet 1998; 35:361–390.
12. Honig PK, Wortham DC, Zoman K, et al. Terfenadine–ketoconazole interaction: pharmacokinetic and electrocardiographic consequences. J Am Med Assoc 1993; 269:1550–1552.
13. Monahan BP, Ferguson CL, Killeavy ES, et al. Torsades de pointes occurring in association with terfenadine use. J Am Med Assoc 1990; 264:2788–2790.
14. Woosley RL, Chen Y, Freiman JP, et al. Mechanism of the cardiotoxic actions of terfenadine. J Am Med Assoc 1993; 269:1532–1536.
15. McConn DJ, Zhao Z. Integrating in vitro kinetic data from compounds exhibiting induction, reversible inhibition and mechanism-based inactivation: in vitro study design. Curr Drug Metab 2004; 5:141–146.
16. Schrag ML, Wienkers LC. Covalent alteration of the CYP3A4 active site: evidence for multiple substrate binding domains. Arch Biochem Biophys 2001; 391:49–55.
17. Ansede JH, Thakker DR. High-throughput screening for stability and inhibitory activity of compounds toward cytochrome P450-mediated metabolism. J Pharm Sci 2004; 93:239–255.
18. Zhou S, Yung Chan S, Cher Goh B, et al. Mechanism-based inhibition of cytochrome P450 3A4 by therapeutic drugs. Clin Pharmacokinet 2005; 44:279–304.
19. Dresser GK, Spence JD, Bailey DG. Pharmacokinetic–pharmacodynamic consequences and clinical relevance of cytochrome P450 3A4 inhibition. Clin Pharmacokinet 2000; 38:41–57.
20. Michalets EL, Williams CR. Drug interactions with cisapride: clinical implications. Clin Pharmacokinet 2000; 39:49–75.
21. Spinler SA, Chang JW, Kindwall KE, et al. Possible inhibition of hepatic metabolism of quinidine by erythromycin. Clin Pharmacol Ther 1995; 57:89–94.
22. Williams D, Feely J. Pharmacokinetic–pharmacodynamic drug interactions with HMG-CoA reductase inhibitors. Clin Pharmacokinet 2002; 41:343–370.

23. Anderson JR, Nawarskas JJ. Cardiovascular drug–drug interactions. Cardiol Clin 2001; 19:215–234.
24. Simonsen U. Interactions between drugs for erectile dysfunction and drugs for cardiovascular disease. Int J Impot Res 2002; 14:178–370.
25. Patsalos PN, Froschor W, Pisani F, et al. The importance of drug interactions in epilepsy therapy. Epilepsia 2002; 43:365–385.
26. Spina E, Pisani F, Perucca E. Clinically significant pharmacokinetic drug interactions with carbamazepine. Clin Pharmacokinet 1996; 31:198–214.
27. Diaz D, Fabre L, Daujat M, et al. Omeprazole is an aryl hydrocarbon-like inducer of human hepatic cytochrome P450. Gastroenterology 1990; 99:737–747.
28. McDonnell WM, Scheiman JM, Traber PG. Induction of cytochrome P450IA genes (CYP1A) by omeprazole in the human alimentary tract. Gastroenterology 1992; 103:1509–1516.
29. Rice JM, Diwan BA, Ward JM, et al. Phenobarbital and related compounds: approaches to interspecies extrapolation. Prog Clin Biol Res 1992; 374:231–249.
30. LeCluyse EL, Madan A, Hamilton G, et al. Expression and regulation of cytochrome P450 enzymes in primary cultures of human hepatocytes. J Biochem Mol Toxicol 2000; 14:177–188.
31. LeCluyse EL. Human hepatocyte culture systems for the in vitro evaluation of cytochrome P450 expression and regulation. Eur J Pharm Sci 2001; 13:343–368.
32. LeCluyse EL, Audus KL, Hochman JH. Formation of extensive canalicular networks by rat hepatocytes cultured in collagen-sandwich configuration. Am J Physiol 1994; 266:C1764–C1774.
33. LeCluyse EL, Bullock PL, Parkinson A, et al. Cultured rat hepatocytes. Pharm Biotechnol 1996; 8:121–159.
34. Sidhu JS, Farin FM, Omiecinski CJ. Influence of extracellular matrix overlay on phenobarbital-mediated induction of CYP2B1, 2B2, and 3A1 genes in primary adult rat hepatocyte culture. Arch Biochem Biophys 1993; 301:103–113.
35. Ferrini JB, Pichard L, Domergue J, et al. Long-term primary cultures of adult human hepatocytes. Chem Biol Interact 1997; 107:31–45.
36. Greuet J, Pichard L, Ourlin JC, et al. Effect of cell density and epidermal growth factor on the inducible expression of CYP3A and CYP1A genes in human hepatocytes in primary culture. Hepatology 1997; 25:1166–1175.
37. Li AP, Gorycki PD, Hengsiter JG, et al. Present status of the application of cryopreserved hepatocytes in the evaluation of xenobiotics: consensus of an international expert panel. Chem Biol Interact 1999; 121:117–123.
38. Roymans D, Annaert P, Van Houdt J, et al. Expression and induction potential of cytochromes P450 in human cryopreserved hepatocytes. Drug Metab Dispos 1999; 33:1004–1016.
39. Alexandre E, Viollon-Abadie C, David P, et al. Cryopreservation of adult human hepatocytes obtained from resected liver biopsies. Cryobiology 2002; 44:103–113.
40. Hengstler JG, Utesch D, Steinberg P, et al. Cryopreserved primary hepatocytes as a constantly available in vitro model for the evaluation of human and animal drug metabolism and enzyme induction. Drug Metab Rev 2000; 32:81–118.

41. Hengstler JG, Ringel M, Biefang K, et al. Cultures with cryopreserved hepato-cytes: applicability for studies of enzyme induction. Chem Biol Interact 2000; 125:51–73.
42. Lin JH, Lu AY. Role of pharmacokinetics and metabolism in drug discovery and development. Pharmacol Rev 1997; 49:403–449.
43. Verhoeven CH, Gloudemansa RH, Groothuisab GM, et al. Excretion balance and metabolism of the progestagen Org 30659 in healthy postmenopausal women. J Steroid Biochem Mol Biol 2000; 73:39–48.
44. Anderson LW, Collins JM, Kleckar RW, et al. Metabolic profile of XK469 (2(R)-[4-(7-chloro-2-quinoxalinyl)oxyphenoxy]-propionic acid; NSC698215) in patients and in vitro: low potential for active or toxic metabolites or for drug–drug interactions. Cancer Chemoth Pharm 2005; 56:351–357.
45. Hegstad S, Lundanes E, Holmes JA, et al. Characterization of metabolites of benz(j)aceanthrylene in faeces, urine and bile from rat. Xenobiotica 1999; 29:1257–1272.
46. Samuel K, Yin W, Stearns RA, et al. Addressing the metabolic activation poten-tial of new leads in drug discovery: a case study using ion trap mass spectrometry and tritium labeling techniques. J Mass Spectrom 2003; 38:211–221.
47. Perkins EJ, Cramer JW, Forid NA, et al. Preclinical characterization of 2-[3-[3-[(5-ethyl-4'-fluoro-2-hydroxy[1,1'-biphenyl]-4-yl)oxy]propoxy]-2-propylphenox-y]benzoic acid metabolism: in vitro species comparison and in vivo disposition in rats. Drug Metab Dispos 2003; 31:1382–1390.
48. Bruning PF. Droloxifene, a new antioestrogen in postmenopausal advanced breast cancer: preliminary results of a double-blind dose-finding phase II trial. Eur J Cancer 1992; 28A:1404–1407.

6

Considerations for Applying Physiologically Based Pharmacokinetic Models in Risk Assessment

Chadwick Thompson and Babasaheb Sonawane

National Center for Environmental Assessment, Office of Research and Development, U.S. Environmental Protection Agency, Washington, D.C., U.S.A.

Andy Nong and Kannan Krishnan

Groupe de Recherche Interdisciplinaire en Santé et Département de Santé Environnementale et Santé au Travail, University of Montreal, Montreal, Quebec, Canada

INTRODUCTION

Pharmacokinetics (*pharmacon* + *kinetics*; *pharmakon* (Greek) = drugs and poisons; *kinetics* = change as a function of time) involves the study of the time course of the concentrations or amounts of a parent chemical or its metabolites in biological fluids, tissues, and excreta as well as the construction of mathematical models to interpret such data (1). The time course of the concentration of a chemical or its metabolite in biota is determined by the rate and extent of absorption, distribution, metabolism, and excretion (ADME). The pharmacokinetics (PK) or ADME of a substance determines the *delivered dose*, i.e., the amount of chemical available for interaction in tissue(s). Relating an adverse response observed in biota to an appropriate measure of delivered dose as opposed to an administered dose (or exposure concentration) improves the characterization of many dose-response relationships.

Dose-response relationships that appear unclear or confusing at the administered dose level may become more understandable when expressed on the basis of the internal dose of the chemical. Figure 1 depicts the case of a hypothetical chemical for which the correlation between dose and response is weak or complex (*Panel A*). However, once the relationship is based on internal dose, there emerges a clear and direct relationship between dose and response (*Panels B and C*). The major advantage of constructing dose-response relationships on the basis of internal or delivered dose is that it can provide a stronger biological basis for conducting extrapolations and for comparing responses across studies, species, routes, and dose levels (2–6).

The use of blood and tissue concentrations for relating dose and response in exposed organisms has long been recognized in pharmacology (1). The target tissue dose closely related to ensuing adverse responses is often referred to as the "dose metric" (7). Dose metrics used for risk assessment applications should reflect the biologically active form of the chemical (parent chemical, metabolites, or adducts), its level (concentration or amount), duration of internal exposure (instantaneous, daily, lifetime, or specific window of susceptibility), intensity (peak, average, or integral) as well as the appropriate biological matrix (e.g., blood, target tissue, surrogate tissues). For assessment of health risks related to lifetime exposure of systemically acting chemicals, in the absence of mode of action (MOA) information to the contrary, the integrated concentration of the toxic form of the chemical over a lifetime [i.e., area under the concentration (AUC) vs. time curve] in target tissue is often considered to be an appropriate dose metric (8–10).

When the toxic moiety is not the parent chemical, but a reactive intermediate, the amount of metabolite produced per unit of time or the amount of metabolite in target tissue over a period of time (e.g., milligram metabolite/L tissue during 24 hours) has been used as the dose metric (7). For developmental effects, the dose surrogate is defined in the context of window of susceptibility

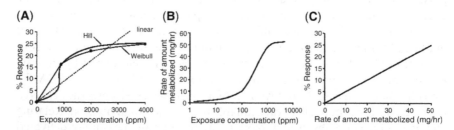

Figure 1 Relationship between the exposure concentration and adverse response for a hypothetical chemical. *Panel A* depicts the case of a chemical for which the correlation between dose and response is weak or complex, along with equally plausible curve fits (linear, Hill, and Weibull). This dose-response relationship is improved when it is based on an appropriate measure of internal dose (*Panels B and C*).

for a particular gestational event (11,12). Although the AUC and rate of metabolite formation are among the most commonly investigated dose metrics, other surrogates of tissue exposure may also be appropriate for risk assessment purposes, depending on the chemical and MOA; examples include maximal concentration of the toxic moiety, duration, and extent of receptor occupancy, macromolecular adduct formation, and depletion of glutathione (10).

Equally important as target tissue dose is the target tissue location relative to the portal of entry (POE). A range of modeling approaches are used to characterize exposures and resulting delivered doses that reflect the differences in chemical and physical characteristics (e.g., stable or reactive gases, particulate matter, lipophilic organics, water soluble compounds) and their ability to cause POE and/or systemic toxicity (13–16). Exposure to many drugs and toxicants occur via the oral route and cause systemic effects; thus noncompartmental and compartmental pharmacokinetic modeling has focused on the characteristics related to such exposures (17,18). Compartmental modeling has since evolved to include more physiologically based approaches (19), which are the focus of this chapter.

The objective of this chapter is to provide a description of approaches for using physiologically based pharmacokinetic (PBPK) models and supporting data in human health risk assessment. Specifically, the chapter focuses on the use of PBPK models in predicting internal dose at target or surrogate tissues.

PHARMACOKINETIC DATA REQUIRED FOR RISK ASSESSMENTS

The dose-response analysis portion of a risk assessment can be used to determine a POD for one or more of the most sensitive critical effects, based on the relationship between administered dose and observed responses in laboratory or field studies. The most robust pharmacokinetic data set needed for risk assessment would consist of the time-course data on the dose metric associated with exposure scenarios and doses used in the critical studies chosen for the assessment (e.g., animal bioassays or human epidemiological studies) and relevant human exposure conditions. This information would be obtained for the window of susceptibility, route and scenario of exposure associated with the critical animal study, as well as for the window of susceptibility, appropriate route, and exposure scenarios in humans. If such a pharmacokinetic data set were available for each scenario of interest, then there would be no real need for any pharmacokinetic models, even though models could facilitate simulation of other potential scenarios of interest and critical determinants of tissue dose. In almost all cases, however, dose metric measurements associated with human exposures to environmental chemicals will not be available. Further, the available animal pharmacokinetic data may not correspond to the active toxic moiety, relevant route, or appropriate dose levels. In the absence of experimental data on the biologically active form of a chemical in target tissues, data on blood concentration of parent chemical, urinary metabolite levels, or fraction absorbed may be used as surrogates of dose metrics. These data can be used to develop a PBPK model to estimate the level of the toxic moiety of interest, and the uncertainty in those estimates can be formally characterized.

CHOOSING PBPK MODELS APPROPRIATE FOR USE IN RISK ASSESSMENT

Whether or not a PBPK model was initially intended for risk assessment purposes, it can be useful for risk assessment if it permits simulations of the tissue and blood concentrations of the toxic moiety (parent chemical or metabolite) associated with the animal toxicity or human epidemiological study serving as the basis for the derivation of health protective values [e.g., reference concentration (RfC), reference dose (RfD), cancer slope factors]. Specifically, the model should be able to simulate the dose metrics in the test species and/or in humans for the exposure route and exposure scenario of relevance. For most applications in risk assessment, a PBPK model:

- Should have been developed or calibrated for species and life stage of relevance for the endpoints of interest
- Should consist of parameters essential for simulating uptake via routes associated with human exposures as well as the critical study chosen for the assessment
- Should be able to provide predictions of the time-course of concentration of toxic moiety (parent chemical or metabolite) in the target organ or a surrogate compartment
- Should be peer-reviewed and evaluated for its structure and parameters

Figure 2 is a generic decision tree for applying the above criteria to PBPK model selection. In general, only peer-reviewed PBPK models for the relevant species and life stage consisting of parameters for simulating relevant routes of exposure and potentially relevant dose metrics are appropriate for use in risk assessment. The first criterion, though appearing self-evident, is quite fundamental, because the models available in the literature sometimes may not have been developed for the specific life stage and species used in the critical toxicological study forming the basis of a risk assessment. For example, although PBPK models for methanol have been developed in rats, monkeys, and humans, literature suggests that developmental toxicities in mice may be the critical effects (20–22) although a newer two-generation study in rats may also be important (23). When the PBPK model has not been developed for the life stage or species used in the study forming the basis of the POD for an assessment, additional work may have to be undertaken to resolve the situation before the PBPK model can be used for that particular assessment.

The PBPK model for the relevant species and life stage should be peer-reviewed to determine the adequacy of the structure, parameter estimation

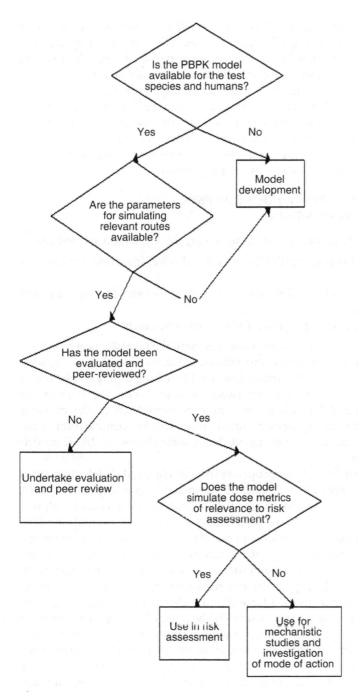

Figure 2 Decision tree for selecting physiologically based pharmacokinetic models appropriate for use in risk assessment.

methods, and the evaluation efforts. If the PBPK model for a chemical is available in the species and life stage corresponding to the study forming the basis of an assessment but it has not been peer-reviewed, then efforts may be directed toward such a review. Table 1 lists some fundamental questions and issues to be considered while conducting a peer-review evaluation of the adequacy of PBPK models intended for risk assessment applications. Finally, the peer-reviewed model(s) chosen for risk assessment applications should be able to provide simulations of the tissue dose of the toxic moiety or an appropriate dose metric for exposure scenarios and routes associated with the critical study as well as human exposures.

EVALUATION OF DOSE METRICS FOR PBPK MODEL-BASED ASSESSMENTS

When using PBPK models in risk assessment, the basic data needed are:

1. Point of departure (POD) and critical effect from one or more key studies
2. Peer-reviewed PBPK model for the relevant test species and humans
3. Dose metric appropriate for the risk assessment.

The methods and challenges associated with the identification of critical effects and PODs for an assessment remain the same regardless of whether one uses a PBPK model. The approaches for identifying PODs can be found elsewhere (15,24). The criteria and issues associated with the selection of PBPK models useful for risk assessment were considered in the previous section. The third data requirement noted above, i.e., the identification of the appropriate dose metric, is a key aspect determining the use of PBPK models in risk assessment.

The dose metric or the appropriate form of chemical closely associated with the toxicity varies from chemical to chemical, depending on the MOA and critical effect. The dose metric for PBPK-based risk assessment is chosen following the identification of the potential toxic moiety and evaluation of the relationship with the end point of concern. Analysis of available information on MOA of a chemical, along with descriptive toxicity data from various studies, including those using inhibitors or inducers of hepatic metabolism, should be useful in identifying the potential toxic moiety (parent chemical or metabolite). Further, the available data on closely related chemicals may be used to infer the likely toxic moiety. If the toxic effects of a chemical are enhanced following treatment with a specific inhibitor of hepatic metabolism, then it would implicate the parent chemical as the toxic moiety. Alternatively, if the effects are ablated by such inhibition or enhanced by the induction of the rate of metabolism, it would indicate that a metabolite is the toxic moiety. Similarly, the toxicity data for various exposure routes and modes of administration may be compared to infer the potential toxic moiety (25).

Table 1 Questions and Issues to Be Considered While Evaluating the Adequacy of a Physiologically Based Pharmacokinetic Model

Is an appropriate target organ or a surrogate tissue identified as one of the model compartments?

Are the known major sites of storage, transformation, and clearance included in the model structure?

Is the form of equation used to describe chemical uptake justified on the basis of the hypothesized mechanism of tissue uptake of the chemical?

Have enzymatic processes been described appropriately? (If simulated exposure levels are well below saturation, use of first-order kinetics is acceptable.)

Are the units throughout differential equations consistent (e.g., concentration should not be in mg/mL in one place and M in another)?

Are the input parameters related to the characteristics of the host, chemical, or environment?

Is the sum total of compartment volumes within 100% of the body weight?

Do the tissue blood flow rates add up to the cardiac output?

Is the ventilation:perfusion ratio specified in the model within physiological limits?

Is the volume of each tissue compartment within known physiological limits?

Is the approach used to establish partition coefficients valid?

Is the method used for estimating biochemical parameters adequate?

Is the allometric scaling done appropriately?

Is the integration algorithm used in the study known for solving differential equations and appropriate?

Does the shape of the pharmacokinetic curve generated by the model match that obtained experimentally?

Is the model, with its set of parameters, able to consistently describe available pharmacokinetic data?

After the potential toxic moiety has been identified, the appropriate measure of tissue exposure to the toxic moiety should be chosen. For example, the peak concentration has been related to neurotoxic effects of solvents (5,26–28), and concentration of trichloroethylene at the time of testing correlated with effects on behavioral and visual functions (29). Tissue concentrations of 2,3,7,8-tetrachlorodibenzo-*p*-dioxin measured during a critical period of gestation have been reported to predict the intensity of developmental responses (30). Gender-specific genotoxic effects of benzene in mice are related to differences in the rate of oxidative metabolism (31).

For chronic effects of chemicals, the integrated concentration of the toxic form of the chemical in target tissue over time (i.e., AUC) is often considered a reasonable dose metric (2,8–10). For carcinogens producing reactive intermediates, the amount of metabolite produced per unit of time and the amount of metabolite in target tissue over a period of time (e.g., milligram metabolite/L tissue during 24 hours) have been used as dose metrics (7). For developmental effects, the dose surrogate is defined in the context of window of susceptibility for a particular gestational event (11). Although the AUC and rate of metabolite formation figure among the most commonly investigated dose metrics, other surrogates of tissue exposure may also be appropriate for risk assessment purposes, depending on the chemical and its MOA (e.g., maximal concentration of the toxic moiety, duration, and extent of receptor occupancy, macromolecular adduct formation, or depletion of glutathione) (10).

When the appropriate dose metric cannot be identified readily, an evaluation of the relationship with the end point of concern should be undertaken with each of the dose metrics in order to identify the one that exhibits the closest or the best association (2,32). This becomes particularly important when there are little or confusing data on the plausible MOA of the chemical. At a minimum, the appropriate dose metric can be identified as the one that demonstrates a consistent relationship with positive and negative responses observed at various dose levels, routes, and scenarios in a given species. In other words, the level of the dose metric should be lower for exposure conditions with no effect and higher for toxic exposures, regardless of the route and exposure scenario (10). Where there is inadequate basis for giving priority to one dose metric over another, the most conservative one (the dose metric producing the highest risk or lowest acceptable exposure level) should be used if the objective is to be health protective (10). The use of the appropriate dose metric helps to reconcile route and species differences in cancer and noncancer responses, provided there are no pharmacodynamic differences.

GENERIC EXAMPLES OF THE USE OF PBPK MODELS IN RISK ASSESSMENT

High-Dose to Low-Dose Extrapolation

PBPK models facilitate high-dose to low-dose extrapolation of tissue dosimetry by accounting for the dose-dependency of relevant processes (e.g., saturable metabolism, enzyme induction, enzyme inactivation, protein binding, and depletion of glutathione reserves) (3). The description of metabolism in PBPK models has frequently included a capacity-limited metabolic process that becomes saturated at high doses. Nonlinearity arising from mechanisms other than saturable metabolism, such as enzyme induction, enzyme inactivation, depletion of glutathione reserves, and binding to macromolecules, have also been described with PBPK models (3,33). A PBPK model intended for use in high-dose to low-dose extrapolation should have the equations and parameters describing dose-dependent phenomena if dose-dependence occurs in the range of interest or assessment. Because the determinants of nonlinear behavior may not be identical across species and age groups (e.g., maximal velocity for metabolism, glutathione concentrations), these parameters are required for each species/age group. During the conduct of high-dose to low-dose extrapolation, no change in the parameters of the PBPK model is required except for the administered dose or exposure concentration.

An example of high-dose to low-dose extrapolation is presented in Figure 3. In this figure, both the blood AUC and the amount metabolized over a period of time (12 hours) are plotted as a function of exposure concentrations of toluene. For conducting high-dose to low-dose simulation in this particular example, only the numerical value of the exposure concentration (which is an input parameter for the PBPK model) was changed during every model run. All other model parameters remained the same. The model simulations shown in Figure 3 indicate the nonlinear nature of blood AUC, as well as the amount of toluene metabolized per unit time in the exposure concentration range simulated. In such cases, the high-dose to low-dose

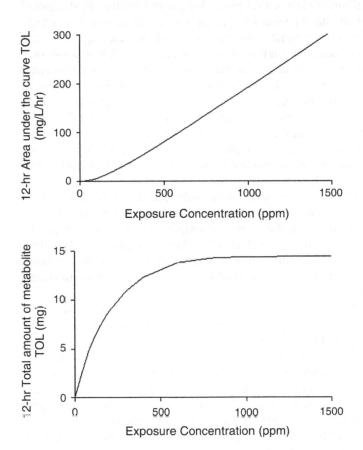

Figure 3 High-dose to low-dose extrapolation of dose metrics using physiologically based pharmacokinetic model for toluene. Inhalation exposures were for 4 hours and area under the concentrations and amount metabolized were calculated for 12 hours. *Source*: From Ref. 34.

extrapolation of tissue dosimetry should not be conducted assuming linearity; rather it should be performed using tools such as PBPK models.

Interspecies Extrapolation

The application of PBPK models for interspecies extrapolation of tissue dosimetry is performed in several steps. First, a rodent model is developed to describe the uptake and disposition of the chemical in question by integrating information on the physiological, biochemical, and physicochemical parameters. Then, a priori predictions of the PBPK model are compared with experimental observations to evaluate the adequacy of the structure and the parameter estimates of the rodent model. The next step involves using species-specific or allometrically scaled physiological parameters in the model and replacement of the chemical-specific parameters (e.g., partition coefficients) with appropriate estimates for the species of interest (e.g., humans). Thus, in this approach, the qualitative determinants of PK are considered to be invariant among the various mammalian species. Qualitative differences between species, if any, can also be factored into the existing structure of PBPK models (e.g., if different metabolic pathways existed among species).

For conducting interspecies extrapolation of pharmacokinetic behavior of a chemical, quantitative estimates of model parameter values (i.e., partition coefficients, physiological parameters, and metabolic rate constants) in the second species are required. The tissue:air partition coefficients of chemicals appear to be relatively constant across species, whereas blood:air partition coefficients show some species-dependent variability. Therefore, the tissue:blood partition coefficients for human PBPK models have been calculated by dividing the rodent tissue:air partition coefficients by the human blood:air partition values (19). The tissue:air and blood:air partition coefficients for volatile organic chemicals may also be predicted using appropriate data on the content of lipids and water in human tissues and blood (35,36).

Although the adult physiological parameters vary coherently across species, the kinetic constants for metabolizing enzymes do not necessarily follow any type of readily predictable pattern, making the interspecies extrapolation of xenobiotic metabolism difficult. Therefore, the metabolic rate constants for xenobiotics should be obtained in the species of interest. In vivo approaches for determining metabolic rate constants are not always feasible for application in humans. The alternative is to obtain such data under in vitro conditions (37,38). A parallelogram approach may also be used to predict values for the human PBPK model on the basis of metabolic rate constants obtained in vivo in rodents as well as in vitro using rodent and human tissue fractions (37,39). Alternatively, for chemicals exhibiting high affinity (low K_m) for metabolizing enzymes, V_{max} has been scaled to the 0.75 power of body weight, keeping the K_m species invariant. This approach may be useful as a crude approximation, but it should be used only when other direct measurements of metabolic parameters are not available or feasible. Whenever the human data for a particular chemical are not available for evaluation purposes, a corollary approach permitting the use of human data on similar chemicals may be attempted (40).

Route-to-Route Extrapolation

There are two different approaches to route extrapolation involving PBPK models. The first one is to use an animal model to extrapolate a POD for one route to a POD for another route on the basis of equivalent dose metric. The second approach involves the estimation of the human POD for one route from the available animal POD for another route on the basis of equivalent dose metric.

The extrapolation of the kinetic behavior of a chemical from one exposure route to another is performed by adding appropriate equations to represent each exposure pathway. For simulating the intravenous administration of a chemical, a single input representing the dose administered to the animal is included in the equation for mixed venous concentration. Oral gavage of a chemical dissolved in a carrier solvent may be modeled by introducing a first-order or a zero-order uptake rate constant, and dermal absorption has been modeled by including a diffusion-limited compartment to represent skin as a POE (19). After the equations describing the route-specific entry of chemicals into systemic circulation are included in the model, it is possible to conduct extrapolations of PK and dose metrics. This approach is illustrated in Figure 4 for oral-to-inhalation extrapolation of the kinetics of chloroform in rats. For simulating the inhalation PK, the oral dose was set to zero, whereas for simulating chloroform kinetics following oral dosing, the inhaled concentration was set to zero (Fig. 4). Accordingly, 4 hours of inhalation exposure to 83.4 ppm chloroform is equal to an oral dose of 1 mg/kg, as determined with PBPK models on the basis of equivalent dose metric (i.e., parent chemical AUC in blood) (Fig. 4).

Duration Adjustment

On the basis of equivalent dose metric, a duration-adjusted exposure value can be obtained with PBPK models (2,42). Accordingly, the AUC of a chemical for the exposure duration of the critical study is determined initially using the PBPK model, and then the atmospheric concentration for a continuous exposure (during a day, window of exposure, or lifetime), yielding the same AUC is determined by iterative simulation. Figure 5 depicts an example of 4- to 24-hour extrapolation of the PK of toluene in rats, based on equivalent 24-hour AUC (2.4 mg/L-hr). The rats exposed to 50 ppm for 4 hours and 9.7 ppm for 24 hours of toluene would receive the same dose metric.

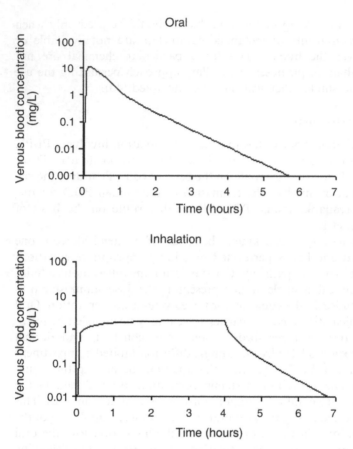

Figure 4 Oral-to-inhalation extrapolation of the pharmacokinetics of chloroform on the basis of same area under the concentration in blood (7.06 mg/L-hr). The oral dose was 1 mg/kg, and the inhaled concentration was 83.4 ppm (4 hours). *Source*: From Ref. 41.

Intraspecies Extrapolation

Intraspecies extrapolation of the dose metric is conducted using PBPK models with the sole intent of estimating the magnitude of the interindividual variability. In this regard, the population distributions of parameters, particularly those relating to physiology and metabolizing enzymes, are specified in a Monte Carlo approach, such that the PBPK model output corresponds to distributions of dose metric in a population. Using the Monte Carlo approach, repeated computations based on inputs selected at random from statistical distributions for each input parameter (physiological parameters, enzyme content/activity with or without the consideration of polymorphism) are conducted to provide a statistical distribution of the output, i.e., tissue

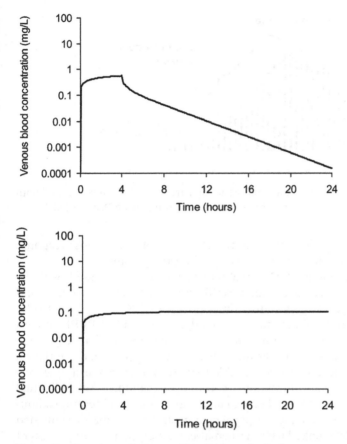

Figure 5 Duration adjustment (4 to 24 hours) of toluene exposures in rats, based on equivalent area under the concentration (2.4 mg/L-hr). The rats were exposed to 50 ppm toluene for 4 hours and 9.7 ppm for 24 hours. *Source*: From Ref. 34.

dose. Using the information on the dose metric corresponding to the 95th percentile and 50th percentile for unimodel, normal distribution (43), the magnitude of the interindividual variability can be computed (Fig. 6).

Even though several past efforts have characterized the impact of the distributions of parameters in the adult population, such variability analyses should also account for the life stage-specific changes in physiology, tissue composition, and metabolic activity (44,45).

PERSPECTIVES AND CONCLUDING REMARKS

Risk assessments based on the use of PBPK models account for only the pharmacokinetic aspect or, more specifically, target tissue exposure to a toxic moiety. If these tissue exposure simulations are combined with

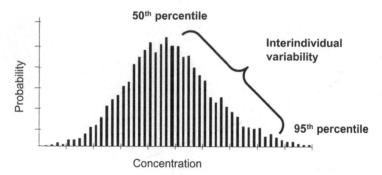

Figure 6 Estimation of interindividual variability from the 50th (median) and 95th percentile values of a dose metric simulated with a probabilistic PBPK model.

pharmacodynamic models, then better characterization of dose-response relationships and prediction of PODs may become possible.

The identification of PODs by simulation may become possible with the use of biologically based dose-response (BBDR) models. These models would require the linkage of quantitative descriptions of PK and pharmacodynamics via mechanism of action. Accordingly, the output of PBPK models is linked to the pharmacodynamic model using an equation that reflects the researchers' hypothesis of how the toxic chemical participates in the initiation of cellular changes leading to measurable toxic responses. For example, certain DNA adducts cause mutations, cytotoxic metabolites kill individual cells, and expression of growth factors can act as a direct proliferation stimulus. In each of these cases, the temporal change in the dose metric simulated by the PBPK model is linked with mathematical descriptions of the process of adduct formation, cytotoxicity, or proliferation in the BBDR models to simulate the quantitative influence of these processes on tumor outcome (46). A characteristic of several of these pharmacodynamic models is that they are able to simulate the normal physiological processes (e.g., cell proliferation rates, hormonal cycle) and additionally account for the ways in which chemicals perturb such phenomena, leading to the onset and progression of injury. The pharmacodynamic models, for linkage with PBPK models, are not available for a number of toxic effects and modes of action. This situation is a result, in part, of the complex nature of these models and the extensive data requirements for development and evaluation of these models for various exposure and physiological conditions.

With the availability of integrated pharmacokinetic–pharmacodynamic models, the scientific basis of the process of estimating PODs and characterizing the dose-response curve will be significantly enhanced. Additionally, such a modeling framework will facilitate a quantitative analysis of the impact of pharmacodynamic determinants on the toxicity outcome, such that the magnitude of the pharmacodynamic component of the interspecies and intraspecies factors can be characterized more confidently. Even though some PBPK models have been used in quantitative dose-response analysis for a few chemicals, the need for applying such models (where possible) in risk assessment should be continuously explored, and the concerns of cost/benefit of PBPK model development addressed.

ACKNOWLEDGMENTS

This chapter represents an abbreviated version of the draft document "Approaches for the Application of PBPK Models and Supporting Data in Risk Assessment." The following are acknowledged for their important contributions to this work: Hugh Barton, Jerry Blancato, Weihsueh Chiu, Rob DeWoskin, Joyce Donohue, Hisham El-Masri, Marina Evans, Lynn Flowers, Karen Hammerstrom, John Lipscomb, Allen Marcus, Dierdre Murphy, Alberto Protzel, Paul Schlosser, and Woodrow Setzer.

REFERENCES

1. Wagner JG. History of pharmacokinetics. Pharmacol Ther 1981; 12:537–562.
2. Andersen ME, Clewell HJ III, Gargas ML, et al. Physiologically-based pharmacokinetics and risk assessment process for methylene chloride. Toxicol Appl Pharmacol 1987; 87:185–205.
3. Clewell HJ III, Andersen ME. Dose, species and route extrapolation using physiologically-based pharmacokinetic models. In: Pharmacokinetics in Risk Assessment: Drinking Water and Health. Washington, DC: National Academy Press, 1987:159–182.
4. Aylward LL, Hays SM, Karch NJ, et al. Relative susceptibility of animals and humans to the cancer hazard posed by 2,3,7,8-tetrachlorodibenzo-p-dioxin using internal measures of dose. Environ Sci Technol 1996; 30:3534–3543.
5. Benignus VA, Boyes WK, Bushnell PJ. A dosimetric analysis of behavioral effects of acute toluene exposure in rats and humans. Toxicol Sci 1998; 43:186–195.
6. Melnick RL, Kohn MC. Dose-response analyses of experimental cancer data. Drug Metab Rev 2000; 32:193–209.
7. Andersen ME, Dennison JE. Mode of action and tissue dosimetry in current and future risk assessments. Sci Total Environ 2001; 274:3–14.
8. Collins J. Prospective predictions and validations in anti-cancer therapy. In: Pharmacokinetics in Risk Assessment Drinking Water and Health. 8. Washington, DC: National Academy Press, 1987:431–440.
9. Voisin E, Ruthsatz M, Collins JM, et al. Extrapolation of animal toxicity to humans: interspecies comparisons in drug development. Regul Toxicol Pharmacol 1990; 12:107–116.
10. Clewell HJ III, Andersen ME, Barton HA. A consistent approach for the application of pharmacokinetic modeling in cancer and noncancer risk assessment. Environ Health Perspect 2002; 110:85–93.
11. Welsch F, Blumenthal GM, Conolly RB. Physiologically based pharmacokinetic models applicable to organogenesis: extrapolation between species and potential use in prenatal toxicity risk assessments. Toxicol Lett 1995; 82–83:539–547.

12. Luecke RH, Wosilait WD, Young JF. Mathematical analysis for teratogenic sensitivity. Teratology 1997; 55:373–380.
13. Andersen ME, Jarabek AM. Nasal tissue dosimetry-issues and approaches for "Category 1" gases: a report on a meeting held in Research Triangle Park. NC, February 11–12, 1998. Inhal Toxicol 2001; 13:415–435.
14. Overton JH Jr. Dosimetry modeling of highly soluble reactive gases in the respiratory tract. Inhal Toxicol 2001; 13:347–357.
15. U.S. EPA. Methods for Derivation of Inhalation Reference Concentrations and Application of Inhalation Dosimetry, EPA/600/8-90/066F, U.S. Environmental Protection Agency, Office of Research and Development, Washington, DC, 1994. Available at http://nepis.epa.gov/pubtitleORD.htm.
16. U.S. EPA. Air Quality Criteria for Particulate Matter, EPA/600/P-99/002aF, U.S. Environmental Protection Agency, Washington, DC, 2004. Available at http://cfpub.epa.gov/ncea/cfm/partmatt.cfm.
17. O'Flaherty EJ. Toxicants and Drugs: Kinetics and Dynamics. New York, NY: John Wiley & Sons, 1981.
18. Renwick AG. Toxicokinetics-pharmacokinetics in toxicology. In: Hayes AW, ed. Principles and Methods of Toxicology. 4th ed. Philadelphia, PA: Taylor & Francis, 2001:137–192.
19. Krishnan K, Andersen ME. Physiologically based pharmacokinetic modeling in toxicology. In: Hayes AW, ed. Principles and Methods of Toxicology. 4th ed. Philadelphia, PA: Taylor & Francis, 2001:193–241.
20. Horton VL, Higuchi MA, Rickert DE. Physiologically based pharmacokinetic model for methanol in rats, monkeys and humans. Toxicol Appl Pharmacol 1992; 117:26–36.
21. Rogers JM, Mole ML, Chernoff N, et al. The developmental toxicity of inhaled methanol in the CD-1 mouse, with quantitative dose-response modeling for estimation of benchmark doses. Teratology 1993; 47:175–188.
22. Bouchard M, Brunet RC, Droz PO, et al. A biologically based dynamic model for predicting the disposition of methanol and its metabolites in animals and humans. Toxicol Sci 2001; 64:169–184.
23. Clark LH, Setzer RW, Barton HA. Framework for evaluation of physiologically-based pharmacokinetic models for use in safety or risk assessment. Risk Anal 2004; 24:1697–1717.
24. U.S. EPA. Guidelines for Carcinogen Risk Assessment, SAB Review Draft, NCEA-F-0644, U.S. Environmental Protection Agency, Office of Research and Development, Washington, DC, 1999.
25. IPCS (International Programme on Chemical Safety). Guidance Document for the Use of Data in Development of Chemical-specific Adjustment Factors (CSAF) for Interspecies Differences and Human Variability in Dose/concentration Response Assessment, World Health Organization, Geneva, 2001. Available at www.who.int/entity/ipcs/publications/methods/harmonization/en/csafs_guidance_doc.pdf.
26. Bushnell PJ. Concentration time relationships for the effects of inhaled trichloroethylene on signal detection behavior in the rats. Fund Appl Toxicol 1997; 36:30–38.
27. Pierce CH, Dills RL, Morgan MS, et al. Biological monitoring of controlled toluene exposure. Int Arch Occup Environ Health 1998; 71:433–444.
28. MacDonald AJ, Rostami-Hodjegan A, Tucker GT, et al. Analysis of solvent central nervous system toxicity and ethanol interactions using a human population physiologically based kinetic and dynamic model. Regul Toxicol Pharmacol 2002; 35:165–176.

29. Boyes WK, Bassnell PJ, Crofton KM, et al. Neurotoxic and pharmacokinetic responses to trichloroethylene as a function of exposure scenario. Environ Health Perspect 2000; 108:317–322.

30. Hurst CH, Devito MJ, Setzer RW, et al. Acute administration of 2,3,7,8-tetrachlorodibenzo-p-dioxin (TCDD) in pregnant Long Evans rats: association of measured tissue concentrations with developmental effects. Toxicol Sci 2000; 53:411–420.

31. Kenyon EM, Kraichely RE, Hudson KT, et al. Differences in rates of benzene metabolism correlate with observed genotoxicity. Toxicol Appl Pharmacol 1996; 136:49–56.

32. Kirman CR, Hays SM, Kedderis GL, et al. Improving cancer dose-response characterization by using physiologically based pharmacokinetic modeling: an analysis of pooled data for acrylonitrile-induced brain tumors to assess cancer potency in the rat. Risk Anal 2000; 20:135–151.

33. Krishnan K, Gargas ML, Fennell TR, et al. A physiologically based description of ethylene oxide dosimetry in the rat. Toxicol Ind Health 1992; 8:121–140.

34. Tardiff R, Charest-Tardiff G, Brodeur J, et al. Physiologically based pharmacokinetic modeling of a ternary mixture of alkyl benzenes in rats and humans. Toxicol Appl Pharmacol 1997; 144:120–134.

35. Poulin P, Krishnan K. A tissue composition-based algorithm for predicting tissue:air partition coefficients of organic chemicals. Toxicol Appl Pharmacol 1996; 136:126–130.

36. Poulin P, Krishnan K. A mechanistic algorithm for predicting blood:air partition coefficients of organic chemicals with the consideration of reversible binding in hemoglobin. Toxicol Appl Pharmacol 1996; 136:131–137.

37. Lipscomb JC, Fisher JW, Confer PD, et al. In vitro to in vivo extrapolation for trichloroethylene metabolism in humans. Toxicol Appl Pharmacol 1998; 152:376–387.

38. Lipscomb JC, Teuschter LK, Swartout J, et al. The impact of cytochrome P450 2E1-dependent metabolic variance on a risk-relevant pharmacokinetic outcome in humans. Risk Anal 2003; 23:1221–1238.

39. Reitz RH, Mandrela AL, Park CN, et al. Incorporation of in vitro enzyme data into the physiologically-based pharmacokinetic (PBPK) model for methylene chloride: implications for risk assessment. Toxicol Lett 1988; 43:97–116.

40. Jarabek AM. Inhalation RfC methodology: dosimetric adjustments and dose-response estimation of non-cancer toxicity in the upper respiratory tract. Inhal Toxicol 1994; 6:301–325.

41. Corley RA, Mandrela AL, Smith FA, et al. Development of a physiologically based pharmacokinetic model for chloroform. Toxicol Appl Pharmacol 1990; 103:512–527.

42. Brodeur J, Laperé S, Krishnan K, et al. Le probléme de l'ajustement des valeurs limites d'exposition pour des horaires de travail non-conventionnels: utilité de la modélisation pharmacocinétique à base physiologique. Travail et Santé 1990; 6:S11–S16.

43. Naumann BD, Silverman KC, Dixit R, et al. Case studies of categorical data-derived adjustment factors. Human Ecol Risk Assess 2001; 7:61–105.

44. O'Flaherty EJ. Physiologic changes during growth and development. Environ Health Perspect 1994; 102:103–106.

45. Corley RA, Mast TJ, Garney EW, et al. Evaluation of physiologically based models of pregnancy and lactation for their application in children's health risk assessments. Crit Rev Toxicol 2003; 33:137–211.

46. Page NP, Singh DV, Farland W, et al. Implementation of EPA revised cancer assessment guidelines: incorporation of mechanistic and pharmacokinetic data. Fund Appl Toxicol 1997; 37:16–36.

7

Considerations of Design and Data When Developing Physiologically Based Pharmacokinetic Models

Peter J. Robinson, Jeffery M. Gearhart, Deirdre A. Mahle, Elaine A. Merrill, Teresa R. Sterner, and Kyung O. Yu
Air Force Research Laboratory, Wright-Patterson AFB, Dayton, Ohio, U.S.A.

John C. Lipscomb
National Center for Environmental Assessment, Office of Research and Development, U.S. Environmental Protection Agency, Cincinnati, Ohio, U.S.A.

INTRODUCTION

Optimal dose–response analysis relies on *target site dosimetry*, by mathematically describing the four major processes of pharmacokinetics: bioavailability, distribution, metabolism, and elimination. In other words, uncertainty factors for animal to human extrapolation and even human interindividual extrapolation are replaced with efforts to directly compare the dose of chemical in the target tissue under both the experimental conditions and the human exposure conditions of interest. This comparison is then used to assess the likelihood of health effects in humans from the observed responses of the experimental animals. In contrast to the use of uncertainty or "safety" factors, this approach tries to provide a direct estimate of risk, rather than a conservative upper bound.

Physiologically based pharmacokinetic (PBPK) modeling has become the tool of choice to develop estimates of target site dosimetries in both animals and humans. PBPK models have advantages over more traditional kinetic models in that PBPK compartments correspond directly to the tissues and organs in the animal. It is thus possible to meaningfully extrapolate from one animal to another by simply taking into account physiological differences (different organ volumes, blood flows, etc.). The drawbacks of PBPK modeling primarily relate to the time, effort, and cost involved in appropriately developing, validating, and applying a model for the situation at hand. In this chapter, we outline some of the practical issues involved in the appropriate development of a PBPK model, so that such costs may be kept to a minimum. The overall process of model formulation, refinement, and validation is iterative, as shown in Figure 1.

In many cases, PBPK models have become necessarily complicated and, therefore, pose a risk communication problem. Hopefully, the approaches and guidelines presented below will help in the focused development of models that are as simple as possible to explain and extrapolate the data, but not too simple, in that they fail to take into account the

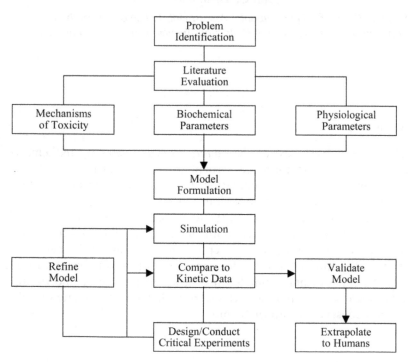

Figure 1 Schematic representation of the model development process in a risk assessment context. *Source*: Adapted from Ref. 1.

appropriate biological processes that underlie a meaningful risk assessment. In terms of their historical development, PBPK models have progressed from the simple to the more complex. Among the first useful models to be developed were those for volatile organics, the so-called Ramsey–Andersen models (2). In this seminal work, a model structure, set of algorithms and key concepts (such as flow-limited diffusion) were established. The value of those contributions serves as the reason why the model structure originally developed for styrene has been successfully applied for so long to so many volatile compounds. Many volatile organic compounds could be modeled to a sufficient degree of physiological detail by making a number of simplifying assumptions. This model structure was based on the concept of parsimony that a model should be developed only to the level of complexity necessary to address the issue at hand. It is possible to develop very complex PBPK models; however, doing so may not represent optimal expenditure of resources, and the addition of superfluous elements may decrease the level of certainty placed in model predictions. Relatively, simple models have been successfully developed for styrene, methylene chloride, and other compounds (2,3).

Certain experimental techniques are central to the successful development of models, regardless of their level of complexity. These include methods to experimentally determine blood and tissue partition coefficients (PCs) (vial equilibration and in vivo methods); gas uptake and in vitro methods to determine the values for metabolic parameters; and appropriate design of exposure studies to optimize model development and validation, including inhalation (whole-body or nose-only), oral exposure, and injection (bolus and infusion) studies. The values for these parameters may vary according to species (even strain), age, health status, and sex.

In developing a PBPK model, organs and tissues are represented as a finite number of individual compartments with their individual values for blood flow, tissue volume, and PC. Clearly, the actual number of organs/tissues in any particular model depend on the needs and circumstances of the model. Multiple tissues may be lumped together and treated as a single compartment (as in the slowly perfused or rapidly perfused tissues). This occurs when the tissues are sufficiently similar for the chemical of interest (in terms of perfusion and partitioning) and it is not necessary to distinguish between them in the application of the model. In many cases, data limitations also influence the decision to lump tissues together. Conversely, it sometimes happens that a particular organ is sufficiently important in the context of the model (and the data are sufficiently rich) that it may be specifically modeled (with organ- or tissue-specific values for flow, volume, and PC). This may occur, for example, with fat (brown and white fat), brain (individual brain regions) or liver. In the liver (or other organ for that matter), if metabolism and clearance is sufficiently great that a significant concentration gradient of the substrate is set up from the arterial to the venous ends of the liver sinusoids, the assumption of a "well-stirred" liver compartment may become untenable, and separate subcompartments for perivenous, periarterial, and intervening zones of the liver may need to be set up. Alternatively, a noncompartmental organ model that takes concentration gradients into account [such as the sinusoidal perfusion model (4)] may be incorporated into the overall PBPK model context. Clearly, if needed, and if the data can support it, the level of modeling detail can even be taken down to the cellular and molecular level.

For some chemicals, however, relatively simple models are insufficient to adequately simulate the data. In such cases, the underlying assumptions need to be changed and additional biological complexity often is introduced into the model. At this point, a great deal of scientific judgement comes into play in deciding the degree of model complexity that is appropriate. Issues may include more detailed modeling of metabolic processes and specific organs, such as the liver and fat; changes in physiology due to development, pregnancy, or aging (life-stage modeling); and interactions between more than one chemical. In many cases, it may also be necessary to interface the pharmacokinetic models with models of the interaction of the chemical with the target tissue [pharmacodynamic (PD) models] in order to provide an appropriately complete description of the overall process. A sensitivity analysis [see Evans and Andersen (5)] can be conducted to demonstrate whether certain parameters are deterministic toward a given pharmacokinetic outcome. Parameters (and compartments) that are sensitive should be given careful consideration. We illustrate some of these processes by considering a recently developed PBPK/PD model for perchlorate and iodide in both rats and humans. It should be noted, however, that the development process for PBPK models outlined here is presented as guidance only, and that there is no substitute for sound scientific judgement regarding the structure and level of detail required in each case. In addition, this process is usually iterative in nature, in that an initial model is developed, tested against available data, and if necessary, appropriately revised (either by changing the value of key parameters, or by altering the structure of the model itself) so as to fit the data better.

Figure 2 shows a schematic for a typical PBPK model for a volatile organic chemical (in this case styrene). As described earlier, PBPK models, as opposed to classical compartmental kinetic models, contain compartments analogous to anatomic and physiologic organization. They include parameter values that quantitatively describe actual biological systems. These include physiological and anatomical parameters such as organ weights and blood flows, and physiochemical and biochemical parameters such as tissue PCs, transport parameters, and metabolic rate constants. Within a species, physiologic parameter values should be fairly consistent, but the values for physiochemical and biochemical parameters will vary according to the unique properties of the chemical.

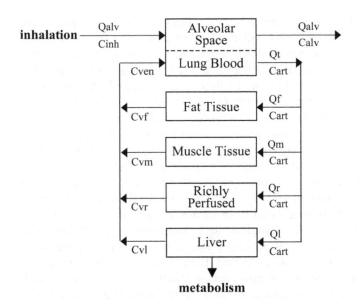

Figure 2 Schematic representation of a simple Ramsey–Andersen PBPK model. *Source*: Adapted from Ref. 2.

The mass balance of the chemical in each (nonmetabolizing) tissue t, for use in the model, is in general given by:

$$V_t\left(\frac{dC_t}{dt}\right) = Q_t\left(C_a - \frac{C_t}{(PC)_t}\right) \tag{1}$$

where V_t is the tissue volume (in units of L, for example), Q_t is the tissue blood flow (L/minutes or L/hour), C_a is the arterial concentration (mg/L), C_t is the tissue concentration (mg/L), and $(PC)_t$ is the blood–tissue PC for the chemical (Fig. 3). Note that values for tissue volumes and flows are given

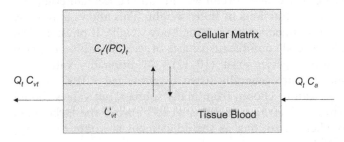

Figure 3 Schematic of a tissue compartment where Q_t is the tissue blood flow rate, C_a is the arterial concentration, C_{vt} is the venous blood leaving the tissue, C_t is the tissue concentration, and PC_t is the tissue–blood partition coefficient. *Source*: Adapted from Ref. 6.

here for the whole animal; in many cases, values for these parameters are given per kg body weight, and will need to be converted. In some cases, too, values may not be reported for a specific species or strain; in such cases, they can be estimated from their relative body weights (see also discussions of scaling below). For organ flows, values are often reported as fractions of the total cardiac output Q_{tot}. This in turn may be reported not as a total value for the animal as a whole, but also per kg body weight.

The most commonly used parameters in PBPK models are listed in Table 1. They consist of anatomical and physiological parameters (such as tissue blood flows and volumes) that are in principle common to all chemicals being modeled, and chemical-specific parameters. Anatomical and physiological parameters are usually available from the literature for a particular species, or may be calculated from values for similar species based on scaling considerations. Physiological and anatomical parameters also change during the life span of an organism, as well as during pregnancy, and attempts have been made to collect these kinds of data as well, and to incorporate these differences into specific PBPK models (7). Chemical-specific parameters, such as blood–air and tissue–blood PCs, may need to be determined experimentally for the chemical of interest, although methods are being developed to estimate these using quantitative structure–activity relationships [see, for example, Beliveau et al. (8,9)].

PHYSIOLOGICAL PARAMETERS

Whenever pharmacokinetic studies are performed for developing a PBPK model, the actual body and organ weights of the animals studied should be used to represent model compartments. Some compartments, which involve active cellular transport mechanisms of the chemical of concern, may require subcompartments; for example, the thyroid compartment in a model may include follicle, lumen, and stroma subcompartments. If possible, actual values of these cellular subcompartments should be used. If such kinetic studies were not performed, organ weights and subcellular compartments can be described as fractions of body weight. This allows automatic adjustment of the compartments with varying body weight. If physiological data are unavailable, several comprehensive lists of values for physiological parameters for multiple species exist (10–12).[a] Additionally, values for physiological parameters are generally listed in published PBPK models. Values for parameters can be chosen from many sources, but values should be well-justified. These values may be adjusted to improve the fit between observation and prediction, so long as the adjustment is within the range of values deemed valid.

---

[a] Arms and Travis did not intend to capture variability, only to mention the mean or central tendency from the many data sets that they summarized.

Table 1 Commonly used Physiological Parameters in Physiologically Based Pharmacokinetic Models

Description	Symbol	Unit (if any)
Body weight	BW	kg
Cardiac output or total systemic blood flow (also equal to the blood flow to the lungs). May be available as flow per kg of body weight	Q_{tot}	L/h
Fraction of cardiac output going to a specific organ, tissue or (lumped) group of organs/tissues, t. Typically, t includes both tissues that are of intrinsic interest (as target sites for toxicity, for example) and those that have significant impact on kinetics (typically liver, fat, muscle, often kidney, and lung). Remaining tissues/organs are lumped as slowly perfused or rapidly perfused tissues so that the total flow adds up to the cardiac output	f_t	–
Blood flow to organ/tissue t, $Q_t = f_t \times Q$	Q_t	L/h
Fraction of body weight that is associated with specific organ/tissue	f_{vt}	–
Organ/tissue volume, $V_t = f_{vt} \times BW$	V_t	L
Alveolar ventilation rate, or total volume of air exchanged with the blood per unit of time. May be given as breathing rate times, the effective exchangeable volume of a single breath, but a correction for the "dead space" in the lungs should be included	Q_{alv}	L/h
Blood/air partition coefficient for compound; usually measured experimentally using, for example, the vial equilibration method	PB	–
Tissue/blood partition coefficient for tissue t ($t = l$ for liver, for example); usually measured experimentally using, for example, the vial equilibration method	$(PC)_t$	–
Arterial concentration of compound (model calculated)	C_a	mg/L
Concentration of compound in venous blood from organ/tissue (model calculated)	C_{vt}	mg/L
Concentration of compound in organ tissue (model calculated)	C_t	mg/L
Maximum metabolic rate (for metabolism in the liver, for example); usually measured experimentally using, for example, the gas uptake method or by fitting IV kinetics	V_{max}	mg/h
Michaelis constant (for metabolism in the liver, for example); usually measured experimentally using, for	K_m	mg/L

(Continued)

Table 1 Commonly used Physiological Parameters in Physiologically Based Pharmacokinetic Models (*Continued*)

Description	Symbol	Unit (if any)
example, the gas uptake method or by fitting IV kinetics. Expressed in venous blood at equilibrium with liver		
The following parameters are used in the case of diffusion—limitation in tissue x for a particular compound		
Volume of cellular matrix in tissue *t* available to the compound for diffusion	V_{mt}	L
Concentration of compound in cellular matrix in tissue *t*	C_{mt}	mg/L
Effective permeability–surface area product for diffusion into the cellular matrix of tissue *t*	$(PA)_t$	L/h
Volume of blood subcompartment of tissue *t*	V_{bt}	L
Concentration of chemical in blood subcompartment of tissue *t*	C_{bt}	mg/L

Note: These are often available in compilations for particular species, such as rat and human. In many cases, changes in parameter values with development and aging are also available.
Abbreviations: IV, intravenous; PC, partial coefficient.

Mean or central tendency values are often used to represent the tissue volumes and blood flows used in PBPK models. However, considerable variability in these parameters may exist for some tissues. For example, blood flow to the stomach can increase 10-fold in response to enhanced functional activity (secretion and digestion) (13). In such a case, an estimate of resting blood flow may be used (11,12).

Parameters for perinatal and lifetime models needs to be considered relative to age-related growth, which varies across tissues. Allometric scaling cannot sufficiently describe the changes in tissue growth, blood flows, or fetal/neonatal growth taking place during the perinatal period. As opposed to the typical growth scenario, organs and tissues cannot be assumed to increase at the same rate in this dynamic system. During gestation, the placenta, uterus, mammary tissue, and fetal volume are growing at an accelerated rate in comparison to the other organs. Likewise, during lactation, the mammary gland and maternal and neonatal fat content show the most dramatic changes. Body weight of the neonate also changes dramatically. Maternal weight gain is mainly due to the growth of these select tissues rather than growth of other organs. Therefore, the total change in the maternal body weight can be simply described as the change in these specific tissue volumes added to the initial prepregnancy body weight. Likewise, the temporal changes in maternal cardiac output during gestation and lactation can be described as the sum of the initial cardiac output and the change in blood flow to the mammary gland, fat, placenta, and uterine tissues.

Growth equations for many perinatal physiological parameters are provided in Gentry et al. (14) and O'Flaherty et al. (15). To date, organ weights and blood flow are not available for early time points in gestation. Clewell et al. (16) extrapolated these to early period of growth by using a best-fit exponential curve to organ-specific data measured later in gestation.

Significant sex differences in some tissues can also be a source of variability. The distribution of a chemical is affected by multiple factors, including fractional content of adipose tissue (reflected in body mass index), body composition, plasma volume, organ blood flow, sex-dependent differences in metabolism, and the extent of tissue and plasma protein binding of the chemical. Women have a higher body fat percentage than men [21.0–32.7%, respectively (11)]. Such a large difference would result in disparities in the rate and extent of the chemical's distribution. Women also have a lower average body weight, a smaller average plasma volume, and lower average organ blood flows. Gender-specific values are recommended when incorporating a fat compartment, especially for humans. Other major sex differences exist in several protein groups responsible for binding drugs in human plasma. These differences are influenced by concentrations of sex hormones. Sex-based differences in drug metabolism seem to play a greater role in intergender pharmacokinetic variability than any of the other pharmacokinetic parameters (17). When data pertaining to humans or other species of concern are not available, values measured from other species, such as mice or rats, can sometimes be allometrically scaled as described later.

CHEMICAL-SPECIFIC PARAMETERS

Development of a PBPK model requires certain essential chemical-specific components. These are (i) experimentally measured or theoretically calculated PCs, (ii) estimated metabolic rate constants, and (iii) measured excretion rates of parent and metabolite.

Partition Coefficients

PC is an essential physiological parameter in a PBPK model. The PC value quantitatively represents the ratio of the concentrations at equilibrium of unbound chemical between the two compartments being compared. Variability in a PC value can adversely impact the predictive accuracy of the model. Quite often, in the absence of experimental data, the value of this parameter is estimated based on physical properties of the chemical in question. Algorithmic calculations have been developed, which predict a chemical-specific PC value based on physiochemical properties of the chemical (e.g., octanol:water PC), water and lipid composition of the tissue investigated, and/or experimentally derived PCs from oil and water tissue surrogates (18–20). Use of these algorithms can be beneficial, particularly when there is a need to reduce costs associated with obtaining tissues from animals. To increase the certainty in model predictions (when pharmacokinetic outcomes are sensitive to PC values), PC values should be experimentally determined.

The vial equilibration method is the most common in vitro method for determining chemical-specific PCs (18,21). This method has been modified to improve reproducibility (22) and to increase throughput by quantifying multiple chemical PCs as a mixture (23). The vial equilibration method described below is ideal for volatile and semivolatile compounds and has been used most successfully for volatile organic solvents. Tissues are harvested from the species of interest, particularly tissues that are the target of the chemical or site of toxicologic effect, and incubated with the test compound until equilibrium is reached between the tissue and the headspace. A number of operational equations have been derived to calculate these ratios under specific experimental conditions (18,21–23). Time to steady state is critical and should be optimized for the test compound. Metabolism of the compound in exposed tissue samples must be controlled for. Analysis is done by gas chromatography in a verified linear range. Human tissues can be obtained from tissue bank organizations to provide species specificity to models developed with human data.

To measure blood:air or tissue:air PCs, blood or minced tissue is smeared onto the sides of a 20 mL headspace vial [based on modifications by Gearhart et al. (22)]. Saline homogenization is unnecessary if tissues are finely minced and eliminates the need to measure saline:air PCs. The vial is sealed with a crimp cap. Reference vials receive no tissue and are crimp sealed, as well. One milliliter of headspace from a Tedlar® bag of the volatile chemical of interest at 10,000 ppm is drawn out with a gastight syringe and injected into each headspace vial (preparation of gas bag is described later). Vials are incubated at 37°C until the chemical is at equilibrium. One milliliter of headspace from each vial is drawn off and injected onto a gas chromatograph (GC). Peak areas of the chemical of interest are recorded for reference and sample vials. The PC value can be calculated using the equation:

$$PC = (A_{ref} V_{ref}) - [A_s(V_{ref} - V_S)]/(A_s V_s) \qquad (2)$$

where A_{ref} is the peak area of the reference vial as determined by GC, V_{ref} is the volume of the reference vial, A_s is the peak area of the sample vial as determined by GC, and V_s is the volume of headspace in the sample vial. This equation gives the blood:air or tissue:air PC. The tissue:blood PC is mathematically derived using the equation:

$$(PC)_{tissue:blood} = \frac{(PC)_{tissue:air}}{(PC)_{blood:air}} \qquad (3)$$

Calculating Bag Concentrations

Calculating PC values for volatile compounds requires control and reproducibility of concentrations. This is accomplished through dilution of a
"bag concentration" into vials of known volume. Tedlar® bags are
equipped with both an air port and an injection port and can be inflated
with known amounts of either room air or specialty gas by using a wet test
meter to deliver the gas into the vacuum-evacuated bag. The chemicals of
interest that are best suited for this preparation are those chemicals that
are liquid but readily volatilize at room temperature or with mild heating.
To determine the amount (in milliliters, mL) of chemical to be injected into
the Tedlar bag, Eq. (4) is used:

$$\text{mL of chemical to be added}$$
$$= \text{bag volume, L} \times (1/\text{molar gas constant})$$
$$\times (1/\text{density of liquid, g/L}) \times (\text{desired ppm}/1,000,000) \quad (4)$$

where molar gas constant is 24.45 mol/L at 25°C.

Volatilization of the chemical adds volume to the bag and must be
taken into account by subtracting the volume expansion from the total volume desired.

$$\text{Expansion volume, L} = \text{mL of chemical to be added}$$
$$\times \text{density} \times (1/\text{MW}) \times \text{molar gas constant}$$
$$(5)$$

where MW is the molecular weight of the chemical in g/mol. The expansion
volume must be subtracted from the total bag volume to determine the volume of air to be added.

Once the liquid chemical is injected into the Tedlar bag, it is advisable
to heat and mechanically manipulate the bag gently to insure complete volatilization and homogeneous mixing.

To estimate PCs for compounds of low or nonvolatility, the method of
Jepson et al. (24) can be used. This in vitro filtration method has been validated in several biological tissues with compounds having a vapor pressure
as low as 9×10^{-6} mmHg and as high as 14.2 mmHg. PCs for nonvolatile,
water-soluble compounds can also be determined in vivo after establishing
steady state concentrations in tissues. While PCs have been estimated in
vivo after a single dose, infusion of the compound to steady-state provides
a more accurate estimation of the PC (25). The in vivo PC can be calculated
as the ratio C_t/C_B, where C_t denotes the concentration in tissue and C_B
denotes concentration in blood, both at steady state. In principle, the tissue
concentration needs to be corrected for the amount of the chemical still
residing in the blood volume B (as a fraction of tissue volume) within the

tissue at the time of measurement. The "true" tissue concentration C_{t*} is then given by

$$C_t^* = (C_t - BC_B)/(1 - B) \tag{6}$$

so that

$$PC = C_t^*/C_B = (C_t/C_B - B)/(1 - B) \tag{7}$$

For an eliminating organ such as liver or kidney, this ratio may underestimate the PC value and should be modified to include the effect of blood flow and clearance (26).

When incorporating PC values into PBPK models, it is important to remember that variability between species and between age groups within a species may exist for the same matrix. Aged Sprague–Dawley rats have been shown to have higher PC values for blood, liver, kidney, fat, and brain when compared to postnatal rats of the same strain (27). Another consideration is the effects of freezing on tissue quality and measured PC value. The availability of human tissues from tissue banks has allowed for human-specific PC determination; however, tissues are often shipped frozen and may result in under- or overestimated values for some chemicals (28). The need for a correcting factor can be determined by comparing PC values determined using previously frozen and fresh rat tissues.

In many cases, the investigator should be prepared to approximate the values for chemical-specific parameters from very limited available data. For example, in a recently developed PBPK model for perchlorate, PCs for the thyroid subcompartments were based on electrical potentials measured within the thyroid stroma, follicular membrane, and lumen after perchlorate (ClO_4^-) dosing (29). Electrical potential differences can be interpreted as effective PCs for charged moieties, such as ClO_4^- and iodide (I^-). These electrical potentials in the thyroid hinder entrance of negatively charged ions from the stroma into the follicle, while the opposite potential from the follicle to the lumen enhances passage of negatively charged species into the lumen and indicates an effective PC greater than one. Using Chow and Woodbury's measured electrical potentials at the stroma:follicle and follicle:lumen interfaces, effective partitions were calculated (30).

METABOLISM: GAS UPTAKE AND IN VITRO METHODS

Metabolism is discussed below as taking place in the liver, and this is usually the case. However, metabolism may in fact take place in other tissues and organs, such as the lung or the skin. In such cases, a metabolic elimination term can be added to the mathematical equation for the mass balance of chemical in each metabolizing tissue (although some experimental techniques such as the gas uptake method described below still estimate only *total* metabolism in the organism). Metabolism is a saturable process usually described by means of the Michaelis–Menten equation in terms of a maximum elimination rate V_{max} (units of mg/minutes or mg/hour) and a Michaelis constant K_m (mg/L). The latter is the concentration of substrate in the liver at which half the maximum elimination capacity is achieved, although PBPK models express K_m values as concentration in venous blood at equilibrium with liver. Metabolism in the liver in the context of PBPK modeling is usually described by the so-called venous equilibration (or "well-stirred") model for hepatic elimination. In such a model, elimination is described by Eqs. (10) and (11).

The metabolic parameters for a particular volatile compound are typically determined by means of closed-chamber gas uptake system. The physical apparatus for this is shown in Figure 4. The animals are placed in the chamber and the lid is sealed using vacuum grease. Oxygen level in the chamber is maintained at 21% to 22%, and is allowed to equilibrate for 15 minutes before gas injection. The chemical is introduced to the chamber via the injection port on the chamber lid. The gas chromatograph is programmed to determine the concentration of the chemical in air within the chamber at predetermined intervals (typically every 5–15 minutes). The decrease in concentration is indicative of the rate of metabolism of the chemical by the animal. V_{max} and K_m are determined by PBPK modeling

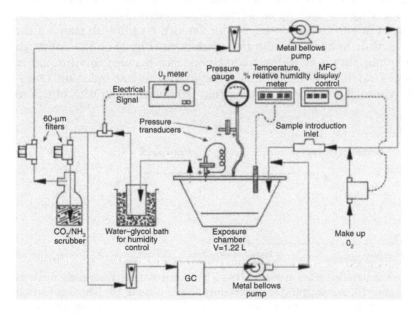

Figure 4 Schematic of the closed-chamber gas uptake and monitoring apparatus. *Source*: Adapted from Ref. 31. *Abbreviations*: MFC, mass flow controller; GC, gas chromatography.

of the rate of disappearance of the chemical. The same procedure is repeated without animals to determine the rate of loss of the chemical from the chamber itself (loss run). For some chemicals, it may be necessary to repeat the procedure with a dead animal in the chamber to take into account adsorption of the chemical onto the surface (fur and skin) of the animal.

For volatile compounds, the gas uptake technique has been extensively used for determination of in vivo metabolism (32,33). In the case of nonvolatile chemicals, an approach taken to determine in vivo metabolism is to dose animals at varying concentrations, preferably by intravenous (IV) injection, and sample the blood over time for parent and/or metabolite concentrations until there is significant disappearance of the chemical. Then, with the parameter values for PCs and excretory pathways set at experimentally determined values, values in the PBPK model for Michaelis–Menten parameters, V_{max} and K_m, can be adjusted simultaneously to reproduce the blood concentration time course of the chemical of interest over the course of the study.

Excretion rates for parent and metabolite(s) are determined by collecting urine, feces, and exhaled breath samples from animals dosed at different concentrations and then left in metabolism cages long enough to account for the majority of the mass of parent chemical initially administered. The values for model parameters responsible for excretion of parent or metabolite are then adjusted to account for the mass of parent and metabolite(s) chemical leaving the body by each route of excretion (34).

Use of in vitro results to infer values in vivo is central to many of the inferences made in human health risk assessment (Fig. 5) (1,35). Although there may not always be a direct correspondence between in vitro and in vivo values, careful consideration must be given before abandoning a measured in vitro value in favor of fitting that parameter along with others to

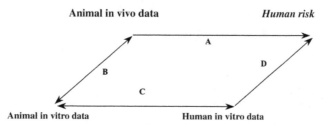

Figure 5 The parallelogram approach. (**A**) This extrapolation is normally made in human health risk assessments; an uncertainty factor is used. (**B**) The comparison between in vivo and in vitro animal data is frequently made. (**C**) This comparison can be made when in vitro systems derived from humans are utilized. (**D**) Extrapolation from in vitro human data, along with extrapolation from in vivo animal data, will improve human health risk evaluation.

the kinetic data as a whole. The latter approach will no doubt produce a better overall fit to the data (and will give an effective value for the parameter that reflects the local biological milieu), but this comes at the expense of the ability to exploit measured differences in vitro between animals and humans to reduce uncertainty in the extrapolation. For a more full treatment of this topic refer to Chapter 9.

Allometric Scaling and Interspecies Extrapolation

To account for the difference in parameter values due to variations in body weight (BW) across species, allometric scaling is commonly employed. Allometry takes the form of power-law equations relating parameter values to body mass. The usual allometric scaling relationship relating metabolic rate (B) to body mass (M) is

$$B = B_0 M^{3/4} \tag{8}$$

where B_0 is the normalization parameter or coefficient (36,37). This formula has been extended to a wide range of organisms from the microbes to large vertebrates and plants. Hence, in pharmacokinetics, values for blood flows, clearance rates, and maximum velocities are most commonly scaled by multiplication with $BW^{3/4}$. It should be noted that body weight in these formulas assumes "normal" body composition. Further adjustments for the prediction of body size differences clearance and volume in obese or very lean people will usually require the use of other covariate information (e.g., height and skin thickness) to predict the weight with "normal" body composition. Allometric scaling within a species is not appropriate. Additional adjustments may be required to account for developmental changes (refer to Chapter 11, by G. Ginsberg) (38).

MODEL DEVELOPMENT AND VALIDATION

Experimental Design Considerations

To develop a model, one should determine what kind of problems occur with chemicals of interest. Information on route of exposure, mechanism of target organ toxicity, and biochemical and physiological constants can be obtained by searching the literature. The mechanism of target organ toxicity will determine whether the parent compound, metabolite(s) and/or reactive intermediate(s) are responsible for tissue toxicity. This will help determine whether to investigate a parent chemical and/or metabolite(s) for model development. In the absence of evidence to the contrary, it is often assumed that toxicity is a function of cumulative exposure (AUC) rather than of peak exposure (C_{max}).

Exposure (Administration)

The four major routes of administration of a chemical are inhalation, oral absorption, dermal exposure, and IV injection. Many conditions of a model, including length of exposure time and concentrations of the chemical, are decided depending upon available kinetic data.

A gas uptake system, as described previously, can be used to expose animals to volatile chemicals in pharmacokinetic studies (33,39). The lungs are the main sites of airborne chemical absorption in either whole-body or nose-only exposure and are also elimination sites for some volatile chemicals, markedly so postexposure. When planning to conduct a gas-uptake study, four to five exposure levels should be used that range from lower concentrations (most likely a first-order uptake process) to higher concentrations, which should produce saturated kinetics (Michaelis–Menten process). Complications, such as dermal absorption during whole-body exposure or ingestion from grooming (licking the fur) following exposure, should be considered if its contribution is significant.

The exchange of gases in the lung, leading to systemic exposure via the inhalation route, is typically described by the following steady-state equation (6):

$$C_a = \frac{Q_{alv} C_{inh} + Q_{tot} C_v}{Q_{tot} + \left(\dfrac{Q_{alv}}{PB}\right)} \tag{9}$$

where C_a and C_v are the arterial and venous blood concentrations (mg/L), Q_{alv} is the alveolar ventilation rate (L/minutes or L/hour), C_{inh} is the concentration in the inhaled air (mg/L), Q_{tot} is the cardiac output (total blood flow through the lung, L/minutes or L/hour), and PB is the blood–air PC. This equation assumes rapid equilibration of the chemical across the alveolar membrane, no significant metabolism in the lung tissue and negligible storage capacity in the lungs. Additional factors that may be taken into account in describing the exchange of gases in the lung include the lung "dead space" in the conducting airways, in which only 70% of the maximum air capacity of the lung is actually exhaled in any one breath, while 30% (the "dead space") remains to be mixed with the next incoming breath; and potential "scrubbing" of chemical in the upper respiratory tract (URT), which prevents it from being immediately absorbed in the lung (though it may be ultimately ingested) (40). The degree of scrubbing varies widely from chemical to chemical and may lead to local effects in the URT. This effect may be important for some chemicals and is covered from a risk assessment perspective in Chapter 1, by L. Haber.

Drinking water ingestion and oral bolus may result in the same AUC, but bolus will give a higher C_{max}. Toxicologically, this may be important— humans ingest water in multiple, typically small volumes throughout the day. In contrast, animal dosing is frequently via gavage (bolus). PBPK models are simpler when they simulate a bolus exposure, though humans and experimental animals typically sip a drinking water dose or encounter an oral contaminant at intervals. Animals on test are frequently gavaged. If C_{max} is the important dose metric, then care must be given to verify absorption rates and gastrointestinal (GI) transit times. It should be recognized, however, that different species have GI tracts with different anatomical and physiological characteristics (41). The rate of absorption of a chemical into the systemic circulation depends on GI blood flow, GI transit time, surface area, contents, chemical solubility, and the vehicle used. Chemicals can be absorbed by passive diffusion as well as carrier-mediated transport through the intestine epithelium. Enterohepatic circulation (which differs from biliary elimination) can affect uptake and excretion of a compound. In addition, some of the chemical will be reabsorbed from bile (enterohepatic recirculation) and result in an increase of parent or metabolite(s) concentration in systemic circulation well after initial uptake. When a chemical is absorbed through oral uptake, the first passage through the liver will produce metabolite(s) that may be excreted into bile before ever reaching systemic circulation. This first pass will lower the chemical's initial systemic concentration in blood (its bioavailability).

Oral uptake of a chemical is greatly influenced by vehicle effects on GI absorption. The chemical properties of the vehicles used in the experiments will result in different uptake patterns when compared with consumption of regular food or drinking water, the normal vehicles of environmental exposure. When absorption of halogenated hydrocarbons in water or corn oil was investigated, for example, the rate of uptake of these chemicals in oil was slower and blood concentrations were lower than those in water (42). Vehicle effects, enterohepatic circulation, and fasting (versus regular feeding of animals before dosing) are important issues to consider when designing experiments for oral gavage studies. Absorption rate constants of chemicals by oral gavage are obtained by fitting data with a preliminary pharmacokinetic model.

IV injection, by a single bolus or infusion, eliminates the absorption process and introduces chemicals directly into systemic circulation. Unlike the process of oral ingestion, chemicals reach the target organ before first-pass elimination by the liver. Elimination kinetics can easily be obtained by IV injection. Water or physiological saline is a frequently used vehicle to dissolve chemicals; oily vehicles should not be used with this route. The constant infusion of a chemical into blood circulation will reach a steady state when the input rate of a chemical is same as its elimination rate and in vivo PCs can be measured by analyzing tissue concentrations once steady state is reached.

Distribution (Translocation) and Sequestration

After entering the blood circulation via inhalation, dermal absorption, oral uptake or IV injection, a chemical is distributed to tissues throughout the body. Its distribution is related to cardiac output and regional blood flow to the target organs. Richly perfused tissues such as kidney, liver, heart, and brain will initially receive the highest portion of the chemical dose. Chemical partitioning from blood into tissues then plays a role in determining the transfer of a chemical into perfused tissues. The free proportion of chemical in blood, not the bound fraction, is usually considered available for distribution to tissues and to exert the toxicological action.

Plasma proteins such as albumin may bind chemicals and act as storage depots. When uptake is rapid relative to rate of protein binding, the equilibrium-free fraction may not be a good estimate for the locally free material that is actually available for uptake. Likewise, when on- and off-rates for binding are "faster" than the duration of time it takes a given volume of blood to course through a tissue, the amount of chemical available for distribution to the tissue will be greater than the free fraction estimated by equilibrium binding. In such cases, more detailed data on the kinetics of binding (such as association and dissociation rate constants) may be required to fully characterize the uptake of the material into tissues [Robinson and Rapoport (43)].

Liver, kidney, and adipose tissues are other storage reservoirs that have a high affinity for many chemicals. Lipophilic compounds are accumulated in fat and slowly released to systemic circulation, which alters a chemical's concentration in the bloodstream and may prolong internal exposures even after the cessation of an external exposure.

Metabolism (Biotransformation)

Liver is the major site for chemical metabolism. Other tissues such as the kidney, skin, GI tract, and lungs also have capacity for metabolism. For some chemicals, metabolites are more toxic than their parent compounds (44–46). Liver metabolism is presented as a sum of the mathematical descriptions for the first-order pathway and/or saturable metabolism, as necessary. Saturable metabolism is usually described in terms of Michaelis–Menten kinetics, so that the metabolism rate (mg/minutes or mg/hour) is given by:

$$\text{Metabolism Rate} = V_{\text{max}} \left(\frac{C_l}{C_l + K_m} \right) \tag{10}$$

where V_{max} (mg/minutes or mg/hour) is the maximum hepatic metabolic rate, K_m (mg/L) is the Michaelis–Menten constant, and C_l (mg/L) is the concentration of the chemical in the liver. The mass balance of the chemical in the liver for use in the model is then given by:

$$V_l \left(\frac{dC_l}{dt} \right) = Q_l \left(C_a - \frac{C_l}{(PC)_L} \right) - V_{\text{max}} \left(\frac{C_l}{C_l + K_m} \right) \tag{11}$$

where V_1 is the liver volume, Q_1 is the liver blood flow, C_a is the arterial concentration, C_1 is the concentration in liver tissue, and $(PC)_L$ is the blood–liver PC for the chemical (Fig. 3). Note that in the steady state, $dC_1/dt = 0$ and

$$\text{Metabolism Rate} = Q_1 \left[C_a - \frac{C_1}{(PC)_L} \right] = V_{max} \left[\frac{C_1}{C_1 + K_m} \right] \tag{12a}$$

At low substrate concentrations relative to K_m ($C_1 \ll K_m$), Eq. (12a) reduces to:

$$C_1 = \frac{C_a}{\dfrac{1}{(PC)_L} + \dfrac{V_{max}}{Q_1 K_m}} \tag{12b}$$

Here, if in addition $(PC)_1 V_{max}/Q_1 K_m \gg 1$, C_1 approaches zero, describing the extreme limit of *flow-limited clearance* in which, from Eq. (12a) *Metabolism Rate* $= Q_1 C_a$. At the opposite extreme in which $C_1 \gg K_m$, Eq. (10) and (12a) reduce to the simple relation that *Metabolic Rate* $= V_{max}$, describing *enzyme-limited clearance* by the liver.

Eq. (11) describes what is known as the *venous equilibration* (or "well-stirred") model for hepatic elimination, in which, like other tissues, the substrate in the liver is in equilibrium with the venous outflow, so that all the enzyme in the liver is assumed to be working at the same substrate concentration C_1. A more realistic model for the liver takes into account the physiological fact that there is always a concentration gradient in the liver from its arterial to its venous ends, and that the enzyme in the zones of the liver near its arterial end are exposed to higher substrate concentrations than enzyme(s) toward the venous end, and are in fact responsible for this substrate concentration gradient. A mathematical description of liver metabolism that takes this into account is the so-called *sinusoidal perfusion model* for hepatic elimination [Robinson (4)], and this mathematical description may be substituted for Eq. (12) in such cases in which there is a significant arteriovenous substrate concentration difference in the liver.

Elimination

An important consideration when developing a PBPK model is the elimination of the parent compound or metabolites through the urine, feces, or exhaled air. Biotransformation of xenobiotics often results in more water-soluble metabolites that are excreted in urine or feces. After dosing, animals can be maintained in metabolism cages designed to collect urine and feces separately. Analysis for parent compound and metabolites can provide a time-dependent excretion rate. Volatile organic solvents are also largely exhaled during and following exposure, especially when the dose route is inhalation. Metabolism of many organic compounds yields CO_2, which is also exhaled. Specialized metabolism cages have been designed to trap exhaled compounds, particularly exhaled CO_2. By quantifying the loss of parent or metabolite through all possible elimination pathways, a mass balance can be achieved.

Monte Carlo Analysis and Parameter Sensitivity

Statistical dependencies of model outputs on variations in model inputs are often determined using Monte Carlo techniques. In this technique, the effects of parameter variability on the simulation are assessed by assuming a statistical distribution for each parameter of interest. Whenever possible, the actual distribution of parameter values is utilized. Input quantities can be varied by sampling from a number of probability distribution functions, including Uniform, Normal, Exponential, Beta, Chi-Square, F, and Gamma distributions. A random value is then chosen from each of these distributions (e.g., liver blood flow, fat:blood PC, VFC; assuming statistical independence when warranted) and a single simulation is performed. This process is repeated until a statistical distribution of the output (e.g., AUC of parent chemical in brain tissue) is produced (typically thousands of iterations). The distribution of the model output gives an indication of how much uncertainty (or variability) is due to the uncertainty (or variability) in the values of the input parameters.

If only one input parameter is varied in this way (with the others held constant), the magnitude of change in value of the output—relative to the magnitude of change in the input parameter, indicates the sensitivity of the system to the specific input parameter. Typically, the relative impact of each parameter on model predictions is assessed first. This is done after finalizing all model parameter values. The model is run at a dose low enough to not saturate its nonlinear metabolic mechanisms for a period long enough to ensure equilibrium. A common dose metric, such as serum concentration area under the curve (AUC) or the dose metric most closely associated with toxicity, is predicted. The model is then repeatedly rerun, typically using a 1% increase for the value of each evaluated input parameter to determine the resulting change in predicted biomarker. Sensitivity coefficients for each parameter are calculated as shown in Eq. (13), where A represents the value for the dose metric resulting from a 1% increase in the parameter value, B represents the value for the dose metric using the original parameter value, C represents the value of the parameter increased 1% from original value, and D represents the original value of the parameter. In cases where specific parameters drive most of the output variation (i.e., the parameters have the highest sensitivity coefficients), it is particularly important to ensure that an effort is made to define those parameters more precisely than the other, less consequential parameters [Evans and Andersen (5)]. Parameters with absolute values for sensitivity coefficients less than or equal to 0.1 are generally considered insensitive—variations in their values are relatively unimportant in determining the dose metric simulated.

$$\text{Sensitivity Coefficient} = \frac{(A - B)/B}{(C - D)/D} \tag{13}$$

Model Validation

A major advantage of a PBPK model over classical kinetic compartmental models is that a PBPK model will ideally reproduce the underlying structure and processes of the biological system. In order to see whether or not this is in fact the case, the model is validated (evaluated). Once a model has been developed and optimized so that it adequately reproduces the data at hand, it is then required to satisfactorily predict the behavior of the biological system under quite different conditions (i.e., adequately simulate data from additional studies). In some cases, particularly where relevant data are scarce, a choice may need to be made as to which portion of data from a study is used for model development and which part is retained for model validation. The most convincing argument for the utility of a model is if it makes clear predictions, which are subsequently verified through the use of different and diverse data sets.

When the model verification process fails, it may mean that the model structure or the level of physiological detail is inappropriate. In the model development process, decisions have probably been made (based in large part on the investigators' experience) as to what biological processes to include and which to exclude from the model as (probably) irrelevant or insensitive. At this stage, these decisions need to be reconsidered in order that the model more adequately describe the biological reality. The overall model development process thus takes on an iterative structure as shown in Figure 1.

ADVANCED PBPK MODELS

By their very nature, PBPK models are specific to the compound or compounds in question and need to take into account those processes that are particularly important for those compounds. While no model should be developed beyond the simplest tool that adequately describes a compound's distribution, some models may require an appreciable level of complexity. Thus, their structure can become much more complex than the structure of the Ramsey and Andersen (2) model. In this section, we consider some of the processes that may need to be considered in the development of more complex PBPK models.

Diffusion Limitation

In many cases, the simple description of a compound instantaneously equilibrating between tissue and (venous) blood according to a characteristic PC

(Eq. 1) is insufficient to describe the tissue kinetics. One reason for this may be that the tissue compartment is sufficiently large and poorly perfused with capillaries so that the material takes a finite time to diffuse from the interface with the blood into the depths of the tissue itself. This may be the case for a lipophilic material distributing (slowly) in the fat or liver compartment, for example. This process can be described with a diffusion term as is traditional in compartmental analysis. Note that such a diffusion limitation term is often expressed in terms of a "*PA*" or a permeability–surface area product (units of L/hour or cm^3/hour). This term may be misleading, as it is usually not suggested that there is an actual membrane barrier to the diffusion process. However, uptake through the tissue subcompartments (Fig. 3) must be considered. This requires that tissue blood and cellular matrix be described separately (6). The rate of change in the amount of chemical in the cellular matrix (subscript m) is equal to the product of the diffusion rate constant and the net flux from tissue blood:

$$V_{mt}\, dC_{mt}/dt = (PA)_t(C_{vt} - C_t/(PC)_t) \tag{14}$$

The rate of change in the tissue blood subcompartment equals the sum of the net retention from blood flow plus the net flux from cellular matrix:

$$V_{bt}\, dC_{bt} = Q_t(C_a - C_{vt}) + (PA)_t(C_t/(PC)_t - C_{vt}) \tag{15}$$

If the diffusion of a chemical from tissue blood to cellular matrix is slow with respect to total tissue blood flow, both equations are necessary (6). On the other hand, if tissue blood flow (i.e., perfusion) is slow with respect to diffusion, tissues are described as homogeneous, well-mixed compartments such that the rate of change in the amount of chemical in the tissue is described with a single equation for the whole tissue mass (cellular matrix plus tissue blood) given above as Eq. (1). Note that in this equation, the effluent venous blood concentration (C_{vt}) is in equilibrium with the tissue concentration (C_t) as specified by the tissue–blood PC [$(PC)_t$] such that $C_t = (PC)_t C_{vt}$.

PBPK/PD

These models include a quantitative, mechanism-based description, not just of the kinetic processes (absorption, distribution, metabolism, and elimination), but also the relevant interactions of the chemical(s) with the target site(s) and biological responses. There is a natural transition between what is "kinetic" and what is "dynamic." Some models seamlessly cover both areas in order to describe the biology as a whole. In those cases, a specific biological response is measured as a result of a specific exposure. This response, as well as the kinetics of the exposed material, needs to be modeled in order to provide a complete description of the process and to be able to predict the response under different exposure conditions.

As an example, we consider a recent set of models for the biological effect of perchlorate on thyroid function (16,30,47). Detection of ClO_4^- in several drinking water sources across the United States has led to public concern over health effects from chronic low-level exposures (48). Perchlorate inhibits thyroid iodide (I^-) uptake at the sodium-iodide symporter (NIS), thereby disrupting the initial stage of thyroid hormone synthesis. A PBPK model was developed to describe the kinetics and distribution of both radioactive I^- and ClO_4^- in both humans and rats. The model also simulates the subsequent inhibition of thyroid uptake of radioactive I^- by ClO_4^-, as well as the response of the system to upregulate NIS in the presence of sustained levels of perchlorate. Although thyroid hormones and their regulatory feedback are not incorporated in the model structure, the model's successful prediction of free and bound radioactive I^- and perchlorate's interaction with free radioactive I^- provide a basis for extending the structure to address the complex hypothalamic–pituitary–thyroid feedback system and, ultimately, predict the effects of iodide deficiency and perchlorate exposure. This progressive development of the model structure in order to describe greater levels of detail of the biological system is a major advantage of the PBPK/PD approach to data analysis.

CONCLUSIONS: APPLICATION TO RISK ASSESSMENT

To use a PBPK model in risk assessment (rather than as a generator of ideas and hypotheses), we must have a degree of confidence in the model. How do we know we have an adequate model? Generally, as a rule of thumb, a "good" PBPK model will

- Fit or adequately describe diverse data sets, often in multiple species and routes of exposure.
- Use independently estimated (in vivo, in vitro, or even theoretical) values for parameters whenever possible, rather than fitting many parameters to the kinetic data.
- Be physiologically realistic and as simple as possible, with few, if any, ad hoc inclusions.

In general, there is a trade-off between model detail and ability to independently estimate key model parameters.

At present, interspecies extrapolation is perhaps the major success story for PBPK modeling. It serves to reduce the uncertainties associated with the extrapolations A and D in Figure 5 by explicitly providing a quantitative framework, based on species-specific physiology, in which to incorporate parameters (physiological, in vitro biochemical and theoretical) for both animals and humans. Present successes in intraspecies extrapolations have been limited in part due to a relative lack of data describing human anatomic, physiologic, and biochemical variability, as well as a relative paucity of adequately reported human data sets.

REFERENCES

1. Clewell HJ III, Andersen ME. Biologically motivated models for chemical risk assessment. Health Phys 1989; 57:129–137.
2. Ramsey JC, Andersen ME. A physiologically based description of the inhalation pharmacokinetics of styrene in rats and humans. Toxicol Appl Pharmacol 1984; 73:159–175.
3. Andersen ME, Clewell HJ, Gargas ML, et al. Physiologically-based pharmaco-kinetics and risk assessment process for methylene chloride. Toxicol Appl Pharmacol 1987; 87:185–205.
4. Robinson PJ. Hepatic modeling and risk assessment: compartmental versus tube models and interspecies scaling. Drug Metab Rev 1991; 23:381–398.
5. Evans MV, Andersen ME. Sensitivity analysis of a physiological model for 2,3,7, 8-tetrachlorodibenzo-p-dioxin (TCDD): assessing the impact of specific model parameters on sequestration in liver and fat in the rat. Toxicol Sci 2000; 54:71–80.
6. Krishnan K, Andersen ME. Physiologically based pharmacokinetic modeling in 'toxicology'. In: Hayes AW, ed. Principles and Methods of Toxicology. 3rd ed. New York, 1994:149–188.
7. Clewell RA, Merrill EA, Robinson PJ. The use of physiologically based models to integrate diverse data sets and reduce uncertainty in the prediction of per-chlorate and iodide kinetics across life stages and species. Toxicol Ind Health 2001; 17:210–222.
8. Beliveau M, Tardiff R, Krishnan K. Quantitative structure-property relation-ships for interspecies extrapolation of the inhalation pharmacokinetics of organic chemicals. Toxicol Appl Pharmacol 2003; 189:221–232.
9. Beliveau M, Lipscomb JC, Tordiff R, et al. Quantitative structure-property rela-tionships for interspecies extrapolation of the inhalation pharmacokinetics of organic chemicals. Chem Res Toxicol 2005; 18:475–485.
10. Arms AD, Travis CC. Reference Physiological Parameters in Pharmacokinetic Modeling. EPA/600/6-88/004, NTIS PB88–196019, U.S. Environmental Pro-tection Agency, Washington, DC, 1988.
11. Brown RP, Delp MD, Lindstedt SL, et al. Physiological parameter values for physiologically based pharmacokinetic modelsPhysiological parameter values for physiologically based pharmacokinetic models. Toxicol Ind Health 1997; 13:407–484.
12. Davies B, Morris T. Physiological parameters in laboratory animals and humans. Pharm Res 1993; 10:1093–1095.
13. Granger DN, Barrowman JA, Kvietys PA. Clinical Gastrointestinal Physiology. Philadelphia, PA: WB Sauders Co., 1985.
14. Gentry PR, Covington TR, Anderson ME, Clewell HJ. Application of a physiolo-gically based pharmacokinetic model for isopropanol in the derivation of a reference dose and reference concentration. Regul Toxicol Pharmacol 2002; 36:51–68.
15. O'Flaherty EJ, Scott W, Schreiner C, Beliles RP. A physiologically based kinetic model of rat and mouse gestation: disposition of a weak acid. Toxicol Appl Pharmacol 1992; 112:245–256.

16. Clewell HJ III, Merill EM, Yu KO, et al. Predicting fetal perchlorate dose and inhibition of iodide kinetics during gestation: a physiologically-based pharmacokinetic analysis of perchlorate and iodide kinetics in the rat. Toxicol Sci 2003; 73:235–255.
17. Gandhi M, Aweeka F, Greenblat RM, Blaschke TF. Sex differences in pharmacokinetics and pharmacodynamics. Ann Rev Pharmacol Toxicol 2003; 44:499–523.
18. Gargas ML, Burgess RJ, Voisard DE, et al. Partition coefficients of low-molecular-weight volatile chemicals in various liquids and tissues. Toxicol Appl Pharmacol 1989; 98:87–99.
19. Meulenberg CJW, Vijverberg HP. Empirical relations predicting human and rat tissue: air partition coefficients of volatile organic compounds. Toxicol Appl Pharmacol 2000; 165:206–216.
20. Poulin P, Krishnan K. Molecular structure-based prediction of the partition coefficients of organic chemicals for physiological pharmacokinetic models. Toxicol Methods 1996; 6:117–137.
21. Sato A, Nakajima T. Partition coefficients of some aromatic hydrocarbons and ketones in water, blood and oil. Br J Ind Med 1979; 36:231–234.
22. Gearhart JM, Mahle DA, Grene RJ, et al. Variability of physiologically-based pharmacokinetic (PB-PK) model parameters and their effects on PB-PK model predictions in a risk assessment for perchloroethylene (PCE). Toxicol Lett 1993; 68:131–144.
23. Fisher J, Mahle D, Bankston L, Greene R, Gearhart J. Lactational transfer of volatile chemicals in breast milk. Am Ind Hyg Assoc J 1997; 58:425–431.
24. Jepson GW, Hoover DK, Black RK, et al. A partition coefficient determination method for nonvolatile chemicals in biological tissues. Fund Appl Toxicol 1994; 22:519–524.
25. Chen HS, Gross JF. Estimation of tissue-to-plasma partition coefficients used in physiological pharmacokinetic models. J Pharmacokinet Biop 1979; 7:117–125.
26. Gibaldi M. Physiologic pharmacokinetic models. In: Pharmacokinetics. New York, NY: Marcel Dekker, 1982; 9:355–384.
27. Mahle DA, Grisgsby CC, Godfrey RL, et al. Comparison of partition coefficients for a mixture of volatile organic compounds in rats and humans at different life stages Toxicol Sci 2004; 78:22.
28. Mahle DA, Gearhart JM, Godfrey RJ, et al. Determination of partition coefficients for a mixture of volatile organic compounds in rats and humans at different life stages. AFRL-HE-WP-TR-2005–0012, Mantech Environmental Technology Inc. Air Force Research Laboratory, Human Effectiveness Directorate, Wright-Patterson Air Force Base, Dayton, OH, 2005.
29. Chow SY, Woodbury DM. Kinetics of distribution of radioactive perchlorate in rat and guinea-pig thyroid glands. J Endocrinol 1970; 47:207–218.
30. Merrill EA, Clewell RA, Gearhart JM, et al. PBPK predictions of perchlorate distribution and its effect on thyroid uptake of radioiodide in the male rat Toxicol Sci 2003; 73:256–269.
31. Gardner SY, Sarah Y, McGee KJ, et al. Emission-particle induced ventilatory abnormalities in a rat model of pulmonary hypertension. Environ Health Perspect 2004; 112:872–878.
32. Filser JG. The closed chamber technique—uptake, endogenous production, excretion, steady-state kinetics and rates of metabolism of gases and vapors. Arch Toxicol 1992; 66:1–10.

33. Gargas ML, Andersen ME, Clewell HJ III. A physiologically-based simulation approach for determining metabolic constants from gas uptake data. Toxicol Appl Pharmacol 1986; 86:341–352.
34. Andersen ME, Clewell HJ III, Gargas ML, et al. Inhalation pharmacokinetics: evaluating systemic extraction, total in vivo metabolism and the time course of enzyme induction for inhaled styrene in rats based on arterial blood: inhaled air concentration ratios. Toxicol Appl Pharmacol 1984; 73:176–187.
35. Clark LH, Setzer RW, Barton HA, et al. Framework for evaluation of physiologically-based pharmacokinetic models for use in safety or risk assessment. Risk Anal 2004; 24:1697–1717.
36. Knaak JB, al Bayati MA, Raabe OG. Development of partition coefficients, V_{max} and K_m values, and allometric relationships. Toxicol Lett 1995; 79:87–98.
37. Lindstedt SL. Allometry: body size constraints in animal design. In: Pharmacokinetics in Risk Assessment: Drinking Water and Health. Vol. 8. Washington, DC: National Academy Press, 1987:65–79.
38. Anderson BJ, Woollard G, Holford NHG. A model for size and age changes in the pharmacokinetics of paracetamol in neonates, infants and children. Br J Clin Pharmacol 2000; 50:125–134.
39. Gargas ML, Andersen ME. Physiologically based approaches for examining the pharmacokinetics of inhaled vapors. In: Gardner DE, Crapo JD, Massaro EJ, eds. Toxicology of the Lung. New York, NY: Raven Press, 1988:449–476.
40. Clewell HJ III, Gentry PR, Gearhart JM, et al. Development of a physiologically based pharmacokinetic model of isopropanol and its metabolite acetone. Toxicol Sci 2001; 63:160–172.
41. Kararli TT. Comparison of the gastrointestinal anatomy, physiology, and biochemistry of humans and commonly used laboratory animals. Biopharm Drug Dispos 1995; 16:351–380.
42. Withey JR, Collins BT, Collins PG. Effect of vehicle on the pharmacokinetics and uptake of four halogenated hydrocarbons from the gastrointestinal tract of the rat. J Appl Toxicol 1983; 9:173–177.
43. Robinson PJ, Rapoport SI. Kinetics of protein binding determine rates of uptake of drugs by brain. Am J Physiol 1986; 251:R1212–R1220.
44. Bruckner JV, David BD, Blancato JN. Metabolism, toxicity and carcinogenicity of trichloroethylene. Crit Rev Toxicol 1989; 20:31–50.
45. Henningsen GM, Yu KO, Salomon RA, et al. The metabolism of t-butylcyclohexane in Fischer-344 male rats with hyaline droplet nephropathy. Toxicol Lett 1987; 39:313–318.
46. Medinsky MA, Sabourin PJ, Lucier G, et al. A physiological model for simulation of benzene metabolism by rats and mice. Toxicol Appl Pharmacol 1989; 99:193–206.
47. Clewell HJ III, Merrill EM, Yu KO, et al. Predicting neonatal perchlorate dose and inhibition of iodide uptake in the rat during lactation using physiologically-based pharmacokinetic modeling. Toxicol Sci 2003; 74:416–436.
48. Motzer WE. Perchlorate: problems, detection, and solutions. Environ Forensics 2001; 2:301–311.

8

In Silico Predictions of Partition Coefficients for Physiologically Based Pharmacokinetic Models

Kannan Krishnan and Andy Nong
Groupe de Recherche Interdisciplinaire en Santé et Département de Santé Environnementale et Santé au Travail, University of Montreal, Montreal, Quebec, Canada

Sami Haddad
Department of Biological Sciences, University of Quebec at Montreal, Montreal, Quebec, Canada

INTRODUCTION

The health risk assessment approaches for noncarcinogens (reference concentration and reference dose) generally account for interindividual differences in pharmacokinetics and pharmacodynamics, by way of a composite factor of 10 (1–5). There is increasing effort to understand and quantify the pharmacokinetic and pharmacodynamic components of the interindividual variability factor (6–8). The intraspecific or interindividual variability factor, reflecting the pharmacokinetic component, can be estimated either using measured data on biomarker concentrations collected in human studies or using pharmacokinetic models (8,9). The pharmacokinetic models frequently used in this regard are the compartmental models—commonly known as the physiologically based pharmacokinetic (PBPK) models (10).

The PBPK models facilitate the estimation of the magnitude of interindividual variation in tissue dose resulting from the interindividual variability in mechanistic determinants or input parameters, namely physiological parameters (e.g., cardiac output, breathing rate, tissue volumes, and tis sue blood flow rates), partition coefficients (PCs) (e.g., blood:air PC and tissue:blood PC), and metabolic constants (maximal velocity and Michaelis constant) (11–19). Frequently, the Monte Carlo simulation methodology is applied to perform repeated computations of internal dose (based on inputs selected at random from statistical distributions representing population variability of each input parameter). Based on the data on internal dose corresponding to the 95th percentile and 50th percentile (in the case of unimodel, normal distribution), the magnitude of the interindividual variability factor can be computed for risk assessment purposes (Fig. 1). For estimating the magnitude of the interindividual (pharmacokinetic) variability factor using PBPK models, information on the population distribution and variability of the PCs is required. These PCs primarily relate to blood:air (P_{ba}) and tissue:blood (P_{tb}). In the case of volatile organic substances, tissue:air (P_{ta}) is often obtained and it is then divided by P_{ba} to get P_{tb}. However, it is not feasible to undertake experimental studies of P_{ba}, P_{ta}, and P_{tb} for each chemical in each population or individual of interest. Alternatively, it may be feasible to estimate the population variability of P_{ba}, P_{ta}, and P_{tb} using in silico approaches. The objective of this chapter is to provide an overview of the in silico approaches for estimating interindividual variability in PCs for use in PBPK modeling and risk assessment.

IN SILICO APPROACHES FOR ESTIMATING PCs

The in silico approaches for estimating PCs are one of the three types: empirical, mechanistic, and semiempirical (20,21). The mechanistic and semiempirical methods are based on the biological and chemical properties that determine the PCs. On the contrary, the empirical approaches relate the PCs to structural features of chemicals through a mathematical function (f):

$$PC = f(\text{structural feature}) \tag{1}$$

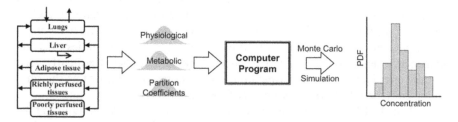

Figure 1 Schematic representation of the development and use of physiologically based pharmacokinetic models to generate population distributions of internal dose based on Monte Carlo simulation methodology. *Abbreviation*: PDF, probability density function.

The empirical approach is essentially data based, meaning that the resulting model and parameter estimates are derived from the training data set. Two common types of empirical quantitative structure-activity relationship (QSAR) approaches used to estimate PCs are: the linear free energy (LFE) approach and the Free-Wilson approach (21). Whereas the LFE approach relates PCs to electrostatic, steric, or hydrophobic features of chemicals, the Free-Wilson approach calculates PCs as the additive function of the contributions by fragments present in a molecule. The empirical QSARs, even though relatively easy to develop, cannot be used in their present form for evaluating interindividual differences in PCs and tissue dose. The major reason is that these QSARs are not comprised of parameter(s) that vary among individuals of a population; rather they only attempt to explain the variation of PCs among chemicals on the basis of the variation in structure/properties of chemicals (22–24).

Mechanistic and semiempirical in silico approaches, on the other hand, are based on properties that are specific to chemicals as well as characteristics that are specific to an individual or a population (25–33). Thus, these approaches account for the determinants and processes responsible for blood and tissue partitioning of chemicals, amenable for estimating the magnitude of the interindividual variation in P_{ba}, P_{ta}, and P_{tb}. Mechanistic in silico approaches have been successfully applied to calculate P_{ba}, P_{ta}, and P_{tb} of low-molecular-weight volatile organic substances (alkyl benzenes, chloroethanes, chloroethylenes, ketones, acetate esters, and alcohols) as well as some highly lipophilic organochlorine compounds in rats and humans [polychlorinated biphenyls (PCBs), polybrominated biphenyls (PBBs), dieldrin, dichlorodiphenyl trichloroethane (DDT), and dioxins] (25–28,32–34).

MECHANISTIC IN SILICO APPROACHES FOR PREDICTING INTERINDIVIDUAL DIFFERENCES IN PCs

Blood:Air

The mechanistic in silico approaches for predicting P_{ba} are based on the hypothesis that the blood solubility and protein binding are the key processes of the partitioning process. While chemical solubility is likely to be determined by the neutral lipid, phospholipid, and water contents of blood, the binding would appear to be associated with plasma proteins and/or hemoglobin. In initial efforts (25), the human blood:air PCs of low-molecular-weight volatile organic compounds (VOCs) have been calculated on the basis of the consideration of solubility alone, as follows:

$$P_{ba} = [P_{ow} \times P_{wa} \times (F_{nb} + 0.3 \times F_{pb})] + [P_{wa} \times (F_{wb} + 0.7 \times F_{pb})] \quad (2)$$

where $P_{wa} = n$-octanol (or vegetable oil): water PC

P_{wa} = water:air PC

F_{nb} = Fraction of neutral lipids in blood

F_{pb} = Fraction of phospholipids in blood, and

F_{wb} = Fraction of water in blood

In the above approach, based on the lipophilicity–hydrophilicity characteristics of phospholipids, the solubility in this fraction can be approximated to that of a mixture of 30% neutral lipid and 70% water (27). As a result, Eq. (2) can be rewritten as follows:

$$P_{ba} = [P_{ow} \times P_{wa} \times F_{neb}] + [P_{wa} \times F_{web}] \tag{3}$$

where F_{neb} = volume fraction of neutral lipid equivalents in blood, calculated as the sum of neutral lipids plus $0.3 \times$ phospholipid content, and F_{web} = volume fraction of water equivalents in blood, calculated as the sum of blood water content plus $0.7 \times$ phospholipid content.

Figure 2 illustrates the calculation of blood:air PC using this methodology, with 1,1,1-trichloroethane as the example. In this case, the chemical structure information is used for computing P_{ow} and P_{wa}, which are then used along with the average value for lipid and water contents in human blood for predicting P_{ba} of 1,1,1-trichloroethane. Using individual-specific values on lipid and water contents in this algorithm, predictions of interindividual differences in P_{ba} can be obtained. The impact of the variation of blood lipid and water levels on P_{ba} depends on the P_{ow} of the chemical. The more lipophilic the chemical, the greater the influence of the variation in lipid content. As shown in Figure 3, for chemicals with P_{ow} less than 20, the variation in lipid content is not expected to significantly influence P_{ba}. However, for chemicals with P_{ow} greater than 2000, the interindividual variation in lipid content is likely to translate into a proportional influence on the P_{ba}. The sensitivity ratio for the neutral lipid component is close to 1 for such cases, as shown in Figure 3. The interindividual variation in blood water content might have a significant impact only for chemicals with low P_{ow}. The above methodology has recently been extended to include the impact of protein binding on the human blood:air PCs of organic chemicals, thus facilitating the consideration of the interindividual variability in the binding affinity and concentration of binding proteins (32).

Tissue:Air

The tissue:air PCs can be predicted on the basis of the consideration of chemical solubility in tissue neural lipids, phospholipids, and water. The algorithm for computing P_{ta} according to this approach is as follows:

$$P_{ta} = [P_{ow} \times P_{wa} \times F_{net}] + [P_{wa} \times F_{wet}] \tag{4}$$

where F_{net} = volume fraction of neutral lipid equivalents in tissues, calculated as the sum of neutral lipids plus $0.3 \times$ phospholipid content, and F_{wet} = volume fraction of water equivalents in tissues, calculated as the sum of tissue water content plus $0.7 \times$ phospholipid content.

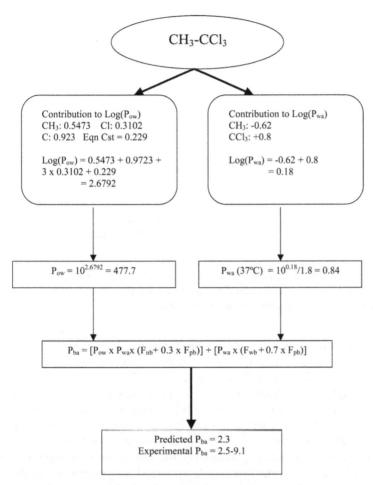

Figure 2 An example of the calculation of blood:air partition coefficient (P_{ba}) for 1,1,1-trichloroethane (CH_3CCl_3) using a mechanistic in silico approach (o: octanol, w: water, a: air, F_{neb}: volume fraction of neutral lipid equivalents in blood, and F_{web}: volume fraction of water equivalents in blood). *Source*: From Ref. 10.

The above algorithm facilitates the computation of tissue:air PCs of chemicals, on the basis of not only their physicochemical properties but also the individual-specific tissue composition data. Compilations of typical and ranges of tissue composition data (lipid and water contents) are available in the literature. Using the range of lipid and water contents in various tissues of adult humans (Table 1) in the above equation, P_{ta} has been calculated for several organic chemicals (Table 2). In effect, the variability of experimental P_{ta} values might arise from method-related differences and/or interindividual

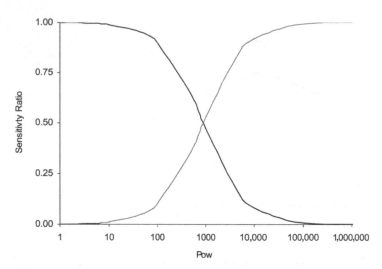

Figure 3 The outcome of the sensitivity analysis of blood water content (*black*) and neutral lipid content (*gray*) on the blood:air partition coefficient as a function of *n*-octanol:water partition coefficients of chemicals ($P_{o:w}$).

variation in tissue composition. Figure 4 depicts the sensitivity ratio (i.e., ratio of change in PCs to change in tissue lipid or water content value) associated with the prediction of liver:air and adipose tissue:air PCs of organic chemicals with P_{ow} ranging from 0.0001 to 100,000. It can be seen that in the case of human liver:air PC, the individual variation in lipid content does not influence the outcome in case of chemicals with $P_{ow} < 1$. However, the adipose tissue:air PC is strongly influenced by the lipid content, which might vary from one individual to another, particularly for chemicals with high P_{ow} values. The results of the sensitivity analysis presented in Figure 4 essentially allow one to appreciate the importance of interindividual variability in lipid and water content on the outcome (i.e., tissue:air PC) as a function of the lipophilicity characteristics of the chemical of interest.

Table 1 Ranges of Neutral Lipids and Water in Human Tissues

Tissue	Neutral lipids[a]	Water[b]
Adipose tissue	0.7100–0.8700	0.23–0.10
Liver	0.0186–0.0853	0.67–0.80
Muscle	0.0271–0.0806	0.70–0.80

[a]Based on Pelekis et al.
[b]Values correspond to the low- and high end of the range of tissue composition data found in the literature. The values correspond to fraction of tissue weight.
Source: From Ref. 35.

Table 2 Comparison of Predicted and Experimental Human Tissue:Air Partition Coefficients Using Low- and High-End Values of Tissue Neutral Lipids and Water[a]

Partition coefficients and chemicals	Predicted range	Experimental values
Adipose tissue:air		
Dichloromethane	70.4–86.8	85 ± 3.5
Chloroform	360.8–443.0	340–394
1,1,1-Trichloroethane	151.2–185.3	251 ± 11
Benzene	275.8–338.1	425 ± 50
Toluene	964.6–1182.2	962 ± 168
Liver:air		
Dichloromethane	5.3–12.5	7.2 ± 0.1
Chloroform	15.1–50.0	16.6–17.4
1,1,1-Trichloroethane	4.5–18.8	16.5 ± 2.8
Benzene	8.5–34.6	22.6 ± 4.5
Toluene	26.6–117.5	48.2 ± 8.4
Muscle:air		
Dichloromethane	6.3–12.1	4.8 ± 0.5
Chloroform	19.7–47.6	9.8–15.8
1,1,1-Trichloroethane	6.3–17.8	6.7 ± 0.5
Benzene	11.8–32.8	16.4 ± 2.5
Toluene	38.2–111.1	34.9 ± 7.6

[a]Experimental values obtained from the literature are expressed as a range (minimum–maximum) or as mean$\pm$standard deviation.
Source: From Ref. 27.

Tissue:Blood

The tissue:blood PCs (P_{tb}) can either be calculated by dividing the computed values of P_{ta} and P_{ba}, or calculated, on the basis of Eqs. (3) and (4), as follows:

$$P_{tb} = \frac{[P_{ow} \times P_{wa} \times F_{net}] + [P_{wa} \times F_{wet}]}{[P_{ow} \times P_{wa} \times F_{neb}] + [P_{wa} \times F_{web}]} \qquad (5)$$

Simplifying the above equation,

$$P_{tb} = \frac{[P_{ow} \times F_{net}] + F_{wet}}{[P_{ow} \times F_{neb}] + F_{web}} \qquad (6)$$

Essentially, this equation is appropriate for predicting tissue:blood PCs of organic chemicals, for which macromolecular binding in tissue and blood is negligible. Contribution of binding to blood and tissue partitioning of chemicals may additionally be accounted for, as necessary (29,32).

The above equation has been used to calculate the human tissue:blood PCs for a number of chemicals (20,26,27). A close examination of this

(A)

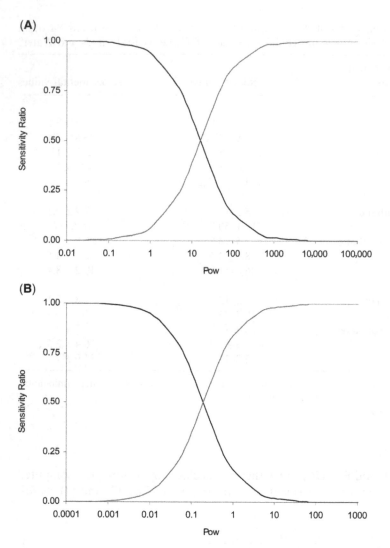

(B)

Figure 4 The outcome of the sensitivity analysis of tissue water content (*black*) and neutral lipid content (*gray*) on the tissue:air partition coefficient (**A**: liver and **B**: adipose tissue) as a function of *n*-octanol:water partition coefficients of chemicals ($P_{o:w}$).

relationship will indicate that, for highly lipophilic chemicals (i.e., log P_{ow} ≥ 4) the first term $[P_{ow} \times F_{net}]$ is very large compared to the second term $[F_{wet}]$ in both the numerator and the denominator, such that Equation (4) can be simplified to the following form:

$$P_{tb} = \frac{[P_{ow} \times F_{net}]}{[P_{ow} \times F_{neb}]} \tag{7}$$

The P_{ow} in the numerator and denominator of the above equation cancels out, such that tissue:blood PC becomes a function of the ratio of neutral lipids in tissues and blood:

$$P_{tb} = \frac{F_{net}}{F_{neb}} \tag{8}$$

In other terms, regardless of the magnitude of P_{ow} of highly lipophilic organic chemicals, their P_{tb} is likely to be a biologically based, individual-specific value as determined by the ratio of lipids in tissues and blood. The assertion has been verified by comparing the experimental and predicted human P_{tb} values for several volatile and nonvolatile organics (34). For example, for VOCs with $P_{o:w}$ values ranging from 2738 to 67,200 (cyclohexane, n-pentane, methylcyclopentane, 2,2-dimethyl butane, n-hexane, 2-methyl pentane, n-heptane and 3-methyl hexane), the average human adipose tissue:blood (P_{atb}) was just about the ratio of neutral lipids in adipose tissues and blood (i.e., 199). Similarly, the average human adipose tissue:serum PC (P_{ats}) reported for PCBs, PBBs, and DDT is 229 (SD: 72), which is close to the average ratio of neutral lipids in human adipose tissue and serum (=173) (34). The literature human P_{ats} values for other nonvolatile organics such as dieldrin (156–179), mirex (264), and hexachlorocyclohexane (300) are also within a factor of 2 of the average ratio of neutral lipid content of human adipose tissue and

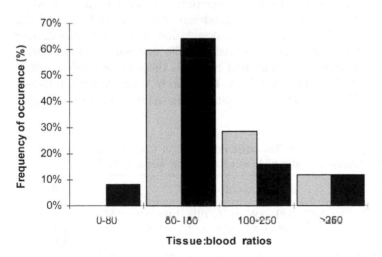

Figure 5 Frequency distribution of the adipose tissue:blood ratio of 2,3,7,8 tetra chlorodibeno-*p*-dioxin concentration (*black bars*) and the ratio of lipid contents of adipose tissues and serum (*grey bars*) in a group of individuals ($n = 50$). *Source:* From Refs. 34 and 36.

Table 3 Average Lipid and Water Contents of Human Tissues and Blood[a]

Tissue	Neutral lipids[b]	Phospholipids[b]	Water[b]	Water equivalents[c]	Neutral lipid equivalents[d]
Adipose	0.7980	0.0020	0.15	0.1947	0.7986
Blood	0.0033	0.0024	0.82	0.8217	0.0040
Plasma	0.0040	0.0021	0.92	0.9214	0.0046
Brain	0.0462	0.0638	0.79	0.8027	0.0653
Kidney	0.0338	0.0182	0.77	0.7772	0.0393
Liver	0.0389	0.0281	0.72	0.7217	0.0473
Lung	0.0030	0.0090	0.75	0.7697	0.0057
Muscle	0.0347	0.0104	0.75	0.7563	0.0378

[a]Data from Poulin and Krishnan.
[b]Expressed as fraction of tissue volume.
[c]Calculated as the sum total of the volume fraction of water and 0.7 times the phopholipid fraction in tissues.
[d]Calculated as the sum total of the volume fraction of neutral lipids and 0.3 times the phospholipid fraction in tissues.
Source: From Ref. 27.

serum (i.e., 173). Even though the average ratio of tissue/serum lipids obtained from the literature corresponds to 173, the reported individual human P_{ats} for highly lipophilic compounds range from 140 to 336, indicating that the variation in PCs is likely due to interindividual variation in lipid levels. Figure 5 shows the positive association between the interindividual differences in P_{ats} of 2,3,7,8-tetrachlorodibenz-*p*-dioxin (TCDD) and the ratio of tissue/blood lipids in a human population (36). These observations emphasize that for nonreactive, superlipophilic compounds, the relative lipid level in two matrices (tissues and blood) is the sole mechanistic determinant of tissue: blood PCs. Similarly, the ratio of water content in tissues and blood would be the key determinant of the tissue:blood PCs of highly

Table 4 Comparison of Rat Tissue:Blood Partition Coefficients of Some Hydrophilic Chemicals to Ratio of Water Content in Tissues and Blood

Tissue:blood PCs	Experimental PCs[a]	RW
Liver	0.90, 0.81, 0.96, 0.76, 1.08, 1.00, 0.97, 0.89	0.86
Muscle	1.16, 0.80, 0.85, 0.85, 0.78, 0.97, 0.90, 0.65	0.90
Adipose tissue	0.06, 0.11	0.15

[a]Data for alcohols, acetate esters, and ketones with P_{ow} of ≤ 1 (liver and muscle) or ≤ 0.3 (adipose tissue), obtained from a compilation by Poulin and Krishnan.
Abbreviations: PCs, partition coefficients; RW, ratio of water content.
Source: From Ref. 27.

hydrophilic chemicals (Table 3). This is further illustrated by the comparison of the ratio of tissue/water content with the human tissue:blood PCs for alcohols, acetate esters, and ketones (Table 4).

Thus, the tissue composition-based algorithms, as supported by the experimental data and human observations, suggest that P_{tb}s of organic chemicals will not increase proportionally as a function of their P_{ow}; instead the P_{tb} reaches a plateau after a certain "cutoff" value of P_{ow}, because chemicals with high P_{ow} dissolve mainly in tissue lipids (and their levels in tissue water component is negligible). In other words, the general perception of a proportional increase in tissue:blood PC as a function of lipophilicity or P_{ow} of chemicals is unlikely to be entirely true above a certain P_{ow} value— based on mechanistic considerations. Figure 6 shows the nonlinearity of the relationship between P_{tb} and P_{ow} of organic chemicals (Table 5). The influence of the variation of tissue content of lipids and water on predicted P_{tb} depends on where the chemical is located in the curves shown in Figures 6 and 7.

If we consider a highly lipophilic chemical, such as styrene ($P_{ow} > 5000$), the lipid solubility is the major determinant of its tissue: blood partitioning process. For styrene, the sensitivity coefficients— related to tissue water and blood water content are essentially negligible (Fig. 8A). This implies that interindividual variation in water content is unlikely to influence the tissue:blood PCs (liver, muscle, and fat) of styrene. However, the sensitivity ratio is about +1 for F_{nlt}, implying that the tissue:blood partitions will increase proportional to the interindividual variation in lipid content of tissues. However, such a direct influence cannot be expected for all chemicals—particularly for hydrophilic chemicals

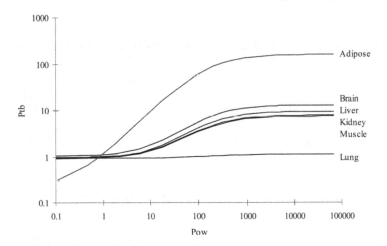

Figure 6 Relationship between tissue:blood partition coefficients (P_{tb}) and *n*-octanol: water partition coefficient (P_{ow}). *Source*: From Ref. 34.

Table 5 List of Chemicals and Their P_{ow} Values Used Deriving the Relationship Presented in Figure 6[a]

Chemicals	P_{ow}
Methanol	0.076
Ethanol	0.479
Acetone	0.575
2-Propanol	0.692
2-Butanone	1.820
1-Propanol	2.188
Isobutanol	4.466
Diethyl ether	5.889
Dichloromethane	17.782
Chloroform	93.330
Benzene	134.914
Trichloroethylene	195.000
1,1,1-Trichloroethane	309.072
Toluene	446.633
Styrene	893.714
Cyclohexane	2738.028
n-Pentane	4131.579
Methylcyclopentane	4562.000
2,2-Dimethyl butane	6500.000
n-Hexane	13157.143
2-Methyl pentane	18605.882
n-Heptane	40433.333
3-Methyl hexane	67200.000

[a]P_{ow} values were obtained from Haddad et al.
Source: From Ref. 34

such as methyl chloride ($P_{ow} = 5$). Figure 8B shows that the lipid content of blood (F_{lb}) is not a sensitive parameter at all, whereas F_{nlt} (neutral lipid content of tissues) is a sensitive parameter only for the adipose tissue. Contrary to the styrene example, it can be seen that the water content of tissues and blood are among the sensitive parameters. The implication is that, for hydrophilic chemicals, the interindividual variation in tissue and blood water content is likely to influence the tissue:blood PCs in a significant manner.

INTERINDIVIDUAL DIFFERENCES IN PCs, TISSUE DOSE, AND RISK ASSESSMENT

Risk assessments are increasingly being based upon tissue dose or internal dose of chemicals rather than their exposure concentrations. For performing

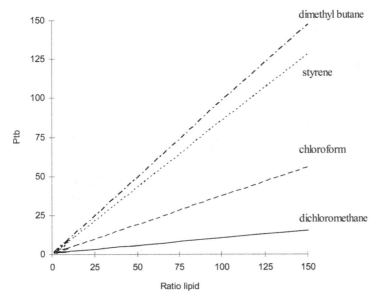

Figure 7 Relationship between the tissue:blood partition coefficients (P_{tb}) and ratios of lipid content of tissues and blood for various chemicals. *Source*: From Ref. 34.

tissue dose-based risk assessments, and more specifically for characterizing the magnitude of interindividual variability in tissue dose of chemicals, the use of PBPK models is inevitable due to the lack of human data. One type of key determinant of tissue dose relates to the PCs. The blood:air PC is a key factor determining the steady-state concentrations of inhaled organics. The tissue:blood PCs, on the other hand, are key determinants of tissue concentrations of chemicals relative to their levels in systemic circulation. The interindividual variations in these PCs can lead to individual differences in tissue dose. As illustrated in this chapter, the interindividual variance in tissue PCs can be predicted with knowledge on interindividual variation and population distribution of tissue and blood lipid and water contents. Predictions of PCs based on tissue and blood lipid and water contents have been obtained successfully for a number of low-molecular-weight organic chemicals and persistent organochlorine contaminants. The information on the interindividual variability of PCs, however, cannot be used in isolation to estimate the individual differences in tissue dose of chemicals. This information should be used along with data on interindividual variance of metabolic parameters and physiological parameters (37). PBPK models provide an appropriate framework for integrating data on interindividual variance of PCs, metabolic parameters, and physiological parameters in order to provide

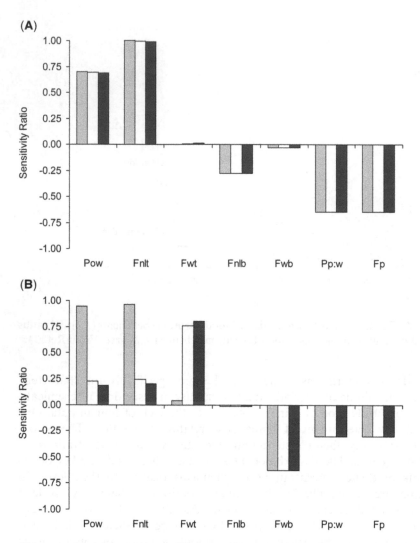

Figure 8 Sensitivity analyses of the input parameters for the estimation of tissue: blood partition coefficients (**A**: lipophilic chemical and **B**: hydrophilic chemical). Sensitivity ratios are based on the calculated change of predicted tissue:blood partition coefficients (fat: gray, liver: white, and muscle: black) in relation to a 10% change in the value of input parameters [n-octanol:water constant (P_{ow}), neutral lipid content of tissues (F_{nlt}), water content of tissues (F_{wt}), water content of blood (F_{wb}), neutral lipid content of blood (F_{nlb}), protein:water partition coefficient (P_{pw}), and blood protein content (F_{p})].

simulations of the population distribution of tissue dose of chemicals and estimation of the interindividual variability factor for use in noncancer risk assessments.

REFERENCES

1. Barnes DG, Dourson ML. Reference dose (RfD): description and use in health risk assessments. Regul Toxicol Pharmacol 1988; 8:471–486.
2. Jarabek AM, Menache MG, Overton JH Jr., et al. The U.S. Environmental Protection Agency's inhalation RfD methodology:risk assessment for air toxics. Toxicol Ind Health 1990; 6:279–301.
3. Dourson ML, Knauf LA, Swartout JC. On reference dose (RfD) and its underlying toxicity data base. Toxicol Ind Health 1992; 8:171–189.
4. U.S. Environmental Protection Agency. Methods for derivation of inhalation reference concentrations and application of inhalation dosimetry, U.S. Environmental Protection Agency, Office of Research and Development, Office of Health and Environmental Assessment, Washington, DC, EPA/600/8-90/066F, 1994.
5. U.S. Environmental Protection Agency. A review of the reference dose and reference concentration processes, U.S. Environmental Protection Agency, Risk Assessment Forum, Washington, DC, EPA/630/P-02/002F, 2002.
6. Naumann BD, Sargent EV. Setting occupational exposure limits for pharmaceuticals. Occup Med 1997; 12:67–86.
7. Renwick AG, Walker R. An analysis of the risk of exceeding the acceptable or tolerable daily intake. Regul Toxicol Pharmacol 1993; 18:463–480.
8. IPCS (International Programme on Chemical Safety), Guidance Document for the Use of Data in Development of Chemical-specific Adjustment Factors (CSAF) for Interspecies Differences and Human Variability in Dose/concentration Response Assessment, World Health Organization, Geneva, 2001. Available at www.who.int/entity/ipcs/publications/methods/harmonization/en/csafs_guidance_doc.pdf.
9. Dourson ML, Felter SP, Robinson D. Evolution of science-based uncertainty factors in noncancer risk assessment. Regul Toxicol Pharmacol 1996; 24: 108–120.
10. Krishnan K, Andersen ME. Physiologically based pharmacokinetic modeling in toxicology. In: Hayes AW, ed. Principles and Methods of Toxicology. 4th. Philadelphia, PA: Taylor & Francis, 2001:193–241.
11. Thomas RS, Lytle WE, Keefe TJ, et al. Incorporating Monte Carlo simulation into physiologically based pharmacokinetic models using advanced continuous simulation language (ACSL): a computational method. Fundam Appl Toxicol 1996; 31:19–28.
12. Bois FY. Analysis of PBPK models for risk characterization. Ann N Y Acad Sci 1999; 895:317–337.
13. Johanson G, Jonsson F, Bois F. Development of new technique for risk assessment using physiologically based toxicokinetic models. Am J Ind Med 1999; Suppl (1), 101–103.
14. El-Masri HA, Bell DA, Portier CJ. Effects of glutathione transferase theta polymorphism on the risk estimates of dichloromethane to humans. Toxicol Appl Pharmacol 1999; 158:221–230.
15. Pelekis M, Nicolich MJ, Gauthier JS. Probabilistic framework for the estimation of the adult and child toxicokinetic intraspecies uncertainty factors. Risk Anal 2003; 23:1239–1255.

16. Bogaards JJ, Freidig AP, van Bladeren PJ. Prediction of isoprene diepoxide levels in vivo in mouse, rat and man using enzyme kinetic data in vitro and physiologically-based pharmacokinetic modelling. Chem Biol Interact 2001; 138:247–265.

17. Haber LT, Maier A. Scientific criteria used for the development of occupational exposure limits for metals and other mining-related chemicals. Regul Toxicol Pharmacol 2002; 36:262–279.

18. Lipscomb JC, Kedderis GL. Incorporating human interindividual biotransformation variance in health risk assessment. Sci Total Environ 2002; 288:13–21.

19. Lipscomb JC, Teuschle LK, Swartont J, et al. The impact of cytochrome P450 2E1-dependent metabolic variance on a risk-relevant pharmacokinetic outcome in humans. Risk Anal 2003; 23:1221–1238.

20. Payne MP, Kenny LC. Comparison of models for the estimation of biological partition coefficients. J Toxicol Environ Health A 2002; 65:897–931.

21. Béliveau M, Krishnan K. In silico approaches for developing physiologically-based pharmacokinetic (PBPK) models. In: Salem H, Katz SA, eds. Alternative Toxicological Methods. Boca Raton, FL: CRC Press, LLC, 2003:479–532.

22. Abraham MH, Kamlet MJ, Taft RW, et al. Solubility properties in polymers and biological media. 2. The correlation and prediction of the solubilities of nonelectrolytes in biological tissues and fluids. J Med Chem 1985; 28:865–870.

23. Abraham MH, Chadha HS, Whiting GS, et al. An analysis of water-octanol and water-alkane partitioning and the delta log P parameter of seiler. J Pharm Sci 1994; 83:1085–1100.

24. Basak SC, Mills D, Hawkins DM, et al. Prediction of human blood: air partition coefficient: a comparison of structure-based and property-based methods. Risk Anal 2003; 23:1173–1184.

25. Poulin P, Krishnan K. A tissue composition-based algorithm for predicting tissue:air partition coefficients of organic chemicals. Toxicol Appl Pharmacol 1996; 136:126–130.

26. Poulin P, Krishnan K. A mechanistic algorithm for predicting blood:air partition coefficients of organic chemicals with the consideration of reversible binding in hemoglobin. Toxicol Appl Pharmacol 1996; 136:131–137.

27. Poulin P, Krishnan K. A biologically-based algorithm for predicting human tissue: blood partition coefficients of organic chemicals. Hum Exp Toxicol 1995; 14:273–280.

28. Poulin P, Krishnan K. Molecular structure-based prediction of the toxicokinetics of inhaled vapors in humans. Int J Toxicol 1999; 18:7–18.

29. De Jongh J, Verhaar HJ, Hermens JL. A quantitative property-property relationship (QPPR) approach to estimate in vitro tissue-blood partition coefficients of organic chemicals in rats and humans. Arch Toxicol 1997; 72:17–25.

30. Balaz S, Lukacova V. A model-based dependence of the human tissue/blood partition coefficients of chemicals on lipophilicity and tissue composition. Quant Struct Act Relat 1999; 18:361–368.

31. Haddad S, Charest-Tardiff G, Krishnan K. Physiologically based modeling of the maximal effect of metabolic interactions on the kinetics of components of complex chemical mixtures. J Toxicol Environ Health A. 2000; 61:209–223.

32. Béliveau M, Lipscomb JC, Tardiff R, et al. Quantitative structure-property relationships for physiologically based pharmacokinetic modeling of volatile organic chemicals in rats. Chem Res Toxicol 2005; 18:475–485.
33. Béliveau M, Tardiff R, Krishnan K. Quantitative structure-property relationships for interspecies extrapolation of the inhalation pharmacokinetics of organic chemicals. Toxicol Appl Pharmacol 2003; 189:221–232.
34. Haddad S, Poulin P, Krishnan K. Relative lipid content as the sole mechanistic determinant of the adipose tissue:blood partition coefficients of highly lipophilic organic chemicals. Chemosphere 2000; 40:839–843.
35. Pelekis M, Poulin P, Krishnan K. An approach for incorporating tissue composition data into physiologically based pharmacokinetic models. Toxicol Ind Health 1995; 11:511–522.
36. Patterson DG, Needham LL, Pirkle JL, et al. Correlation between serum and adipose tissue levels of 2,3,7,8-tetrachlorodibenzo-p-dioxin in 50 persons from Missouri. Arch Environ Contam Toxicol 1988; 17:139–143.
37. Lipscomb JC, et al. The metabolic rate constants and specific activity of human and rat hepatic cytochrome P-450 2E1 toward toluene and chloroform. J Toxicol Environ Health A 2004; 67:537–553.

9

In Vitro to In Vivo Extrapolation of Metabolic Rate Constants for Physiologically Based Pharmacokinetic Models

Gregory L. Kedderis

Independent Investigator, Chapel Hill, North Carolina, U.S.A.

INTRODUCTION

Physiologically based pharmacokinetic (PBPK) modeling has proven a useful approach in quantifying animal to human and human interindividual differences in tissue dosimetry for application in human health risk assessment. Chemical metabolism is often the limiting factor in the manifestation of toxicity, whether metabolism results in a decrease in toxicity (detoxication) or an increase in toxicity (bioactivation). Chemical metabolism can be studied in the whole animal—research species or human—or can be studied under carefully controlled conditions in vitro in the laboratory. Studies of human variability must include enough subjects so that the population distribution can be assessed, which is troubling when examining the number of subjects typically employed in clinical pharmacokinetic investigations. In vitro studies using tissue preparations or genetically expressed enzymes offer the advantage of carefully controlled conditions, and large numbers of samples can easily be evaluated. However, the results must be extrapolated from the experimental setting to the whole organism for inclusion in PBPK models. In order for these extrapolations to be meaningful, the in vitro experiments must be done under the appropriate conditions to accurately determine the kinetic parameters of the metabolic reactions being studied.

This chapter will describe the biochemical and physiological considerations required for accurate extrapolation of in vitro biotransformation kinetics to whole organisms for implementation in PBPK models. The structure and function of enzymes in mammalian cells will be described along with the fundamentals of enzyme kinetics. The choice of in vitro systems to study xenobiotic biotransformation will be discussed in the context of extrapolation of the kinetics of the biotransformation reactions to the intact organism. Aspects of the design of appropriate in vitro systems and experiments for prediction of pharmacokinetics in vivo will be discussed. It will be shown that basic biochemical and physiological principles are the keys to successful extrapolation of in vitro–derived parameters to the in vivo situation. Indeed, understanding the basic biochemistry and physiology underlying xenobiotic biotransformation and target organ toxicity is essential to the soundness of any scientifically based human health risk assessment.

THE STRUCTURE AND FUNCTION OF ENZYMES

Living organisms are composed of various small molecules and macromolecules including proteins, lipids, polysaccharides, and nucleic acids that are termed biomolecules. Biomolecules are organized into three-dimensional structures that make up mammalian cells. Simple biomolecules such as amino acids, sugars, and lipids are the building blocks of macromolecules, which are components of supramolecular complexes such as membranes and ribosomes. Supramolecular complexes are assembled into organelles, which organize into cells (1). Cells are complex integrated entities that combine macromolecular structure and function. Cellular constituents and metabolic pathways are not randomly dispersed in cells, but rather are localized to specific sites within the three-dimensional structure of cells (2). Cells are highly organized such that even the water in cells has a structure that is different from aqueous solutions (3). The carefully controlled environment inside mammalian cells facilitates the efficiency of metabolic processes.

Proteins are macromolecules composed of covalent chains of some 20 different amino acids folded into secondary and tertiary structures via noncovalent forces such as hydrogen bonds and hydrophobic interactions or through disulfide bonds (1). Proteins have both structural and dynamic functions within cells. Enzymes are dynamic proteins that serve as catalysts of biochemical reactions. Enzymes catalyze biochemical reactions in the same way as chemical catalysts, by lowering the activation energy of the reaction and allowing it to occur under biological conditions. Each enzyme has an active site where the substrate specifically binds and catalysis takes place. The enzyme catalyzing the reaction remains unchanged. Enzymes are named on the basis of the types of reactions they catalyze (1). In general, enzymes are named based on the substrate and the type of reaction, with the suffix, ase.

Enzymes have molecular weights in tens of thousands or greater, while substrates are typically molecules of low-molecular weight. Thus, only a small portion of the enzyme, the active site, is directly involved in catalysis. Chemical modification reagents, which react with specific aminoacid residues, can be used to probe the nature of the enzyme active site. Some enzymes possess a nonprotein prosthetic group in the active site that assists in catalysis. Some typical prosthetic groups include metal ions, metalloporphyrins, flavins, pyridine nucleotides, thiamine, biotin, and cobalamine. Some of these moieties are actually cofactors that are not covalently bound to the enzyme. Cofactors typically act as donors or acceptors in group transfer reactions or electron transfer reactions (oxidations and reductions). The substrate binds to the enzyme active site via noncovalent interactions (hydrogen bonds, hydrophobic interactions, and electrostatic interactions) and is oriented such that the catalytic groups in the active site (amino-acid residues or the prosthetic group) can affect a chemical change to the substrate. Enzymes accelerate the rate of chemical reactions by controlling the proximity and orientation of the reactants and the microenvironment of the reaction. The physicochemical microenvironment of the enzyme active site (i.e., pH, hydrophobicity, and hydration state) facilitates chemical reactions that would not take place in free aqueous solution. The amino-acid residues distant from the active site are involved in maintaining the tertiary structure of the enzyme, which provides the appropriate spatial orientation of the active site for efficient catalysis. The specific binding of a substrate to the active site allows for stereospecific catalysis in some cases.

Some enzymes are membrane bound to cellular organelles, such as the endoplasmic reticulum or mitochondria, while others are present in the soluble portion of the cell known as the cytoplasm. However, the aqueous cytoplasm of the cell is highly organized via a group of polymeric proteins called the cytomatrix, and soluble enzymes appear to be associated with this dynamic network (3–5). This intracellular organization can influence the efficiency of enzyme catalysis and promote the coupling of metabolic processes. For example, a chemical that is hydroxylated by endoplasmic reticulum-bound cytochrome P450 (CYP) can be so efficiently conjugated with glucuronic acid by neighboring membrane-bound glucuronosyl transferase that the free alcohol product cannot be detected in the cell. The coupling of metabolic processes can lead to very efficient detoxication of toxicants, but it can also promote toxication processes that can ultimately lead to cellular damage and death.

FUNDAMENTALS OF ENZYME KINETICS

Enzyme kinetics can be defined as the study of the rates at which enzyme-catalyzed reactions occur, all factors that influence the rates of enzyme-catalyzed reactions, and the explanation of the rates in terms of a reaction mechanism (6). Enzyme kinetics is an example of deductive scientific reasoning. Kinetics cannot prove any enzyme mechanism; it can only exclude inconsistent mechanisms. However, kinetic data usually suggest potential enzyme mechanisms that can frequently be distinguished experimentally.

The apparent velocity of an enzyme-catalyzed reaction (v_{app}) is directly proportional to the total enzyme present ($[E]_T$):

$$v_{app} = k_{app}[S][E]_T \tag{1}$$

where k_{app} is an apparent rate constant and $[S]$ is the substrate concentration. Thus, the initial rate of the enzymic reaction should be a linear function of the enzyme concentration (or added protein) and time at a given $[S]$. The apparent rate constant k_{app} in Eq. (1) contains the more complex kinetic mechanism of the reaction. The kinetic mechanism of an enzyme-catalyzed reaction is a mathematical description of the comings and goings of the substrates and products from the enzyme. Kinetic mechanisms can be straightforward or exceedingly complex.

Most enzyme-catalyzed biotransformation reactions follow Michaelis–Menten saturation kinetics. The initial velocity of the reaction increases hyperbolically as a function of substrate concentration (Fig. 1). V_{max} is a horizontal tangent to the top part (zero-order region) of the curve, while the tangent to the linear portion of the curve (the first-order region) is the initial rate of the reaction, V/K. V_{max} is defined as the maximal rate of the reaction at infinite substrate concentration. V/K is the pseudo-first-order

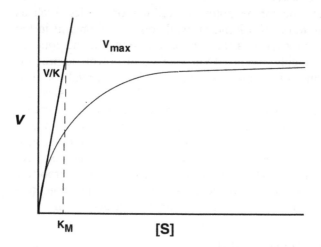

Figure 1   The initial velocity (v) curve for an enzyme-catalyzed reaction following Michaelis–Menten saturation kinetics as a function of substrate concentration ($[S]$). *Source*: From Ref. 7.

rate constant for the reaction at low substrate concentrations. The point where these two tangents intersect corresponds to K_M (7). K_M is defined as the substrate concentration that gives one-half the V_{max}.

Consider the following simple scheme for an enzyme-catalyzed reaction:

$$E + S \underset{k_{-1}}{\overset{k_1}{\rightleftharpoons}} ES \overset{k_p}{\rightarrow} E + P$$

where E is the enzyme, S is the substrate, ES is the enzyme–substrate complex, and P is the product. The mass balance for total enzyme is given by the sum of all enzyme species:

$$[E]_T = [E] + [ES] \tag{2}$$

The initial velocity of the reaction is equal to the concentration of all product-forming species multiplied by the catalytic rate constant:

$$v = k_p[ES] \tag{3}$$

Because [ES] cannot usually be measured directly, Eq. (3) needs to be rearranged into terms that can be measured. Both sides of Eq. (3) are divided by $[E]_T$:

$$\frac{v}{[E]_T} = \frac{K_p[ES]}{[E] + [ES]} \tag{4}$$

The dissociation constant of the ES complex, K_s, is defined in Eq. (5):

$$K_s = k_{-1}/k_1 = [E][S]/[ES] \tag{5}$$

Each enzyme species can be defined in terms of free E. In this example, there is only one equilibrium:

$$[ES] = \frac{[S]}{K_s}[E] \tag{6}$$

Substituting Eq. (6) into Eq. (4) gives Eq. (7):

$$\frac{v}{[E]_T} = \frac{k_p\frac{[S]}{K_s}[E]}{[E] + \frac{[S]}{K_s}[E]} \tag{7}$$

The maximal velocity of the reaction, V_{max}, is defined in Eq. (8):

$$V_{max} = k_p[E]_T \tag{8}$$

Substituting Eq. (8) into Eq. (7) and canceling $[E]$ gives Eq. (9):

$$\frac{v}{V_{\text{max}}} = \frac{[S]/K_s}{1 + [S]/K_s} \tag{9}$$

The numerator in Eq. (9) contains only one term because there is only one product-forming species (ES). The denominator of Eq. (9) contains two terms because there is a total of two different enzyme species (E and ES). Multiplying the numerator and denominator of the right-hand part of Eq. (9) by K_S gives a form of the Henri–Michaelis–Menten Eq. (6):

$$\frac{V}{V_{max}} = \frac{[S]}{K_s + [S]} \tag{10}$$

The derivation presented above gives an equilibrium expression for $[ES]$ in terms of $[E]$, $[S]$, and K_S. For most enzymes, however, k_p is similar in magnitude to k_{-1}; therefore the concentration of ES is not fixed by the concentrations of E and S and the dissociation constant K_S. In this situation, ES is not in equilibrium with E and S, but rather builds up to a near-constant or steady-state level shortly after the reaction begins. The steady-state level of ES would be close to the equilibrium level if k_p is small compared to $k-1$. If k_p is similar to or greater than $k-1$, then the steady-state level of ES would be less than the equilibrium level. The rate at which P is formed is proportional to the steady-state concentration of ES. At steady state, the rate of formation of ES [Eq. (11)] is equal to the rate of decomposition of ES [Eq. (12)], and the overall rate of change in the concentration of ES is zero.

$$\frac{d[ES]}{dt} = k_1[E][S] \tag{11}$$

$$-\frac{d[ES]}{dt} = (k_{-1} + k_p)[ES] \tag{12}$$

$$k_1[E][S] = (k_{-1} + k_p)[ES] \tag{13}$$

Rearranging Eq. (13) gives Eq. (14):

$$[ES] = \frac{k_1[S]}{(k_{-1} + k_p)}[E] \tag{14}$$

Substituting Eq. (14) into the velocity equation [Eq. (4)], rearranging, and substituting the expression for V_{max} [Eq. (8)] gives Eq. (15):

$$\frac{v}{V_{max}} = \frac{k_1[S]/(k_{-1}+k_p)}{1+k_1[S]/(k_{-1}+k_p)} \tag{15}$$

The Michaelis constant, K_M, is defined by the group of rate constants in Eq. (16):

$$K_M = \frac{k_{-1}+k_p}{k_1} \tag{16}$$

Substituting Eq. (16) into Eq. (15) gives the steady-state velocity equation [Eq. (17)]:

$$\frac{V}{V_{max}} = \frac{[S]}{K_M + [S]} \tag{17}$$

Because most in vitro enzyme studies are carried out under conditions where $[S] \gg [E]_T$ and only a small portion of $[S]$ is utilized, Eq. (17) is generally valid. Although the forms of the equilibrium [Eq. (10)] and steady-state [Eq. (17)] velocity equations are the same, the meanings of K_S and K_M are quite different. The equilibrium binding constant, K_S, is a special case of K_M when k_p is small compared with $k-1$. In this case, K_M is the dissociation constant of the ES complex. When $k-1 \ll k_p$, K_M is a kinetic constant. For more complex enzyme mechanisms such as those of xenobiotic-metabolizing enzymes, additional rate constants comprise K_M. Thus, K_M is a binding constant only in the special case in which the rate of dissociation of S from the ES complex is greater than the rate of catalysis (6).

Another important characteristic of enzymes is their inhibition. Many therapeutic agents, endogenous compounds, and xenobiotics are enzyme inhibitors. Enzyme inhibition is an important means of regulating activity in living cells. There are three basic types of enzyme inhibition: competitive, noncompetitive, and uncompetitive.

Competitive inhibitors compete with substrates for the same binding site on the enzyme. Competitive inhibitors can be substrate analogs, alternative substrates, or products of the reaction. Competitive inhibitors that are not metabolized by the enzyme are termed dead-end inhibitors. The steady-state velocity equation in the presence of a competitive inhibitor is as follows:

$$v = \frac{V_{max}[S]}{K_M(1+\frac{[I]}{K_I}) + [S]} \tag{18}$$

where I is the inhibitor and K_I is the inhibition constant. Eq. (18) shows that a competitive inhibitor modifies K_M without affecting V_{max}. K_M value obtained in the presence of a competitive inhibitor is termed the apparent K_M or $K_{M\,app}$. The extent of inhibition produced by a competitive inhibitor depends upon the concentrations of I and S, as well as the values of K_I and K_M. The inhibition produced at a fixed concentration of I can be overcome by increasing the substrate concentration. When initial velocity data obtained in the presence of different concentrations of a competitive inhibitor are plotted in double reciprocal form ($1/v$ versus $1/[S]$), a diagnostic pattern of lines with a common intersection on the $1/v$ axis (at $1/V_{max}$) is obtained (Fig. 2). K_I is defined as the concentration of I that doubles the slope of the double reciprocal plot (6,8). This is distinct from IC_{50}, the concentration of I that yields 50% inhibition of the reaction. Unlike K_I, which is a kinetic constant, IC_{50} is dependent upon substrate concentration (below saturation) and has limited predictive value.

When two chemicals that are substrates for the same enzyme are present at the same time, each will act as a competitive inhibitor of the other. In this case, the inhibition constant for each substrate with respect to the other is its K_M. Essentially, two reactions are being catalyzed by the same enzyme when the enzyme is in the presence of two of its substrates. Competitive inhibition by alternative substrates deserves special consideration in toxicology studies, where insoluble toxicants are often delivered to in vitro or in vivo systems in carrier solvents. Many toxicants and solvents are substrates for the monooxygenase CYP 2E1 (9). Generally, the concentration of the carrier solvent is much higher than the concentration of the toxicant being studied, and the nearly complete inhibition of toxicant biotransformation by the solvent can lead to artifactual observations and erroneous conclusions.

Because noncompetitive inhibitors bind to the enzyme at different sites than the substrate, K_M is not affected. The noncompetitive inhibitor can bind to E or ES, while the substrate can bind to E or EI. The ternary complex of enzyme, inhibitor, and substrate (ESI) is catalytically inactive. Noncompetitive inhibitors decrease V_{max} for an enzyme-catalyzed reaction.

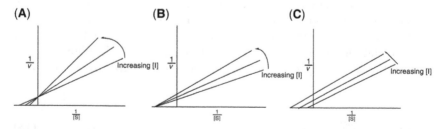

Figure 2 Double reciprocal plots ($1/v$ versus $1/[S]$) showing diagnostic patterns for inhibition types. (**A**) Competitive; (**B**) noncompetitive; (**C**) uncompetitive.

The rapid equilibrium velocity equation in the presence of a noncompetitive inhibitor is shown below:

$$v = \frac{V_{max}[S]}{K_M\left(1 + \frac{[I]}{K_I}\right) + [S]\left(1 + \frac{[I]}{K_1}\right)} \tag{19}$$

Both the K_S and the $[S]$ terms are modified by the inhibition factor. Division of both denominators in Eq. (19) by $(1 + [I]/K_I)$ results in the inhibition factor modifying only the V_{max} term. The noncompetitive inhibitor does not lower V_{max} by affecting k_p, but rather by decreasing the steady-state level of the ES complex. The rapid equilibrium expression for noncompetitive inhibition is presented here because the steady-state treatment results in a complex equation with $[S]^2$ and $[I]^2$ terms (6). Noncompetitive inhibitors affect the slope and $1/v$ axis intercept of the double reciprocal plot. Double reciprocal plots of initial rate data obtained in the presence of different concentrations of a noncompetitive inhibitor have a common intersection on the $1/[S]$ axis at $-1/K_M$ (Fig. 2). For noncompetitve inhibitors, K_I is defined as the inhibitor concentration that yields 50% inhibition of the reaction (6,8).

Many noncompetitive inhibitors affect both V_{max} and K_M of an enzyme-catalyzed reaction. These inhibitors are referred to as mixed-type inhibitors (6). Mixed-type inhibition produces double reciprocal plots that intersect above the $-1/[S]$ axis. Sometimes nonlinear double reciprocal plots are obtained. These effects are due to the presence of an EI complex that can bind the substrate and catalyze product formation at a rate equal to or less than that from the ES complex.

Substances that irreversibly inactivate enzymes are sometimes mistakenly called noncompetitive inhibitors because they decrease V_{max}. Irreversible inhibitors can be distinguished from reversible noncompetitive inhibitors by measuring the effects on V_{max} using different amounts of total enzyme. A plot of V_{max} versus $[E]_T$ for a reversible noncompetitive inhibitor has a different slope, but the same $[E]_T$ intercept as the control (no inhibitor) reaction. For an irreversible inhibitor, the V_{max} versus $[E]_T$ plot has the same slope as the control reaction, but intercepts at a lower $[E]_T$ corresponding to the amount of active enzyme remaining (6).

Uncompetitive inhibitors bind reversibly to the ES complex to form an inactive ESI complex. Uncompetitive inhibitors decrease V_{max} and K_M to the same extent. The steady-state velocity equation for uncompetitive inhibition is shown below:

$$v = \frac{V_{max}[S]}{K_M + [S]\left(1 + \frac{[I]}{K_1}\right)} \tag{20}$$

The inhibition term in Eq. (20) modifies $[S]$. Division of both denominators of Eq. (20) by $(1 + [I]/K_I)$ results in the inhibition term dividing both V_{max} and K_M. Unlike competitive inhibitors, the extent of inhibition by an uncompetitive inhibitor increases with increasing substrate concentration. This is because there is a greater concentration of the ES complex at higher substrate concentrations (6). Double reciprocal plots of initial rate data obtained in the presence of different concentrations of an uncompetitive inhibitor are a series of parallel lines (Fig. 2). At high enough concentrations, uncompetitive inhibitors can drive the reaction velocity to zero (6). Uncompetitive inhibitors are rare in nature. Considering the devastating effects an uncompetitive inhibitor could have on cellular function, it has been suggested that evolution has selected against enzymes that are prone to uncompetitive inhibition (10).

Most of the enzymes involved in the toxication and detoxication of chemicals are multisubstrate enzymes. As the number of substrates required for catalysis increases, the complexity and variety of kinetic mechanisms also increase (11). Enzymes that have two substrates and form two products generally exhibit sequential or ping-pong enzyme mechanisms. Sequential mechanisms require that all substrates bind to the enzyme (in ordered or random fashion) before catalysis takes place. Ping-pong mechanisms involve reaction of the first substrate with the enzyme to form a substituted enzyme and release of the first product before subsequent substrates bind. These enzyme mechanisms can be distinguished from each other by initial velocity and inhibition studies (6).

ENZYMES INVOLVED IN BIOTRANSFORMATION

Many drugs and chemicals are lipophilic and weakly ionizable, so that they are reabsorbed by the kidney and poorly excreted. The biotransformation of chemicals generally leads to the formation of more polar metabolites that are more readily excreted. There are two types of biotransformation pathways, termed by R.T. Williams Phase I and Phase II reactions (12). Phase I reactions include oxidations, reductions, and hydrolyses. Phase II reactions (also termed synthetic reactions) involve the conjugation of chemicals with hydrophilic moieties, such as glutathione, glucuronides, sulfate, or amino acids.

Biotransformation modulates the biological effects of drugs and chemicals. Metabolism might terminate the therapeutic effect of a drug or the toxic effect of a chemical. Conversely, metabolism of a substance may produce a therapeutic or toxic entity. Coadministration of two chemicals can result in exaggerated biological effects due to modulation of the metabolism of one compound by the other. These modulations can occur by inhibition of the biotransformation of the chemical or by induction of an increase in the enzyme system that metabolizes the chemical. Understanding

Table 1 The Major Enzymes Involved in Xenobiotic Metabolism

Phase I	Phase II
Cytochromes P450	Glutathione *S*-transferases
NADPH-cytochrome P450 reductase	UDP-glucuronosyl transferases
Flavin-containing monooxygenases	Sulfotransferases
Alcohol dehydrogenases	*N*-Acetyltransferases
Aldehyde dehydrogenases	*O*-Methyltransferases
Carbonyl reductase	*N*-Methyltransferases
Dihydrodiol dehydrogenase	*S*-Methyltransferases
Glutathione peroxidase	Thiol transferase
Monoamine oxidase	Acetyltransacetylase
Aldehyde oxidase	Amino acid transferases
Xanthine oxidase	
D-Amino acid oxidase	
Quinone reductase	
Epoxide hydrolase	
Esterases	
Amidases	
Rhodanese	

Abbreviation: NADPH, nicotinamide adenine dinucleotide

the properties of the enzymes that catalyze biotransformation reactions is important for accurately predicting the outcomes of chemical metabolism and to effectively diagnose the causes of adverse biological effects due to chemicals.

The major enzymes involved in xenobiotic metabolism are listed in Table 1. This is not intended to be a comprehensive listing of enzymes that might be involved in xenobiotic metabolism, rather a highlight of those enzymes most frequently involved. The major properties of some of the Phase I and Phase II enzymes will be overviewed in this section.

CYP are a superfamily of hemeprotein oxidoreductases with overlapping substrate specificities that utilize electrons from nicotinamide adenine dinucleotide (NADPH) via a flavoprotein reductase to reduce molecular oxygen and oxidize molecules containing C, N, O, S, P, halogens, and other heteroatoms (9,13,14). The CYP system can also catalyze the reduction of a number of moieties, including nitro groups and carbon–halogen bonds. The xenobiotic-metabolizing capability of liver homogenates was first described in the 1940s and was termed a mixed-function oxidase because it utilized both reducing equivalents NADPH and molecular oxygen to hydroxylate substrates (15). The stoichiometry of the CYP–dependent mixed-function oxidase system is shown in Eq. (21):

$$RH + O_2 + NADPH \rightarrow ROH + OH^- + NADP^+ \tag{21}$$

where RH is the substrate and ROH is the product. Studies with ^{18}O have shown that the oxygen atom in the hydroxylated product is from molecular oxygen rather than from water (13–17).

The mechanism of electron transfer in the CYP-dependent mixed-function oxidase system involves electron donation from NADPH to the flavoprotein NADPH-CYP reductase followed by electron transfer to ferric CYP. Ferrous CYP then binds molecular oxygen and the substrate. Transfer of a second electron activates the heme-bound oxygen, and the substrate is hydroxylated. While one high valent iron-oxo species was thought to mediate CYP-catalyzed reactions, recent studies suggest that multiple oxidants are involved in the chemistry taking place at the CYP active site (18). In any case, CYP catalysis takes place from a ternary complex of substrates and enzyme. Steady-state kinetic studies of the hepatic microsomal CYP-catalyzed O-dealkylation of *p*-nitroanisole supported by *t*-butyl hydroperoxide are consistent with catalysis proceeding by an ordered sequential mechanism (19). While many CYP-catalyzed reactions follow Michaelis–Menten saturation kinetics, some reactions exhibit non–Michaelis–Menten kinetics and allosteric behavior, particularly reactions catalyzed by the 3A4 isoform (20–22). This allosteric kinetic behavior could confound in vitro to in vivo extrapolation for certain substances.

Cytochrome P450 enzymes catalyze oxidation and reduction reactions without a strict substrate specificity. While some reactions are preferentially catalyzed by certain isoforms, the metabolism of few xenobiotics is exclusively catalyzed by one isoform of CYP. This is in contrast to CYP–mediated steroid biotransformations, which are often catalyzed by specific enzymes. Cytochromes P450 hydroxylate carbon to form alcohols at methyl, methylene, or methine positions. Oxidation of carbon adjacent to a heteroatom (i.e., *N*-methyl, *S*-methyl, and *O*-methyl) results in the formation of an unstable intermediate that decomposes to form an aldehyde and the desalkyl heterocompound. CYP can also oxygenate heteroatoms to form hydroxylamines, *N*-oxides, sulfones, and haloso compounds. Olefins and aromatic compounds can be epoxidized by CYP. The epoxide may rearrange to a phenol with the migration and retention of adjacent ring substituents such as isotopic hydrogen, halogen, or methyl groups (called the "NIH shift" because it was discovered at the National Institutes of Health). Oxidation of some substrates to reactive intermediates leads to irreversible inactivation of CYP. Cytochromes P450 can also catalyze oxidative dehalogenation and denitrification of some compounds (13,16,17,23).

Cytochromes P450 are membrane-bound enzymes primarily located in the endoplasmic reticulum of hepatocytes. Cytochromes P450 are also present in the inner mitochondrial membrane and the outer nuclear envelope. Heterogeneity of CYP expression has been demonstrated across the liver acinar and the different lobes of the liver (24). Cytochromes P450 are also present in a variety of extrahepatic tissues such as lung, kidney, and adrenal gland (25). One tissue that possesses specific contents of CYP approaching those of the liver is the nasal mucosa (26). Extrahepatic CYP play an important role in chemically induced target organ toxicity.

Another important property of the CYP superfamily is its inducibility by exogenous and endogenous compounds (14,27–29). Enzyme induction can be defined as the increase in the biosynthesis of a catalytically active enzyme in response to a chemical agent or physiological condition. Enzyme induction generally increases the rate and extent of xenobiotic metabolism because the apparent velocity of an enzyme-catalyzed reaction is directly proportional to the total enzyme present [Eq. (1)]. Many chemicals induce their own metabolism, and CYP induction forms the basis for many drug–drug interactions. Rapid induction of CYP P450 enzymes can allow more efficient detoxication of toxic chemicals or can increase the formation of toxic metabolites. For example, CYP 2E1 is rapidly induced in rodents and humans by exposure to ethanol and alcohols, ketones, and heterocyclic compounds (14,30). This enzyme is involved in the bioactivation and detoxication of a variety of xenobiotic carcinogens as well as endogenous ketones and thus is of great relevance in toxicology (9,31,32).

Approximately 57 genes for CYP have been described in humans, with additional genes described in a variety of procaryotic and eucaryotic species (33,34). Various polymorphisms in CYP enzymes have been identified through both metabolism and molecular studies (35). Populations showing bimodal distribution of drug-metabolizing capability have been identified for CYP 1A2 (with phenacetin and caffeine), 2A6 (with coumarin), 2C (with S-mephenytoin and tolbutamide), 2D6 (with debrisoquine), and 3A (with nifedipine). The polymorphism in CYP 2D6 is of great clinical importance because this enzyme metabolizes over 30 different therapeutic agents. A polymorphism in the inducibility of CYP 1A1 by aromatic hydrocarbons has been described. Polymorphisms at the DNA level have been described for CYP 1A1 and 2E1, but these polymorphisms do not appear to have metabolic consequences (35).

The endoplasmic reticulum of liver and other tissues also contains a second mixed-function oxidase system, the flavin-containing monooxygenases. This multigene family of enzymes utilizes electrons from NADPH to reduce molecular oxygen and monooxygenate polarizable heteroatoms, such as N, S, and P (36). Substrates for the flavin-containing monooxygenases include secondary and tertiary amines, hydrazines, thiocarbamates, thioamides, sulfides, disulfides, thiols, and other soft nucleophiles. Many of these substrates are oxidized with stereoselectivity, and this property has been used to distinguish oxidation by flavin-containing monooxygenases from CYP (36). Kinetic studies have shown that release of water from the enzyme–FAD complex is a slow step relative to substrate oxidation, so most substrates are oxidized with the same V_{max} (37). Functional and genetic polymorphisms have been identified in human flavin-containing monooxygenase form 3 (38).

The glutathione *S*-transferases (GST) are a family of enzymes that catalyze the nucleophilic addition of the thiol group of reduced glutathione to electrophilic chemicals (39–41). GST also bind a number of lipophilic chemicals with varying affinity. The active enzymes are homodimers or heterodimers comprised from seven different subunits. Each subunit has an independent active site that is only catalytically active when complexed with another subunit. The active site contains two binding sites, one for glutathione and one for the second substrate. Kinetic studies have shown that these enzymes follow a random sequential mechanism, giving sigmoidal initial velocity plots (39). Normal Michaelis–Menten saturation kinetics have been observed with several substrates, probably because all the reaction flux follows one pathway of the random mechanism.

The GST are located in the cytosol and in the membrane of the endoplasmic reticulum. The enzymes are inducible by a number of drugs and xenobiotics such as phenobarbital. The expression of multiple forms of GST varies among tissues and changes dramatically during growth of the fetus (39–41).

Glutathione (γ-glutamylcysteinylglycine) itself is the major cellular nucleophile involved in the detoxication of reactive chemicals (42). Glutathione is also involved in the regulation of the thiol status of proteins, particularly in mitochondria. Although glutathione conjugation is generally considered a detoxication step, it can also represent bioactivation in that some reversible glutathione conjugates can dissociate to yield reactive metabolites (43) and others can be bioactivated through further metabolism to form reactive toxic species (44). For example, the glutathione conjugate of trichloroethylene, *S*-(1,2-dichlorovinyl) glutathione, is degraded by peptidases to the corresponding cysteine conjugate. The cysteine conjugate is metabolized to the nephrotoxicant 1,2-dichlorovinylthiol by cysteine conjugate β-lyase (44).

Glucuronosyl transferases are a family of enzymes that catalyze the transfer of glucuronic acid from uridine diphospho (UDP)-glucuronic acid to acceptor molecules containing hydroxyl, phenol, carboxylic acid, thiol, or amine groups (45,46). Kinetic studies have shown that these enzymes follow a random sequential mechanism. Glucuronosyl transferases are located in the endoplasmic reticulum, and their biosynthesis can be induced by a number of drugs and xenobiotics. Some forms of glucuronosyl transferases are coordinately induced with CYP. Conjugation with glucuronic acid makes lipophilic molecules much more water soluble and readily excretable and therefore is considered a detoxication pathway. However, acyl glucuronides can undergo structural rearrangements and react with nucleophilic groups in cells (47,48). Thus, glucuronidation of some molecules may represent bioactivation. For example, the acyl glucuronide of the hypolipidemic drug clofibrate acylates cellular macromolecules (47). However, the toxicological significance of this reaction is unclear.

There are a number of sulfotransferase enzymes that catalyze the transfer of sulfate from 3'-phosphoadenosine 5'-phosphosulfate to hydroxyl groups on alcohols, hydroxylamines, steroids, and bile acids (49,50). The formation of N-sulfates from arylamines has also been reported. These enzymes are present in the cytosol. Kinetic studies are consistent with a sequential mechanism, but whether the substrates bind in an ordered or random manner is not clear. There can be competition between sulfotransferase and glucuronosyl transferase in the liver for the same alcohol substrate. The formation of highly charged organic sulfates facilitates the excretion of hydrophobic xenobiotics. However, as with the other conjugation reactions, some sulfate esters are unstable and decompose to form reactive metabolites. This pathway has been shown to be important in the bioactivation of several carcinogens, such as 2-acetylaminofluorene, safrole, and 7,12-dimethylbenz(a)anthracene (23).

IN VITRO SYSTEMS TO STUDY BIOTRANSFORMATION

The enzymes involved in biotransformation can be studied in vitro using isolated perfused organs, intact cells, subcellular fractions, or purified proteins. Genetically expressed enzymes can also be used. From the point of view of toxicology, the in vitro system should bear the closest resemblance possible to the in vivo situation being studied so that the in vitro results will have toxicological relevance. Studying in vitro processes that have no in vivo correlates is of little toxicological value.

Precision-cut liver slices were developed as an integrated system to study metabolism and toxicity that maintains cell–cell interactions between hepatocytes and other hepatic cell types (51,52). However, there appear to be diffusion problems with many drugs such that all cells in the slice do not participate in metabolism (53), confounding the basis for the extrapolation of in vitro data to in vivo. Additionally, different ratios of Phase I and Phase II metabolites were observed in slices compared with isolated hepatocytes and in vivo studies (54), further confounding the interpretation of kinetic data from this in vitro system. More recent studies suggest that these problems may be overcome by using a rotating incubation technique (55).

Freshly isolated hepatocytes in vitro have long been known to qualitatively model xenobiotic biotransformation in vivo in that the same array of metabolites is formed (56). These properties of freshly isolated hepatocytes are in contrast to hepatocytes in monolayer culture, where the cells differentiate rapidly and lose much of their xenobiotic-metabolizing capability (57). Immortalized cell lines such as the liver carcinoma-derived HEP G2 cell line offer attractive properties such as ease of culture and ease of genetic manipulation to express human CYP. While this system can be useful for many aspects of drug metabolism studies, the genetically manipulated cells are not ideal for in vitro prediction of pharmacokinetics because the xenobiotic-metabolizing enzymes are not expressed or regulated in the same manner as in the intact liver.

Subcellular fractions such as microsomes and cytosol have long been used to study the enzymes involved in xenobiotic metabolism. Microsomes are vesicles of endoplasmic reticulum formed in vitro from homogenization of tissue. Microsomes are isolated by differential centrifugation, with the microsomes being the pellet following the final ultracentrifugation and the cytosol being the supernatant (58). Subcellular fractions are easily prepared and can be stored frozen for years as part of a metabolic library. Subcellular fractions are the starting point for the purification of many of the enzymes involved in xenobiotic biotransformation.

DESIGN OF IN VITRO EXPERIMENTS TO PREDICT BIOTRANSFORMATION KINETICS

When using in vitro systems to investigate toxicological problems, care should be taken to insure that the incubation conditions are physiologically meaningful. Whenever possible, in vitro experiments should be done at physiological pH (7.2–7.4) and temperature (37°C). Ionic strength is another experimental variable that can affect the rates of enzyme-catalyzed reactions. As indicated in Eq. (1), the initial rate of the enzymic reaction should be a linear function of the enzyme concentration (or added protein or cell number) and time. When these requirements are satisfied, initial rate conditions have been achieved (6). Comparisons of the rates of enzyme-catalyzed reactions obtained outside the range of initial rate conditions are not valid. This is a pitfall of mining data from some older biomedical literature. Sometimes during the physicochemical characterization of an enzyme or a xenobiotic metabolism pathway, apparent kinetic parameters for the reaction being studied are reported from studies that were improperly or incompletely done. It is imperative to review the experimental details of the study in the experimental procedures section of the publication. Results presented without sufficient experimental detail to understand how they were derived should be treated with skepticism and probably should not be used in PBPK models or quantitative risk assessments.

When using isolated hepatocytes, the cells should be suspended in a nutritive medium containing essential amino acids and cofactors (such as Williams' Medium E), as opposed to the balanced salt solutions or buffers used in many hepatocyte studies. Incubation in a nutritive medium allows the hepatocytes to maintain normal biochemical homeostasis. In the absence of feedback regulation by essential amino acids, hepatocytes actively degrade cellular proteins and glutathione (59), and this autophagy can alter the activities of enzymes involved in xenobiotic metabolism.

Another important factor in all in vitro kinetic studies is knowledge of the actual concentration of the xenobiotic in the cell or microsomal suspension from measurement of the partition coefficient. Techniques have been developed to measure the partition coefficients of both volatile and nonvolatile chemicals (60–62). Most xenobiotics are not exceedingly water soluble, and the nominal concentration added is usually not the actual concentration in the medium containing the enzyme source. Expression of kinetic constants in terms of the nominal or added concentration of xenobiotic substrate is not useful because it is a function of the physical configuration and composition of the in vitro system being used. Any attempt at extrapolation of in vitro data when the actual concentration of the xenobiotic in the medium is not known will be incorrect, especially when the concentration is below the K_M. Additionally, any other kinetically significant component, such as specific protein binding (63), or chemical reactivity (64) also needs to be accounted for in the description of the in vitro system. Care should be taken that volatile chemicals are incubated in sealed flasks and that transfer rates between the vapor phase and the aqueous phase are taken into account. This can be accomplished using a simple two-compartment kinetic model of the incubation system (65).

EXTRAPOLATION OF KINETIC PARAMETERS TO WHOLE ORGANISMS

The theoretical basis for extrapolation of in vitro biotransformation data to whole organisms has been presented in Eq. (1): the overall rate of enzyme-catalyzed reactions is directly proportional to the total enzyme present in the system. Therefore, data generated with subcellular fractions such as microsomes or cytosols can be extrapolated to in vivo based on protein content. Data from intact cellular systems such as hepatocytes can be extrapolated to in vivo based on cell number. There are approximately 128×10^6 hepatocytes per gram of rat or mouse liver (66) and 137×10^6 hepatocytes per gram of human liver (67). The liver is approximately 4% of rat body weight, 5.5% of mouse body weight, and 2.6% of human body weight (68).

Extrapolation of in vitro kinetic data to whole animals requires knowledge (or assumption) of the enzyme kinetic mechanism. As discussed above, many xenobiotic-metabolizing enzymes follow Michaelis–Menten saturation kinetics. Rearrangement of Eq. (17) gives the Michaelis–Menten Equation

$$v = \frac{V_{max}[S]}{K_M + [S]} \qquad (22)$$

where v is the initial velocity of the reaction, V_{max} is the maximal rate of the reaction at infinite substrate concentration, $[S]$ is the substrate concentration, and K_M is the Michaelis constant for the reaction. K_M has units of concentration of the substrate, and provided the true concentration of the substrate in the in vitro system, is known via the appropriate partition coefficient as discussed above, K_M determined in vitro can be directly used in the in vivo pharmacokinetic model. V_{max} for the reaction is extrapolated to the whole organism based on the specific contents of the enzyme catalyzing the reaction, which is directly proportional to the protein content or cell number.

One of the biggest problems with extrapolation of in vitro kinetic data is when biotransformation enzymes follow complex non–Michaelis–Menten kinetics. As discussed above, some CYP-catalyzed reactions do not follow Michaelis–Menten kinetics and exhibit allosteric behavior (20–22) The mechanistic basis for this kinetic behavior is an area of active research. Another problem is the coupling of Phase I oxidations and Phase II conjugations within cells such that the Phase I metabolite is not kinetically observable. This situation is often observed with glucuronidation reactions. Additionally, microsomal UDP-glucuronosyltransferase activity in vitro can be variable due to the membrane-bound nature of the enzyme. Glucuronidation activity can be increased by solubilizing the microsomal membranes with nonionic detergents to allow substrate access. However, this situation clearly does not mimic what happens in intact cells where glucuronosyltransferase is imbedded in membranes. One potential solution is to study the reactions in isolated hepatocytes and treat the coupled metabolic steps as one kinetic process. Computational modeling approaches have also been developed to aid in this problem (69).

In addition to knowledge of the kinetic mechanism of the enzyme involved in the biotransformation reaction, a pharmacokinetic model of the organism is needed to extrapolate in vitro kinetic data. There are two types of compartmental pharmacokinetic models: data based and physiologically based. Data-based pharmacokinetic models are widely used in the analysis of clinical pharmacokinetic data. Data-based compartmental models generally describe individual data sets and cannot usually extrapolate between dose routes or species (70). The mathematical models used do not usually reflect the physiology or anatomy of the organism. This type of analysis of in vivo or in vitro kinetic data is of limited value in understanding species differences in the biotransformation of chemical toxicants and carcinogens. In contrast, physiologically based models represent the physiology and anatomy of the organism and can describe chemical pharmacokinetics in a wide variety of exposure scenarios (70). Physiological models contain both physiological parameters describing the organism and chemical-specific parameters including biotransformation rates and partition coefficients. The goal of PBPK modeling is to describe the behavior of a chemical in an animal, incorporating the necessary degrees of mechanistic detail to ultimately define one set of parameters to describe chemical behavior.

The in vivo pharmacokinetics of a variety of drugs and industrial chemicals have been accurately predicted by biotransformation kinetics determined in isolated hepatocytes and subcellular fractions in vitro. Houston (71) compared the predictions of intrinsic clearance (V/K, in units of flow) from isolated rat hepatocytes and hepatic microsomes for 11 drugs metabolized by CYP in the rat and found that isolated hepatocytes were more consistent than microsomes in accurately predicting intrinsic clearance in vivo. Additional studies (53) have shown that isolated hepatocytes and hepatic microsomes were consistently more accurate in predicting intrinsic clearance than precision-cut liver slices.

One of the first studies using in vitro kinetics determined in isolated hepatocytes and hepatic microsomes to predict pharmacokinetics in vivo using PBPK models was reported by Lin et al (72). They showed that extrapolation of the rat PBPK model for ethoxybenzamide to rabbits accurately simulated blood and tissue time course data, illustrating the usefulness of the in vitro approach for prediction of species-specific pharmacokinetics.

Kedderis et al. (65) showed that the pharmacokinetics of the hepatotoxicant and hepatocarcinogen furan in F-344 rats could be accurately predicted from kinetic studies with isolated hepatocytes. Furan bioactivation involves CYP 2E1 oxidation to a reactive ene-dialdehyde metabolite that mediates furan hepatotoxicity (73). Additional kinetic studies with hepatocytes from B6C3F1 mice and humans allowed comparisons between furan bioactivation in the three species (74). K_M values for furan oxidation were in the low micromolar range for hepatocytes from all three species. Hepatocytes from male mice oxidized furan at a greater rate than rat or human hepatocytes. Hepatocytes from the three different human liver donors oxidized furan at rates equal to or greater than those of rat hepatocytes. These results are not predicted by allometric scaling of the rodent data to humans because allometry predicts slower rates in larger mammals (75). This underscores the importance of using human tissue samples to predict the human bioactivation of toxicants. A greater than two-fold variation in the V_{max} for furan bioactivation was observed among the three preparations of human hepatocytes (74), consistent with CYP 2E1 induction via ethanol ingestion by the liver donors.

The kinetic data from hepatocytes were extrapolated to whole animals based on cell number and used along with physiological parameters from the literature to develop PBPK models for furan in rats, mice, and humans (65,74). Simulation of inhalation exposure to 10 ppm furan for 4 hours indicated significant species differences in the amount of furan absorbed. The absorbed dose of furan (mg/kg; inhaled minus exhaled divided by body weight) and the integrated exposure of the liver to the toxic metabolite were approximately 3.5- and 10-fold greater in rats and mice, respectively, than in humans following the same inhalation exposure. The reason for this species difference is that humans are larger and physiologically slower than mice or rats (75). The volatile toxicant furan is metered into the blood stream via the breathing rate and distributed throughout the organism at rates that are a function of body size. Thus, the inhalation exposure concentration of a toxicant is clearly not an appropriate measure of the dose to the organism. In the case of furan, comparing the absorbed dose (mg/kg) or the target organ (liver) exposure to the toxic metabolite among species is more appropriate. These concepts should be applied when assessing interspecies differences to inhaled toxicants, particularly when animal data are used to estimate human health risks from chemical exposure.

Simulations with the dosimetry models showed that steady-state blood concentrations of furan were achieved by approximately 1 hour after inhalation exposure to 10 ppm and were similar among species (0.7–0.8 μM). Comparison of the initial rates of furan oxidation with the rate of hepatic blood flow for each of the three species indicated that furan oxidation was approximately 13- to 37-fold higher than the rate of furan delivery to the liver via blood flow (74). These results indicate that the bioactivation of furan is limited by hepatic blood flow. Simulations of the liver concentration of the toxic metabolite of furan as a function of furan exposure concentration indicated that furan bioactivation was limited by hepatic blood flow through at least 300 ppm. The hepatic blood flow limitation was also evident following oral bolus administration of furan through approximately 2 mg/kg (76). Despite numerous reports to the contrary (77), the limiting effect of hepatic blood flow on metabolism following oral bolus dosing is not predicted to occur according to the well-stirred two-compartment data-based pharmacokinetic model that is used to analyze most clinical pharmacokinetic data (78). This example further demonstrates the desirability of using PBPK models to analyze pharmacokinetic data for use in risk assessment.

One consequence of the hepatic blood flow limitation of xenobiotic biotransformation is that K_M values cannot be accurately estimated from in vivo studies. Because the initial rate of bioactivation is much greater than the rate of blood flow delivery of the toxicant to the liver, smaller values of K_M (i.e., higher initial rates of oxidation) describe the data just as well as larger values of K_M. Dedrick and Forrester (79) came to a similar conclusion in their studies of the hepatic oxidation of ethanol. Thus, values of K_M estimated with hepatocytes or microsomes in vitro will probably be smaller than K_M estimates obtained from in vivo pharmacokinetic studies because the in vitro estimates are made in systems where other processes such as hepatic blood flow do not limit the apparent rate of metabolism. Another important consequence of the limitation of bioactivation by hepatic blood flow is that increases in V_{max} due to CYP 2E1 induction or interindividual variation will have little or no effect on the amount of the toxic metabolite formed in the liver. Pharmacokinetic analyses of the bioactivation of several other hazardous chemical air pollutants indicated that their CYP–mediated bioactivation is limited by hepatic blood flow and enzyme induction has little effect upon the rate of bioactivation (76).

Studies of the CYP 2E1–mediated oxidation of trichloroethylene provide further illustration of the value of in vitro kinetic studies in risk assessment. Data were developed on the microsomal protein content of human liver (80), the CYP 2E1 content of human liver microsomal protein (81), and the in vitro V_{max} for trichloroethylene oxidation by human liver microsomes (82) from 60 human liver samples. The data were combined and subjected to statistical analysis (80). Analysis of the lognormally distributed data by the method of moments indicated that the 5th and 95th percentiles of the distribution in human liver (trichloroethylene oxidized/min/g liver) differed by approximately six-fold. The units of the data were converted to milligrams trichloroethylene oxidized/hr/kg body weight and incorporated in a human PBPK model for trichloroethylene. Simulations of an 8-hour inhalation exposure to 50 ppm and oral exposure to 5 μg/L in 2 L of drinking water showed that the amount of trichloroethylene oxidized in human liver differed by 2% or less, even though the distribution of metabolic capacity varied six-fold (Table 2). These data indicate that differences in CYP 2E1 enzyme expression among the central 90% of the adult human population account for only approximately 2% of the variance in a risk-relevant pharmacokinetic outcome for trichloroethylene-mediated liver injury (amount of trichloroethylene oxidized in the liver). The results indicate that physiological processes such as hepatic blood flow limit the full impact of interindividual differences in CYP 2E1–mediated oxidation of trichloroethylene to hepatotoxic metabolites (80).

In vitro systems offer the opportunity to study human enzymes and cellular processes without putting a patient at risk through exposure to toxic agents. Data obtained from in vitro systems under the appropriate experimental conditions can be extrapolated to the intact human and used in PBPK models to simulate the exposure of individuals or groups of people to hazardous chemicals using various exposure scenarios. Experiments with

Table 2 Effect of the Distribution of Human Hepatic Cytochrome P450 2E1 Activity on the Bioactivation of Trichloroethylene[a]

	Liver metabolites (g/L)	
Oxidation rate	Inhalation[b]	Oral[c]
5th Percentile (6.6 mg/hr/kg)	258.3	5.4
95th Percentile (39.7 mg/hr/kg)	264.9	5.5
Difference	2%	2%

[a]Data from Lipscomb et al.
[b]Simulation of 8-hour exposure to trichloroethylene (50 ppm) by a 70 kg male human.
[c]Simulation of exposure to trichloroethylene in drinking water (5 μg/L; 2 L) over 24 hr by a 70 kg male human.
Source: From Ref. 80.

human PBPK models can quantitatively characterize the potential outcomes of exposure to toxic chemicals and determine the parameters that affect those outcomes. More complete utilization of the human in vitro systems are available and knowledge of the biochemical characteristics of human enzymes and their regulation will make important contributions to improving human health risk assessment for exposure to potential toxic chemicals.

REFERENCES

1. Lehninger AL. Biochemistry: The Molecular Basis of Cell Structure and Function. 2nd ed. New York, NY: Worth, 1975.
2. Poyton RO. Memory and membranes: the expression of genetic and spatial memory during the assembly of organelle macrocompartments. Mod Cell Biol 1983; 2:15–72.
3. Clegg JS. Intracellular water and the cytomatrix: some methods of study and current views. J Cell Biol 1984; 99:167S–171S.
4. Clegg JS. Properties and metabolism of the aqueous cytoplasm and its boundaries. Am J Physiol 1984; 246:R133–R151.
5. Luby-Phelps K, Lanni F, Taylor DL. The submicroscopic properties of cytoplasm as a determinant of cellular function. Annu Rev Biophys Bio 1988; 17:369–396.
6. Segel IH. Enzyme Kinetics: Behavior and Analysis of Rapid Equilibrium and Steady State Enzyme Systems. New York, NY: John Wiley & Sons, 1975.
7. Northrop DB. Fitting enzyme-kinetic data to V/K. Anal Biochem 1983; 132:457–461.
8. Spector T, Cleland WW. Meanings of K_i for conventional and alternative-substrate inhibitors. Biochem Pharmacol 1981; 30:1–7.
9. Guengerich FP. Reactions and significance of cytochrome P-450 enzymes. J Biol Chem 1991; 266:10019–10022.
10. Cornish-Bowden A. Why is uncompetitive inhibition so rare? A possible explanation, with implications for the design of drugs and pesticides. FEBS Lett 1986; 203:3–6.
11. Fromm HJ. Summary of kinetic reaction mechanisms. Meth Enzymol 1979; 63:42–53.
12. Williams RT. Detoxication mechanisms: the metabolism and detoxication of drugs. In: Toxic Substances and Other Organic Compounds. 2nd ed. New York, NY: John Wiley & Sons, 1959.
13. Guengerich FP, Macdonald TL. Chemical mechanisms of catalysis by cytochromes P-450: a unified view. Acc Chem Res 1984; 17:9–16.
14. Porter TD, Coon MJ. Cytochrome P-450: multiplicity of isoforms, substrates, and catalytic and regulatory mechanisms. J Biol Chem 1991; 266:13469–13472.
15. Kedderis GL. The role of the mixed-function oxidase system in the toxication and detoxication of chemicals: relationship to chemical interactions. In: Goldstein RS, Hewitt WR, Hook JB, eds. Toxic Interactions. San Diego, CA: Academic, 1990:31–60.
16. White RE, Coon MJ. Oxygen activation by cytochrome P-450. Annu Rev Biochem 1980; 49:315–356.

17. Guengerich FP, Macdonald TL. Mechanisms of cytochrome P-450 catalysis. FASEB J 1990; 4:2453–2459.
18. Newcomb M, Hollenberg PF, Coon MJ. Multiple mechanisms and multiple oxidants in P450-catalyzed hydroxylations. Arch Biochem Biophys 2003; 409:72–79.
19. Hollenberg PF. Mechanisms of cytochrome P450 and peroxidase-catalyzed xenobiotic metabolism. FASEB J 1992; 6:686–694.
20. Hlavica P, Lewis DF. Allosteric phenomena in cytochrome P450-catalyzed monooxygenations. Eur J Biochem 2001; 268:4817–4835.
21. Atkins WM. Non–Michaelis-Menten kinetics in cytochrome P450-catalyzed reactions. Annu Rev Pharmacol Toxicol 2005; 45:291–310.
22. Houston JB, Galetin A. Modeling atypical CYP3A4 kinetics: principles and pragmatism. Arch Biochem Biophys 2005; 433:351–360.
23. Guengerich FP. Metabolic activation of carcinogens. Pharmacol Ther 1992; 54:17–61.
24. Gebhardt R. Metabolic zonation of the liver: regulation and implications for liver function. Pharmacol Ther 1992; 53:275–354.
25. Ding X, Kaminsky LS. Human extrahepatic cytochromes P450: function in xenobiotic metabolism and tissue-selective chemical toxicity in the respiratory and gastrointestinal tracts. Annu Rev Pharmacol Toxicol 2003; 43:149–173.
26. Dahl AR, Hadley WM. Nasal cavity enzymes involved in xenobiotic metabolism: effects on the toxicity of inhalants. Crit Rev Toxicol 1991; 21:345–372.
27. Conney AH. Pharmacological implications of microsomal enzyme induction. Pharmacol Rev 1967; 19:317–366.
28. Conney AH. Induction of microsomal enzymes by foreign chemicals and carcinogenesis by polycyclic aromatic hydrocarbons. Cancer Res 1982; 42:4875–4917.
29. Gustafsson JÅ, Mode A, Norstedt G, et al. Sex steroid induced changes in hepatic enzymes. Annu Rev Physiol 1983; 45:51–60.
30. Koop DR. Oxidative and reductive metabolism by cytochrome P450 2E1. FASEB J 1992; 6:724–730.
31. Guengerich FP, Shimada T. Oxidation of toxic and carcinogenic chemicals by human cytochrome P-450 enzymes. Chem Res Toxicol 1991; 4:391–407.
32. Raucy JL, Kraner JC, Lasker JM. Bioactivation of halogenated hydrocarbons by cytochrome P4502E1. Crit Rev Toxicol 1993; 23:1–20.
33. Nelson DR, Koymans L, Kamataki T, et al. P450 superfamily: update on new sequences, gene mapping, accession numbers and nomenclature. Pharmacogenetics 1996; 6:1 42.
34. Lewis DF. 57 varieties: the human cytochromes P450. Pharmacogenetics 2004; 5:305–318.
35. Daly AK, Cholerton S, Gregory W, et al. Metabolic polymorphisms. Pharmacol Ther 1993; 57:129–140.
36. Hines RN, Coshman JR, Philpot RM, et al. The mammalian flavin-containing monooxygenases: molecular characterization and regulation of expression. Toxicol Appl Pharmacol 1994; 125:1–6.
37. Jakoby WB, Zeigler DM. The enzymes of detoxication. J Biol Chem 1990; 265:20715–20718.
38. Cashman JR, Zhang J. Interindividual differences of human flavin-containing monooxygenase 3: genetic polymorphisms and functional variation. Drug Metab Dispos 2002; 30:1043–1052.

39. Mannervik B, Danielson UH. Glutathione transferases—structure and catalytic activity. Crit Rev Biochem 1988; 23:283–337.
40. Pickett CB, Lu AY. Glutathione S-transferases: gene structure, regulation, and biological function. Annu Rev Biochem 1989; 58:743–764.
41. Armstrong RN. Structure, catalytic mechanism, and evolution of the glutathione transferases. Chem Res Toxicol 1997; 10:2–18.
42. Reed DJ. Glutathione: toxicological implications. Annu Rev Pharmacol Toxicol 1990; 30:603–631.
43. Baillie TA, Slatter JG. Glutathione: a vehicle for the transport of chemically reactive metabolites in vivo. Acc Chem Res 1991; 24:264–270.
44. Anders MW, Losh L, Dekant W, et al. Biosynthesis and biotransformation of glutathione S-conjugates to toxic metabolites. Crit Rev Toxicol 1988; 18:311–341.
45. Mulder GJ. Glucuronidation and its role in regulation of biological activity of drugs. Annu Rev Pharmacol Toxicol 1992; 32:25–49.
46. Tukey RH, Strassburg CP. Human UDP-glucuronosyltransferases: metabolism, expression, and disease. Annu Rev Pharmacol Toxicol 2000; 40:581–616.
47. Faed EM. Properties of acyl glucuronides: implications for studies of the pharmacokinetics and metabolism of acidic drugs. Drug Metab Rev 1984; 15:1213–1249.
48. Ritter JK. Roles of glucuronidation and UDP-glucuronosyltransferases in xenobiotic bioactivation reactions. Chem Biol Interact 2000; 129:171–193.
49. Duffel MW et al. Enzymatic aspects of the phenol (aryl) sulfotransferases. Drug Metab Rev 2001; 33:369–395.
50. Negishi M, Pederson LG, Petrotchenko E, et al. Structure and function of sulfotransferases. Arch Biochem Biophys 2001; 390:149.
51. Barr J, Weir AJ, Brendel K, et al. Liver slices in dynamic organ culture. I. An alternative in vitro technique for the study of rat hepatic drug metabolism. Xenobiotica 1991; 21:331–339.
52. Barr J, Weir AJ, Brendel K, et al. Liver slices in dynamic organ culture. II. An in vitro cellular technique for the study of integrated drug metabolism using human tissue. Xenobiotica 1991; 21:341–350.
53. Worboys PD, Brennon B, Bradburg A, et al. Metabolite kinetics of ondansetron in rat. Comparison of hepatic microsomes, isolated hepatocytes, and liver slices, with in vivo disposition. Xenobiotica 1996; 26:897–907.
54. Ekins S, Murray GI, Burke MD, et al. Quantitative differences in phase I and II metabolism between rat precision-cut liver slices and isolated hepatocytes. Drug Metab Dispos 1995; 23:1274–1279.
55. Thohan S, Zurieh MC, Chung H, et al. Tissue slices revisited: evaluation and development of a short-term incubation for integrated drug metabolism. Drug Metab Dispos 2001; 29:1337–1342.
56. Billings RE, McMahon RE, Ashmore J, et al. The metabolism of drugs in isolated rat hepatocytes: a comparison with in vivo drug metabolism and drug metabolism in subcellular liver fractions. Drug Metab Dispos 1977; 5:518–526.
57. Sirica AE, Pitot HC. Drug metabolism and effects of carcinogens in cultured hepatic cells. Pharmacol Rev 1979; 31:205–228.

58. Guengerich FP. Analysis and characterization of enzymes. In: Hayes AW, ed. Principles and Methods of Toxicology. New York, NY: Raven Press, 1989:777–814.
59. Seglen PO, Gordon PB, Poli A. Amino acid inhibition of the autophagic/lysosomal pathway of protein degradation in isolated rat hepatocytes. Biochim Biophys Acta 1980; 630:103–118.
60. Gargas ML, Burgess RJ, Voisard DE, et al. Partition coefficients of low-molecular-weight volatile chemicals in various liquids and tissues. Toxicol Appl Pharmacol 1989; 98:87–99.
61. Jepson GW, Hoover DK, Black RK, et al. A partition coefficient determination method for nonvolatile chemicals in biological tissues. Fund Appl Toxicol 1994; 22:519–524.
62. Artola-Garicano E, Vaes WHJ, Hermens JLM. Validation of negligible depletion solid-phase microextraction as a tool to determine tissue/blood partition coefficients for semivolatile and nonvolatile organic chemicals. Toxicol Appl Pharmacol 2000; 166:138–144.
63. Sjöholm I, Ekakwan B, Kober A, et al. Binding of drugs to human serum albumin: XI. The specificity of three binding sites as studied with albumin immobilized in microparticles. Mol Pharmacol 1979; 16:767–777.
64. Teo SKO, Kedderis GL, Gargas ML. Determination of tissue partition coefficients for volatile tissue-reactive chemicals: acrylonitrile and its metabolite 2-cyanoethylene oxide. Toxicol Appl Pharmacol 1994; 128:92–96.
65. Kedderis GL, Carfagna MA, Held SD, et al. Kinetic analysis of furan biotransformation by F-344 rats in vivo and in vitro. Toxicol Appl Pharmacol 1993; 123:274–282.
66. Seglen PO. Preparation of isolated rat liver cells. Meth Cell Biol 1976; 13:29–83.
67. Arias IM, Popper H, Schachter D, et al. The Liver: Biology and Pathobiology. New York, NY: Raven Press, 1982.
68. Arms AD, Travis CC. Reference Physiological Parameters in Pharmacokinetic Modeling, EPA/600/6-88/004, NTIS PB88–196019. Washington: D.C.: U.S. Environmental Protection Agency, 1988.
69. Miners JO, Smith PA, Sofich MJ, et al. Predicting human drug glucuronidation parameters: application of in vitro and in silico modeling approaches. Annu Rev Pharmacol Toxicol 2004; 44:1–25.
70. Clewell HJ, Andersen ME. Physiologically based pharmacokinetic modeling and bioactivation of xenobiotics. Toxicol Ind Health 1994; 10:1–24.
71. Houston JB. Utility of in vitro drug metabolism data in predicting in vivo metabolic clearance. Biochem Pharmacol 1994; 47:1469–1479.
72. Lin JH, Sugiyama Y, Awazu S, et al. Physiological pharmacokinetics of ethoxybenzamide based on biochemical data obtained in vitro as well as on physiological data. J Pharmacokinet Biop 1982; 10:649–661.
73. Carfagna MA, Held SD, Kedderis GL. Furan induced cytolethality in isolated rat hepatocytes: correspondence with in vivo dosimetry. Toxicol Appl Pharmacol 1993; 123:265–273.
74. Kedderis GL, Held SD. Prediction of furan pharmacokinetics from hepatocyte studies: comparison of bioactivation and hepatic dosimetry in rats, mice, and humans. Toxicol Appl Pharmacol 1996; 140:124–130.

75. Mordenti J. Man versus beast: pharmacokinetic scaling in mammals. J Pharm Sci 1986; 75:1028–1040.
76. Kedderis GL. Extrapolation of in vitro enzyme induction data to humans in vivo. Chem Biol Interact 1997; 107:109–121.
77. Wilkinson GR. Pharmacokinetics of drug disposition: hemodynamic considerations. Annu Rev Pharmacol 1975; 15:11–27.
78. Park BK, Kitteringham NR, Piromohamed M, et al. Relevance of induction of human drug-metabolizing enzymes: pharmacological and toxicological implications. Br J Clin Pharmacol 1996; 41:477–491.
79. Dedrick RL, Forrester DD. Blood flow limitations in interpreting Michaelis constants for ethanol oxidation in vivo. Biochem Pharmacol 1973; 22:1133–1140.
80. Lipscomb JC, Teuscler LK, Swartout J, et al. The impact of cytochrome P450 2E1-dependent metabolic variance on a risk-relevant pharmacokinetic outcome in humans. Risk Anal 2003; 23:1221–1238.
81. Snawder JE, Lipscomb JC. Interindividual variance of cytochrome P450 forms in human hepatic microsomes: correlation of individual forms with xenobiotic metabolism and implications for risk assessment. Regul Toxicol Pharmacol 2000; 32:200–209.
82. Lipscomb JC, Garrett CM, Snawder JE. Cytochrome P450-dependent metabolism of trichloroethylene: interindividual differences in humans. Toxicol Appl Pharmacol 1997; 142:311–318.

10

Use of Physiologically Based Pharmacokinetic Modeling to Evaluate Implications of Human Variability

P. Robinan Gentry

ENVIRON International Corporation, Ruston, Louisiana, U.S.A.

Harvey J. Clewell III

CIIT Centers for Health Research, Research Triangle Park, North Carolina, U.S.A.

INTRODUCTION

One of the more challenging issues that must be considered in performing a human health risk assessment is the heterogeneity of the human population. This heterogeneity is produced by interindividual variations in physiology, biochemistry, and molecular biology, reflecting both the genetic and the environmental factors, and results in differences across individuals in the biologically effective tissue dose associated with a given environmental exposure (pharmacokinetics) as well as in the response to a given tissue dose (pharmacodynamics). Physiologically based pharmacokinetic (PBPK) models have been suggested as a tool to quantify the implications of human variability for the estimation of risk from chemical exposure. Because the parameters in a PBPK model have a direct biological correspondence, they provide a useful framework for determining the impact of observed variations in physiological and biochemical factors on the population variability in dosimetry within the context of a risk assessment for a particular chemical (1).

It is useful to consider the total variability in a human population in terms of three contributing sources: (i) the variation across a population of "typical" individuals at the same age, e.g., young adults; (ii) the variation across the population resulting from their different ages, e.g., infants or the elderly; and (iii) the variation resulting from the existence of subpopulations that differ in some way from the "normal" population, e.g., due to genetic polymorphisms. A fourth source of variability, health status, should also be considered, although it is frequently disregarded in environmental risk assessment. To the extent that the variation in physiological and biochemical parameters across these population dimensions can be elucidated, PBPK models can be used together with Monte Carlo methods to integrate their effects on the in vivo kinetics of a chemical exposure and predict the resulting impact on the distribution of risks (as represented by target tissue doses) across the population.

DETERMINANTS OF IMPACT

There has sometimes been a tendency in risk assessments to use information on the variability of a specific parameter, such as inhalation rate or the in vitro activity of a particular enzyme, as the basis for expectations regarding the variability in dosimetry for in vivo exposures. However, whether or not the variation in a particular physiological or biochemical parameter will have a significant impact on in vivo dosimetry is a complex function of interacting factors. In particular, the structures of physiological and biochemical systems frequently involve parallel processes (e.g., blood flows, metabolic pathways, and excretion processes), leading to compensation for the variation in a single factor. Moreover, physiological constraints may limit the in vivo impact of variability observed in vitro. For instance, high-affinity intrinsic clearance can result in essentially complete metabolism of all the chemical reaching the liver in the blood; under these conditions, variability in amount metabolized in vivo would be more a function of variability in liver blood flow than variability in metabolism in vitro. Thus, it is often true that the whole (the in vivo variability in dosimetry) is less than the sum of its part (the variability in each of the pharmacokinetic factors).

The dosimetric impact of variations in physiological factors also depends on the nature of the chemical causing the toxicity, including such physicochemical properties as reactivity, lipophilicity, water solubility, and volatility. For example, variations in inhalation rate will tend to have more impact on the uptake of a water-soluble chemical such as isopropanol than on a relatively water-insoluble chemical such as vinyl chloride.

In addition, the impact of a particular factor on dosimetry also depends on the mode-of-action of the chemical; that is, how the chemical causes the effect of concern. Of particular importance is whether the toxicity results from exposure to the chemical itself; one of its stable, circulating metabolites; or a reactive intermediate produced during its metabolism.

Another key mode-of-action issue is how the toxicity is occurring. For example, does the toxicity result from direct reaction with tissue constituents, from binding to a receptor, or from physical (e.g., solvent) effects on the tissue. To illustrate these considerations, one can contrast the acute neurotoxicity of many solvents (a physical effect of the chemicals themselves) with their chronic hepatotoxicity (produced by products of their metabolism). The most important pharmacokinetic factors in the acute toxicity of volatile solvents are the blood:air partition coefficient and metabolic clearance, and increasing metabolic clearance typically decreases toxicity. In contrast, the most important pharmacokinetic factors in the chronic toxicity are liver blood flow and metabolism, and increasing metabolic clearance typically increases toxicity.

PHARMACOKINETIC VARIABILITY BETWEEN INDIVIDUALS OF DIFFERENT AGES

PBPK modeling has been routinely used in risk assessment when extrapolating across route and species. The same qualities that make PBPK modeling attractive for these extrapolations also make it a useful platform for predicting age-dependent pharmacokinetics. Specifically, a PBPK model can provide a quantitative structure for incorporating into the risk assessment process information on the age-related variability in pharmacokinetic factors that can impact the relationship between the external (environmental) exposure and the internal (biologically effective) target tissue exposure.

A "life stage" model (2) was specifically developed to evaluate variability across the life span of an individual. With this model, it is possible to investigate how normal changes in pharmacokinetic parameters from birth, through childhood, and across adulthood can affect the internal dosimetry for environmental exposures to chemicals. These changes in parameters can result in different kinetics, depending on the physico/chemical properties of the chemical in question. In this age-dependent model (Fig. 1), all physiological and biochemical parameters change with time, based on data on parameter variability obtained from the literature (3). This model can be used as a tool to examine variability across individuals of different ages following continuous lifetime exposure to various classes of chemicals.

Figure 2 shows the results of using this age-dependent model to simulate continuous inhalation of a volatile compound, isopropanol, at 1 ppb, beginning at birth and continuing for 75 years. The solid line in this figure represents the predicted age-dependent concentration of isopropanol in the blood, while the dashed line represents the corresponding predicted concentration of acetone, the primary metabolite. The model predicts that, for the same inhaled concentration, the blood concentrations achieved during early life are significantly higher than those achieved during adulthood. In the case of the metabolite acetone, however, it should be noted that production

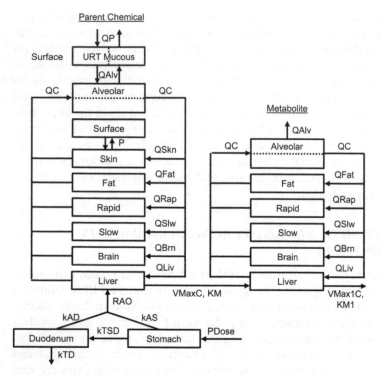

Figure 1   Schematic of the life-stage physiologically based pharmacokinetic model.

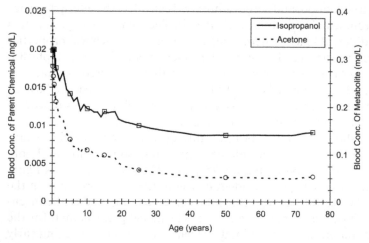

Figure 2   Blood concentrations of isopropanol and its metabolite acetone as a function of age following inhalation exposure.

from isopropanol metabolism would only be a small fraction of endogenous production from ketogenesis. These predicted blood concentrations of isopropanol and acetone vary approximately to a factor of 2 across all life stages, with peak concentrations of both occurring in early life; a second peak for acetone occurs between the ages of 15 and 20 years. The more rapid fluctuations in blood concentrations superimposed over these trends reflect transient changes in hepatic clearance, principally the result of changes in liver clearance associated with growth spurts for that organ.

Quite a different behavior is predicted for daily ingestion of perchloroethylene in drinking water across ages. The exposure in this case is assumed to be a constant intake of perchloroethylene at a rate of 1 µg/kg/day throughout life. As shown in Figure 3, predicted concentrations of perchloroethylene and its major metabolite, trichloroacetic acid (TCA), are much lower during early life than during adulthood. The blood concentrations of perchloroethylene and trichloroacetic acid are predicted to rise consistently from birth to approximately age of 40 years, followed by a plateau between the ages of 40 and 50 years, and a decline from the age of 68 to 75 years. This is in contrast to water-soluble volatiles, such as isopropanol, for which peaks occur early in life. This reflects the ability of perchloroethylene, a poorly metabolized lipophilic volatile, to accumulate in the body, specifically in fat. The variability across ages is also greater with perchloroethylene, with a difference of about seven-fold predicted in the blood concentrations between early life and later ages. The profile of predicted blood concentrations for the persistent compound dioxon 2,3,7,8-tetrachlorodibenzo-*p*-dioxin (TCDD) across ages was

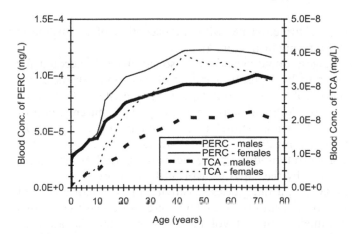

Figure 3 Simulated blood concentrations as a function of age for continuous lifetime oral exposure at a constant daily intake of 1 µg/kg/day for perchloroethylene and its primary metabolite TCA. *Abbreviations*: PERC, perchloroethylene; TCA, trichloroacetic acid.

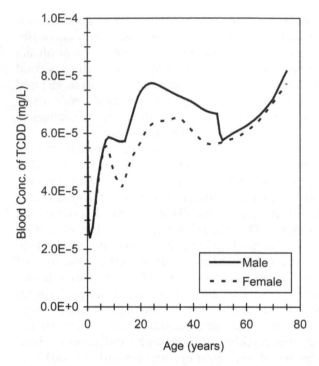

Figure 4   Simulated blood concentrations as a function of age for 2,3,7,8-tetrachloro-dibenzo-*p*-dioxin (TCDD) for continuous oral exposure of males (*solid line*) or females (*dashed line*) at a constant daily intake of 1 µg/kg/day. *Abbreviation*: TCDD, 2,3,7,8-tetrachlorodibenzo-*p*-dioxin.

similar to that for perchloroethylene, with the exception of a substantial decline immediately following birth (Fig. 4).

For TCDD, the gestational period was also simulated to account for bioaccumulation prior to birth, using a model of gestation (4); the rapid clearance of the more volatile compounds, such as perchloroethylene or isopropanol, makes this additional step unnecessary. The concentration of TCDD in the neonate at birth is estimated to be similar to maternal levels, reflecting transplacental exposure to maternal stores of the chemical. The drop in TCDD concentration during the neonatal period results from dilution of TCDD stores by the rapid growth of the neonate.

In the case of TCDD, a time-series sensitivity analysis indicated that the estimated blood concentrations of TCDD were most sensitive to two pharmacokinetic parameters, the fat volume and the metabolic rate constant. In addition, the impact of these parameters on the estimated blood concentrations of TCDD was age dependent (Fig. 5). Blood TCDD concentrations were more sensitive to fat compartment size than clearance during childhood and adolescence, while the rate of clearance exerts more control

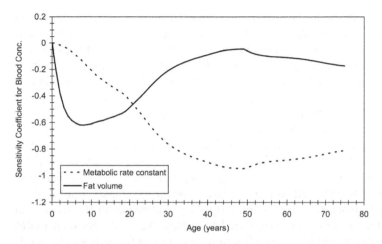

Figure 5 Age-dependent sensitivity coefficients for blood 2,3,7,8-tetrachlorodibenzo-*p*-dioxin (TCDD) concentrations resulting from constant oral exposure. Arterial blood concentration is more sensitive to changes in fat volume (*solid line*) than metabolism for the first several years of life, after which it becomes more sensitive to the metabolic rate constant (*dotted line*). Other parameters showed lower sensitivity.

over blood concentrations during adulthood. This result demonstrates that, where a large number of parameters change with age, the relative importance of the parameters can also change with age, resulting in different expectations regarding risk factors in different age groups.

PHARMACOKINETIC VARIABILITY AMONG INDIVIDUALS AT THE SAME AGE

Even at the same level of exposure, because of the heterogeneity of the human population, it is generally expected that there will be a broad range of observed susceptibilities to the biological effects of exposure to chemicals. PBPK models can provide a quantitative structure for determining the effect of the variation in pharmacokinetic factors in a population at the same age on the relationship between the external (environmental) exposure and the internal (biologically effective) target tissue exposure. For example, PBPK models can be used to determine the impact of differences in key metabolic enzymes due to normal variation in enzyme activities, against a background of variations in physiological parameters that may also affect pharmacokinetics across the general population. This application of PBPK modeling will be discussed later, in the section on polymorphisms.

The variation in pharmacokinetics among individuals of the same age can be demonstrated by comparing the pharmacokinetics of perchloroethylene in males versus females.

Figure 3 provides the simulated concentrations of perchloroethylene and its major metabolite, trichloroacetic acid, in males and females from birth to the age of 75 years. There is a noticeable increase in the predicted blood concentrations of perchloroethylene in females compared to those predicted for males at the same age, starting approximately at the age of 10 years. The concentrations of perchloroethylene are predicted to be as much as 30% higher in the females than in males, with predicted TCA concentrations as much as 50% higher.

The differences in predicted blood concentrations between males and females of the same age are related mainly to the lipophilicity of perchloroethylene. Because it has a much higher lipophilicity than most volatile compounds, it is predicted to have greater storage in fat and other tissues. With females, around the age of puberty, larger increases in the fractional volume of fat are observed, compared to males. Not only between males and females, but variability in the fractional volume of fat across individuals would affect the pharmacokinetics of lipophilic compounds. This variability could impact blood concentrations across individuals, as demonstrated in Figure 3, and therefore impact the potential risk of adverse effects across individuals.

PHARMACOKINETIC VARIABILITY RESULTING FROM DIFFERENCES BETWEEN SUBPOPULATIONS

The increasing recognition that there is a genetic basis for variability in response to chemical exposure has led to a surge in polymorphism-related research in the field of toxicology. The results of this research indicate that the number of identified alleles for a gene of interest may range from two [e.g., *GSTT1*, reviewed in Eaton and Bammler (5)] to more than 70 (*CYP2D6* alleles), some with more than one nucleotide change (6). In particular, many (if not all) of the genes that give rise to enzymes that metabolize environmentally relevant toxicants are polymorphic.

The degree to which polymorphisms increase human variability in toxic response is not well-characterized, although it has been discussed in numerous reviews (7–10). A wide range in activity between different alleles, or between a null allele and a wild type, might lead to the expectation of large differences in tissue dose arising from similar exposures. This conclusion is supported by epidemiological comparisons of cancer risk between populations with the wild-type and variant alleles that show an increased risk (or decreased risk) among populations harboring different alleles (11), and observed variability in blood or tissue levels of pharmaceuticals in patients receiving similar administered doses (12). On the other hand, a genetic polymorphism may have minimal or no impact on toxicity. Overall, the key question for evaluating the effects of polymorphisms in genes encoding metabolic enzymes is how the polymorphism affects the interindividual variability in the tissue dose of active agent resulting from a given administered dose of the parent compound.

A recent study of polymorphisms in the metabolism of warfarin and parathion provide a useful comparison of the potential impact of polymorphisms on risk assessment (13). These compounds are metabolized by two very different metabolic pathways, involving polymorphisms in two unrelated metabolic enzymes, with different biological implications resulting from the presence of the polymorphisms. In the case of warfarin, its stereoisomers [(S)-warfarin and (R)-warfarin] are metabolized by a variety of cytochrome P450 isozymes, but metabolism of (S)-warfarin is primarily by CYP2C9, and metabolism of (R)-warfarin is primarily by CYP2C19 and CYP1A2. The toxic and therapeutic effects of warfarin are due to the parent compound, and the activity of the (S)-enantiomer is approximately three times the activity of the (R)-enantiomer. There are currently three known human alleles of *CYP2C9* for which the effects on the metabolism of (S)-warfarin have been characterized phenotypically (*CYP2C9*1*, *CYP2C9*2*, and *CYP2C9*3*). The protein encoded by the wild-type (*CYP2C9*1*) allele is the most active, and the *CYP2C9*3* allele results in clinically significant reduction in warfarin activity, while the *CYP2C9*2* allele appears to result in a smaller reduction. Information on the variation in the in vitro metabolism of (S) -warfarin by the enzymes encoded by the three human alleles (*CYP2C9*1*, *CYP2C9*2*, and *CYP2C9*3*) is provided in Haining et al. (14), Rettie et al. (15,16), Sullivan-Klose et al. (17), and Takahashi et al. (18,19).

By combining PBPK modeling for (S)-warfarin with Monte Carlo analysis, simulations could be conducted to examine the variability in the area under the blood or plasma concentration curve (AUC) for (S)-warfarin following oral exposure to 1 mg/kg, a concentration in the therapeutic range. In evaluating the potential kinetic variability that may be associated with a polymorphism, the variation in the metabolic parameters associated with that polymorphism should not be considered in isolation; overall variability in all physiological parameters impacting pharmacokinetics should be considered. If the variability in the metabolic parameters is considered in isolation, the impact of the polymorphism on variability may be misstated.

To illustrate this point, the potential effect of the different homozygous genotypes (i.e., the metabolic variability alone) on the distribution of the AUC was evaluated as if there were no other variability in pharmacokinetic parameters. Using an extension of the warfarin model reported by Luecke and Wosilait (20), which includes the metabolism of both the (S)- and the (R)-enantiomers of warfarin, as well as the mutual inhibition of the metabolism of (S)-warfarin and (R)-warfarin (21), and the binding of warfarin in the plasma (22), separate simulations were performed considering the reported variability of the metabolic parameters for each homozygous genotype. The pharmacokinetic parameters (K_m and V_{max}) for homozygous

Table 1 Metabolic Parameters for (S)-Warfarin for Three CYP2C9 Alleles

Allele	References	V_{max} (mg/hour/kg[a,b]) Mean	CV	K_m (mg/L) Mean	CV	Intrinsic Clearance ($V_{max}C/K_m$)
CYP2C9*1	Haining et al.[c] (14)	1.61		1.85		0.87
	Takahashi et al.[d] (19)	2.13	0.0036	0.81	0.12	2.6
	Sullivan-Klose et al.[d] (17)	1.01	0.046	3.57	0.078	0.28
	Rettie et al.[a] (15)	3.20		1.26		2.5
	Rettie et al.[b] (15)	3.20		1.05		3.0
CYP2C9*2	Sullivan-Klose et al.[d] (17)	1.26	0.031	3.85	0.056	0.33
	Rettie et al.[a] (15)	0.21		0.52		0.40
	Rettie et al.[b] (15)	0.36		0.65		0.55
	Rettie et al.[e] (16)	1.10	0.036	1.85	0.17	0.59
CYP2C9*3	Haining et al.[c] (14)	0.31		9.24		0.034
	Takahashi et al.[d] (19)	0.51	0.22	3.20	0.16	0.16
	Sullivan-Klose et al.[d] (17)	1.37	0.044	28.4	0.059	0.048

[a]Hep G2 cells, cell lysate.
[b]Hep G2 cells, particulate preparation.
[c]Baculovirus/insect cell system, purified enzyme.
[d]Yeast expression, microsomes.
[e]Expressed in insect cells, purified enzymes.
Abbreviations: CYP, cytochrome P450; CV, coefficient of variation.

individuals were randomly generated from the distributions for the allele of interest (Table 1), with all other model parameters held fixed at their mean values.

Figure 6 shows the results for consideration of metabolic variability alone for (S)-warfarin (Case 1), in which the plasma concentration over time was simulated for each of the three homozygous genotypes. Each line in the figure represents one Monte Carlo simulation. Inspection of the figures demonstrates that the presence of the *CYP2C9*3* allele has a strong impact

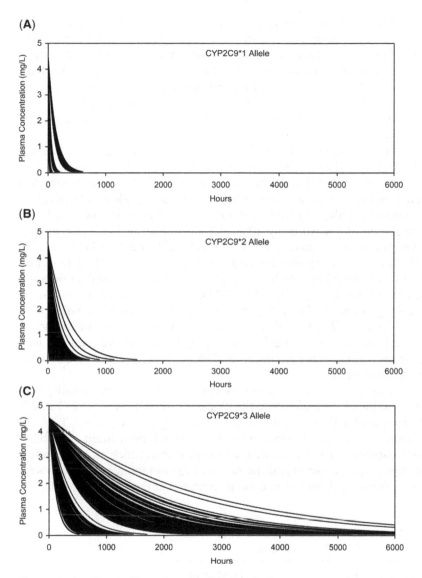

Figure 6   Predicted effect of metabolism by the homozygous alleles of CYP2C9 on the concentration of (S)-warfarin in human plasma. *Abbreviation*: CYP, cytochrome P450.

on the metabolism and clearance of the (S)-enantiomer of warfarin, which is consistent with clinical observations (18,19,23–25).

In reviewing the distributions presented in Figure 6, there appear to be gaps present. These are a reflection of the available information on the distributions of the metabolic parameters associated with each homozygous genotype. Multiple distributions were sampled for each homozygous genotype (Table 1). In some cases, the Coefficient of Variation (CV) associated with a mean for a given parameter was very small, resulting in a tight distribution with little overlap across distributions. The distributions were sampled equally, resulting in gaps between distributions of metabolic parameters, as well as gaps in the estimation of the associated plasma concentrations.

While the Case 1 simulation does demonstrate what appears to be a significant impact on the metabolism of warfarin by the polymorphism, it does not consider the impact that the variability in other physiological parameters may have on the AUC. It also does not consider the prevalence of each allele in the general population (Table 2). *CYP2C9*3* is present in less than 1% of the population. Approximately 12% of the population is a heterozygous genotype possessing a combination of *CYP2C9*1* and *CYP2C9*2* alleles, but this heterozygous genotype would not be expected to contribute significantly to the variability of the dose metric among the total population because the protein encoded by *CYP2C9*2* is nearly as active as the wild-type CYP2C9*1 protein. The heterozygous genotype possessing a combination of *CYP2C9*1* and *CYP2C9*3* alleles, however, is present in 9% of the population, and because the CYP2C9*3 protein is far less active than the wild type, the presence of this heterozygous allele may increase the variability in the AUC dose metric.

To evaluate the impact a polymorphism can have on overall population variability, the variability in the "typical" population (Case 2, Table 3) can be compared to the variability in the overall population (Case 3, Table 3), considering the sensitive subgroups. The normal population variability would be estimated using a Monte Carlo approach in which all of the model parameters would be varied and the values of the metabolic parameters were generated from the distribution chosen to represent the normal population.

Table 2 Average Prevalence of CYP2C9 and PON1 Alleles in the U.S. Population

CYP2C9	Prevalence
S1 homozygous	78%
S1/S2 heterozygous	12%
S1/S3 heterozygous	9%
S2 homozygous	1%
S2/S3 heterozygous	1%
S3 homozygous	0.5%

Abbreviation: CYP, cytochrome P450.
Source: From Ref. 26.

Table 3 Descriptive Statistics of the Area Under the Curve (mg-hour/L) Distribution for (S)-Warfarin

	Case 1 (CYP2-C9*1)	Case 1 (CYP2-C9*2)	Case 1 (CYP2-C9*3)	Case 2 (normal population)	Case 3 (total population)
Mean	157	273	2670	202	311
Standard error	5.28	4.73	52.7	11.1	21.9
Median	58.9	252	2680	83.6	104
Standard deviation	167	149	1670	351	693
Sample variance	27900	22300	2770000	123000	480000
Kurtosis	0.159	2.48	2.09	99.9	92.1
Skewness	1.34	0.822	1.03	7.44	8.21
Range	578	1180	11700	6280	9650
Minimum	22.8	20.1	333	10.5	10.1
95th percentile	509	465	5610	731	1170
99th percentile	555	692	7670	1380	2670
Maximum	601	1200	12100	6290	9660
Count	1000	1000	1000	1000	1000

Note: Case 1, varying only the metabolism parameters defining the polymorphism, using the allele indicated; Case 2, varying all parameters except those defining the polymorphism; Case 3, varying all parameters, using U.S. population frequencies of each allele.
Abbreviation: CYP, cytochrome P450.

For (S)-warfarin metabolism, the most prevalent genotype, homozygous *CYP2C9*1*, was used to represent the normal population because 78% of the population possesses this genotype. This simulation was used to provide a benchmark normal population variability for which meaningful comparisons to the general population, considering the impact of the sensitive subpopulation on variability, could be made. The distribution of the dose metrics for the general population, including all sensitive subpopulations, was obtained by sampling distributions for all of the model parameters as well as for all genotypes, with the frequency of sampling from each genotype distribution determined by the prevalence of that genotype in the U.S. population. In the case of heterozygous alleles, activity was assumed to be intermediate between the two homozygous alleles.

Figure 7 compares the distribution of the (S)-warfarin AUCs for the general population (Case 3) (i.e., varying all parameters, with the metabolic parameters sampled from the distributions for all genotypes according to the prevalence of each genotype in the U.S. population) with the AUC distribution from the "normal" population (Case 2) (i.e., varying all parameters and generating values for the metabolic parameters from the distributions of the homozygous *CYP2C9*1* genotype). In other words, Figure 7 compares the total population variability (including the

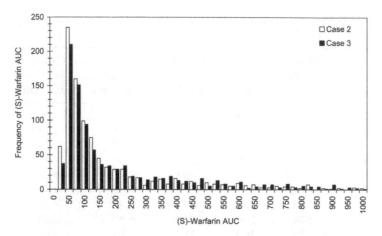

Figure 7 Area under the curve (AUC) for total population (Case 3) compared to AUC for normal population (Case 2). *Abbreviation*: AUC, area under the curve.

polymorphisms) with the variability that would be estimated without accounting for the polymorphism. Consideration of all alleles involved in the polymorphism shifts the normal distribution to the right, demonstrating an increase in the AUC, and reflecting the contribution of the less active allelic forms. As shown in Table 3, the polymorphism also extends the right tail of the distribution, producing a greater upper bound on the AUC and greater overall variability. For example, introducing the polymorphism increases the median AUC value from 84 to 104 mg-hour/L, and increases the 95th percentile from 731 to 1170 mg-hour/L. Thus, the polymorphism in *CYP2C9* does increase the overall population variability in warfarin tissue dose. Even though the *CYP2C9*3* allele occurs at a low prevalence, the difference in activity, coupled with a 9% prevalence of *CYP2C9*1/ CYP2C9*3* heterozygotes, is sufficient to increase the population variability.

It is possible that even though a metabolic enzyme has a polymorphism, the impact on the overall variability for the population may be minimal. This possibility has been suggested in the case of parathion (26). Parathion is metabolized to paraoxon, which can inhibit acetylcholinesterase. Paraoxonase (PON1) is one of the enzymes responsible for the metabolic clearance or detoxification of paraoxon. PON1 is a polymorphic enzyme, with the high-activity PON1 homozygotes accounting for approximately 41% of the U.S. population, the low-activity homozygotes accounting for approximately 15% of the U.S. population, and low/high-activity heterozygotes accounting for approximately 45% of the U.S. population (26–31). Because paraoxon is the active agent, low PON1 activity might be expected to result in increased sensitivity. In vitro data on the two human alleles of paraoxonase (low and high activity) were used to develop distributions for the metabolism parameters to be used in a published PBPK model for parathion (32). As with the warfarin example, Monte Carlo simulations were performed to generate the resulting distribution of predicted blood concentrations of paraoxon across a "normal" or general population (Case 2), considering the variability in other pharmacokinetic parameters.

In contrast to warfarin, for paraoxon metabolism there is no clear wild-type genotype, because the homozygous high-activity genotype is present in 15% of the population, the homozygous low-activity genotype is present in 40% of the population, and 45% of the population is heterozygous. However, the low-activity genotype would represent the sensitive subpopulation, because slower elimination of paraoxon would result in increased toxicity. To provide a benchmark population for which meaningful comparisons to a sensitive subpopulation could be made, the "normal" population was arbitrarily defined as the high-activity homozygotes plus the heterozygotes, which accounts for 60% of the population.

Similarly to warfarin, evaluating the variability in the pharmacokinetic parameters for parathion in isolation demonstrates a change in the distribution of the AUC for the low-activity allele versus the high-activity allele (Case 4) (Fig. 8). However, for parathion, comparison of the variability in the "typical" population (Case 5) to the overall population (Case 6) demonstrates little difference in variability in the predicted distribution for the time-integrated (area under the curve) blood concentrations of paraoxon (mg-hour/L), following exposure to parathion at a dose of 0.033 mg/kg (Fig. 9). While the polymorphism does impact the distribution of blood concentrations, particularly at the higher internal exposures, the overall effect is relatively small when put in the perspective of the variability in other physiological and biochemical factors across the same population.

CONCLUSIONS

The overall pharmacokinetic variability across a population is a function of many chemical-specific, genetic, and physiological factors. Due to the complex interactions among these factors, speculation regarding the extent of population variability on the basis of the observed variation in a single factor can be highly misleading. Analysis using PBPK modeling and Monte Carlo techniques provides a more reliable approach for estimating population pharmacokinetic variability. Analyses such as those described here can be used to develop quantitative Chemical-Specific Adjustment Factors (CSAFs) to replace default uncertainty factors for human pharmacokinetic variability (13). While this approach captures the pharmacokinetic variability, it may not capture the total uncertainty in parameter estimates.

Recent guidance from the International Programme on Chemical Safety (33) provides an approach for replacing default uncertainty factors with CSAFs. This approach divides the animal-to-human interspecies and

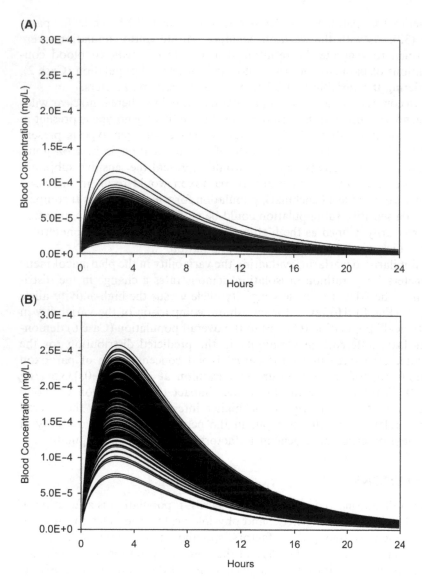

Figure 8 (A) High- and (B) low-activity alleles of paraoxonase on the concentration of paraoxon in human blood.

human intraspecies uncertainty factors into pharmacokinetic and pharmacodynamic components, each of which can be replaced by a CSAF if data are available. For example, the magnitude of the factor for human variability in toxicokinetics (HK_{AF}) may be calculated based on an evaluation of human variability in the area under the tissue concentration–time curve.

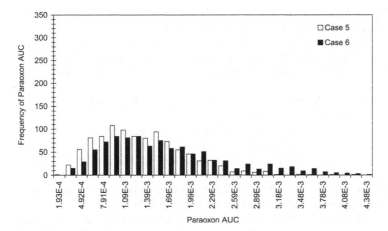

Figure 9 Area under the curve for total population (Case 6) compared to area under the curve for the "typical" population (Case 5, using high-activity and heterozygous alleles).

As demonstrated here, a PBPK model can provide an excellent basis for performing such an evaluation.

PBPK modeling can also be useful in a more qualitative sense, to determine whether there is reason for concern regarding a particular age group that might be more sensitive due to pharmacokinetic differences. Similar analyses can be performed to determine whether exposure during special life stages, such as gestation or lactation, represents a significant concern (4,34). PBPK modeling of this nature, coupled with parameter estimation using quantitative structure–activity relationship (QSAR) techniques and mechanistic information from genomic arrays, may prove particularly valuable in prioritizing testing requirements for new chemicals.

REFERENCES

1. Clewell HJ III, Andersen ME. Use of physiologically based pharmacokinetic modeling to investigate individual versus population risk. Toxicology 1996; 111:315–329.
2. Clewell HJ, Gentry PR, Covington TR, et al. Evaluation of the potential impact of age- and gender-specific pharmacokinetic differences on tissue dosimetry. Toxicol Sci 2004; 79(2):381–393.
3. Clewell HJ, Teeguarden J, McDonald T, et al. Review and evaluation of the potential impact of age and gender-specific pharmacokinetic differences on tissue dosimetry. Crit. Rev. Toxicol 2002; 32(5):329–389.
4. Gentry PR, Covington TR, Clewell HJ III. Evaluation of the potential impact of pharmacokinetic differences on tissue dosimetry in offspring during pregnancy and lactation. Regul Toxicol Pharmacol 2003; 38:1–16.

5. Eaton DL, Bammler TK. Concise review of the glutathione S-transferases and their significance to toxicology. Toxicol Sci 1999; 49:156–164.
6. Ingelman-Sundberg M, Daly AK, Nebert DW, eds. Home Page of the Human Cytochrome P450 (CYP) Allele Nomenclature Committee, Human Cytochrome P450 (CYP) Allele Nomenclature Committee, Feb. 7, 2002 at http://www.imm. ki.se/CYPalleles/.
7. Ingelman-Sundberg M. Genetic variability in susceptibility and response to toxicants. Toxicol Lett 2001; 120:259–268.
8. Knudsen LE, Loft SH, Autrup H. Risk assessment: the importance of genetic polymorphisms in man. Mutat Res 2001; 482:83–88.
9. Linder MW, Valdes R, Jr. Genetic mechanisms for variability in drug response and toxicity. J Anal Toxicol 2001; 25:405–413.
10. Miller MC III, Mohrenweiser HW, Bell DA. Genetic variability in susceptibility and response to toxicants. Toxicol Lett 2001; 120:269–280.
11. Uematsu F, Kikuchi H, Motomiya M, et al. Association between restriction fragment length polymorphism of the human cytochrome P450IIE1 gene and susceptibility to lung cancer. Jpn J Cancer Res 1991; 82:254–256.
12. Furuya H, Fernandez-Salguero P, Gregory W, et al. Genetic polymorphism of CYP2C9 and its effect on warfarin maintenance dose requirement in patients undergoing anticoagulation therapy. Pharmacogenetics 1995; 5:389–392.
13. Gentry PR, Hack CE, Haber L, et al. An approach for the quantitative consideration of genetic polymorphism data in chemical risk assessment: examples with warfarin and parathion. Toxicol Sci 2002; 70(1):120–139.
14. Haining RL, Hunter AP, Veronese ME, et al. Allelic variants of human cytochrome P450 2C9: baculovirus-mediated expression, purification, structural characterization, substrate stereoselectivity, and prochiral selectivity of the wild-type and I359L mutant forms. Arch Biochem Biophys 1996; 333:447–458.
15. Rettie AE, Wienkers LC, Gonzalez FJ, et al. Impaired (S)-warfarin metabolism catalysed by the R144C allelic variant of CYP2C9. Pharmacogenetics 1994; 4:39–42.
16. Rettie AE, Haining RL, Bajpai M, et al. A common genetic basis for idiosyncratic toxicity of warfarin and phenytoin. Epilepsy Res 1999; 35:253–255.
17. Sullivan-Klose TH, Ghanayem BI, Bell DA, et al. The role of the CYP2C9-Leu359 allelic variant in the tolbutamide polymorphism. Pharmacogenetics 1996; 6:341–349.
18. Takahashi H, Kashima T, Nomizo Y, et al. Metabolism of warfarin enantiomers in Japanese patients with heart disease having different CYP2C9 and CYP2C19 genotypes. Clin Pharmacol Ther 1998; 63:519–528.
19. Takahashi H, Kashima T, Nomoto S, et al. Comparisons between in vitro and in vivo metabolism of (S)-warfarin: catalytic activities of cDNA-expressed CYP2C9, its Leu359 variant and their mixture versus unbound clearance in patients with the corresponding CYP2C9 genotypes. Pharmacogenetics 1998; 8:365–373.
20. Luecke R, Wosilait W. Drug elimination interactions: analysis using a mathematical model. J Pharmacokinet Biop 1979; 7:629–641.
21. Kunze KL, Eddy AC, Gibaldi M, et al. Metabolic enantiomeric interactions: the inhibition of human (S)-warfarin-7-hydroxylase by (R)-warfarin. Chirality 1991; 3:24–29.

22. Chan E, McLachlan A, O'Reilly R, et al. Sterochemical aspects of warfarin drug interactions: use of a combined pharmacokinentic-pharmacodynamic model. Clin Pharmacol Ther 1994; 56:286–294.
23. Aithal GP, Day CP, Kesteven PJ, et al. Association of polymorphisms in the cytochrome P450 CYP2C9 with warfarin dose requirement and risk of bleeding complications. Lancet 1999; 353:717–719.
24. Steward DJ, Haining RL, Henne KR, et al. Genetic association between sensitivity to warfarin and expression of CYP2C9*3. Pharmacogenetics 1997; 7: 361–367.
25. Taube J, Halsall D, Baglin T. Influence of cytochrome P-450 CYP2C9 polymorphisms on warfarin sensitivity and risk of over-anticoagulation in patients on long-term treatment. Blood 2000; 96:1816–1819.
26. Haber LT, Maier A, Gentry PR, et al. Genetic polymorphisms in assessing inter-individual variability in delivered dose. Reg Toxi Pharmacol 2002; 35: 177–197.
27. Eckerson HW, Wyte CM, La Du BN. The human serum paraoxonase/arylesterase polymorphism. Am J Hum Genet 1983; 35:1126–1138.
28. Mueller RF, Hornung S, Furlong CE, et al. Plasma paraoxonase polymorphism: a new enzyme assay, population, family, biochemical, and linkage studies. Am J Hum Genet 1983; 35:393–408.
29. Diepgen TL, Geldmacher-von Mallinckrodt M. Interethnic differences in the detoxification of organophosphates: the human serum paraoxonase polymorphism. Arch Toxicol Suppl 1986; 9:154–158.
30. Davies HG, Richter RJ, Keifer M, et al. The effect of the human serum paraoxonase polymorphism is reversed with diazoxon, soman and sarin. Nat Genet 1996; 14:334–336.
31. Sanghera DK, Saha N, Kamboh MI. The codon 55 polymorphism in the paraoxonase 1 gene is not associated with the risk of coronary heart disease in Asian Indians and Chinese. Atherosclerosis 1998; 136:217–223.
32. Gearhart JM, Jepson GW, Clewell HJ, et al. Physiologically based pharmacokinetic model for the inhibition of acetylcholinesterase by organophosphate esters. Environ Health Perspect 1994; 102(11):51–60.
33. IPCS (International Programme on Chemical Safety), Guidance Document for the Use of Data in Development of Chemical-specific Adjustment Factors (CSAF) for Interspecies Differences and Human Variability in Dose/concentration Response Assessment, World Health Organization, Geneva, 2001. Available at www.who.int/entity/ipcs/publications/methods/harmonization/en/csafs_guidance_doc.pdf.
34. Clewell RA, Gearhart JM. Pharmacokinetics of toxic chemicals in breast milk: using PBPK models to predict infant exposure. Environ Health Perspect 2002; 110:A333–A337.

11

Developmental Aspects of Children's Pharmacokinetics

Gary Ginsberg
Connecticut Department of Public Health, Hartford, Connecticut, U.S.A.

Dale Hattis
George Perkins Marsh Institute, Clark University, Worcester, Massachusetts, U.S.A.

Babasaheb Sonawane
National Center for Environmental Assessment, Office of Research and Development, U.S. Environmental Protection Agency, Washington, D.C., U.S.A.

CHILDREN'S PHARMACOKINETICS: OVERVIEW

There are a number of behavioral, dietary, anatomical, and physiological factors that can cause children to be more highly exposed to environmental toxicants than adults (1). Among these factors are pharmacokinetic differences that can cause children, especially early in life, to have increased uptake and reduced clearance of chemicals (2–4). An overview of physiologic and developmental factors that can affect pharmacokinetics in children is presented in Table 1. These factors can influence all aspects of chemical fate (absorption, distribution, metabolism, excretion) with the greatest change in pharmacokinetics relative to adults likely to occur in the first weeks to months of life. Some of the pharmacokinetic differences can be attributed to body size and composition while others are due to functional

Table 1 Overview of Developmental Features that can Affect Pharmacokinetics

Developmental feature	Relevant age period	Implications
Chemical absorption		
Increased oral absorption of certain agents (e.g., metals)	Birth through weaning	Potential for greater chemical uptake
Greater dermal absorption	Primarily in premature neonates	Potential for greater chemical uptake
Greater inhalation rate per respiratory surface area	Birth through several years	Potential for greater local dose in respiratory tract
Body composition Lower lipid content Greater water content	Birth to 3 mo	Less partitioning and retention of lipid-soluble chemicals; larger Vd for water-soluble chemicals
Larger liver weight/body weight	Birth to 6 yrs but largest factor in first 2 yrs	Greater opportunity for hepatic extraction and metabolic clearance; however, also greater potential for activation to toxic metabolites
Immature enzyme function Phase I reactions Phase II reactions	Birth to 1 yr but largest factor in first 2 mo	Slower metabolic clearance of many drugs and environmental chemicals; less metabolic activation but also less removal of activated metabolites
Larger brain weight/body weight; greater blood flow to CNS; higher BBB permeability	Birth to 6 yrs but largest factor in first 2 yrs	Greater CNS exposure, particularly for water-soluble chemicals that are normally impeded by BBB; larger Vd
Immature renal function	Birth to 2 mo	Slower elimination of renally cleared chemicals and metabolites
Limited serum protein binding capacity	Birth to 3 mo	Potential for greater amount of free toxicant and more extensive distribution

Abbreviations: BBB, blood–brain barrier; CNS, central nervous system; Vd, volume of distribution.

immaturity of critical clearance systems. These unique early life factors are further described in subsequent sections.

While pharmacokinetic differences may create important dosimetric differences in children, they can be challenging to analyze because:

1. Numerous factors are immature in young children with the net effect on internal dose of key toxicant (parent compound or metabolites) often difficult to predict. For example, immaturities in chemical activation pathways may be counterbalanced by immaturities in detoxification and clearance.
2. Children represent a wide age range from early postnatal life through puberty. This requires consideration of numerous developmental stages, some of which may be important to consider for pharmacokinetic reasons and others because they represent periods of high intake (exposure factors) or high vulnerability (pharmacodynamic factors).
3. Interindividual variability within any single age group may be considerable due to more rapid growth in some individuals than in others.
4. Not all pathways are equally immature at birth and they develop at different rates. Some enzymes are unique to early development being remnants of fetal life.
5. Pharmacokinetic data for environmental toxicants are generally not available for children.

This chapter provides approaches for addressing the issues identified above, thus making pharmacokinetic factors in early life amenable to being included in human health risk assessments. Much of the focus is on neonatal and infant pharmacokinetic factors, because this period is most different from adults. Therapeutic drug trials demonstrate that allometric scaling of dose based upon body weight or surface area is the least reliable in neonates due to the immaturity of clearance pathways leading to the greatest potential for drug overdose (5). This chapter relies upon developmental physiology data, (6–8) in vitro liver bank enzymology data (9–12) analyses of pediatric therapeutic drug databases (3,13–15), and physiologically based pharmacokinetic (PBPK) modeling analyses (16–19) to describe pharmacokinetics in early life stages. A number of excellent reviews have been published on the ontogeny of pharmacokinetic factors (2,4,6–10,20–22).

ABSORPTION, DISTRIBUTION, METABOLISM, AND EXCRETION IN CHILDREN

Chemical Absorption

Oral Absorption

Oral absorption of certain metals and organic compounds can be greater in early life than in older children or adults. A prime example is that of lead in which a variety of oral uptake studies in humans, monkeys, and rodents have led the U.S. Environmental Protection Agency to select default oral absorption factors for young children of 50% for diet/drinking water and 30% for lead in soil/house dust for the uptake/biokinetic lead model (23). In contrast, gastrointestinal tract absorption of lead in adults is approximately 10% (24,25). The greater oral uptake of lead in neonates has been attributed, at least in part, to greater pinocytic activity of intestinal epithelium prior to closure or maturation of the gastrointestinal tract (24,26). This nonselective uptake mechanism may also increase the absorption of other metals and organic compounds. In animals, gastrointestinal tract closure occurs around the time of weaning although the timing of this event can vary greatly across species (27). The time course for gastrointestinal tract closure in humans is not well defined, although evidence regarding the absorption of large sugars or proteins suggests that it can occur in the first weeks of life and that breast feeding can speed up the closure process (28,29). This is consistent with pediatric trials with zidovudine in children exposed to HIV at birth (30). Further, oral absorption in neonates is erratic due to irregular gastric emptying and peristalsis (24). Thus, the timing of gastrointestinal tract closure and its influence on bioavailability may be difficult to discern.

The possibility of greater uptake of ingested chemicals early in life should be evaluated within the context of the chemical's behavior in the gut. If it is generally well absorbed in rodents and adult humans by the oral route (e.g., small organic molecules), then any increase in absorption during early life stages may not create a substantiative difference in uptake (e.g., methyl mercury). However, for chemicals that are poorly absorbed (e.g., inorganic mercury—Walsh (31); lead, other metals), increased uptake in neonates may be an important factor in exposure assessment. It is prudent to consider this bioavailability increase to exist through the first six months of life, a time period during which the diet is predominated by milk that may increase the absorption of metals (24,32).

While enhanced pinocytosis and milk in the gut may be factors that promote greater oral bioavailability in early life, other factors may also be involved. These include higher requirements for absorption of nutrients, which may also enhance the absorption of environmental agents. Consistent with this are data showing that low iron or calcium intake increases lead absorption from the gut (23,33). Other factors that may affect oral absorption in children are gastric secretion and thus higher stomach pH up to two years of age, immaturity in bile acid metabolism and secretory function that can affect the absorption of lipids from the gastrointestinal tract (34), and differences in blood flow and surface area of gastrointestinal tract absorptive regions, as compared to adults (24).

Dermal Absorption

Children's potential for dermal exposure is generally considered to be greater than adults because of activity patterns, body surface area, and differences in skin structure.

Due to a well-developed stratum corneum, the dermal permeability differs little between full-term newborns and adults (35–39). In contrast, the skin of premature neonates can be substantially more permeable than that of full-term neonates due to immaturity of the stratum corneum (35,40); the difference may subside within two weeks postnatally (39).

Retention of Inhaled Doses

Respiratory tract physiology and morphometry change during development with child/adult differences greatest in neonates. The dosimetry implications of these age-related differences can depend upon the type of material inhaled and the portion of the respiratory tract where deposition occurs. Particles deposit according to their diameter, with large particles greater than $10 \mu M$ depositing primarily in the nose via impaction and small particles depositing in the bronchioles and alveoli via sedimentation and diffusion. Highly reactive or water-soluble gases (e.g., chlorine) are taken out of the inspired airstream in extrathoracic regions (nose, larynx, pharynx); less reactive gases (e.g., ozone) can penetrate to deeper lung regions (tracheobronchial and bronchiolar), while gases that are not reactive or water soluble penetrate to the alveolar region (pulmonary) where they can be taken up into the systemic circulation. A region's airflow and surface area are two factors that determine the percent extraction of inspired chemical as it passes through the region. There can be substantial differences in these factors between children and adults. The ratio of ventilation rate per lung surface area is approximately two-fold greater in neonates as compared to adults over the entire respiratory tract, with the largest difference in the pulmonary region.

Recent analyzes have modeled child and adult respiratory dosimetry for a range of particles and gases based upon known differences in regional airflow and surface area (41,42). These models are based upon depositional studies in adults and extrapolation to children based upon known developmental features of the respiratory tract. They demonstrate that: (i) fine particle deposition in the pulmonary region can be greater in children than in adults, (ii) deposition of reactive gases does not appear to differ markedly across age groups, and (iii) the systemic dose of nonreactive gases can be greater in children depending upon the gas's blood:air partition coefficient and rate of hepatic extraction (42). Blood:air partition coefficient is assumed to be equal in children and adults based upon a limited data set for four volatile anesthetic gases (43). These models can also be affected by activity pattern such that higher activity can lead to greater ventilation rate and deeper penetration in the respiratory tract. The uncertainties with these findings are that the dosimetry models use default assumptions for respiratory tract morphology and they average dose over relatively large areas. More refined deposition models are becoming available based upon imaging techniques that document respiratory tract morphology in combination with computational fluid dynamic modeling approaches (44).

Factors that Affect Chemical Distribution in Children

Newborns and infants have different percent of body content of water and lipid than older children and adults (2,6,7). At birth, there is a greater percent of body water and less body lipid, which can decrease the partitioning and thus retention of lipid-soluble chemicals. In contrast, water-soluble chemicals may tend to have a larger volume of distribution (Vd, expressed in L/kg body weight) because of the expanded water volume in early life. For example, newborns take twice as long as adults to eliminate lidocaine because of the large volume that must be cleared of the drug (45). Body lipid rises steadily after birth for the first nine months of life but then decreases steadily until about age four to seven, when a second period of increasing body lipid begins (6,13). These changes in body composition can alter chemical partitioning into lipid and thus half-life and Vd of lipophilic chemicals. Tissue distribution of chemicals can also be affected by differences in organ size (per body weight) across age groups. Liver mass per body weight is greatest in early life (46–48), and tissues such as liver, kidney, and lung undergo rapid growth during the first two years of life (24). In contrast, reproductive tissues are generally small and not highly active during this period. The brain is disproportionately large in young children. This factor combined with the immaturity of the blood–brain barrier (BBB) leads to a significant additional volume for chemical partitioning (14), thus increasing Vd. This early life differential would apply primarily to those chemicals that do not readily cross the mature BBB. Immaturity of the BBB can also lead to higher brain concentrations of some chemicals and the potential for neurotoxic effects (49).

Another factor that can affect the distribution of chemicals is the binding capacity of plasma proteins. A large number of drugs and certain environmental chemicals (e.g., organic acids such as trichloroacetic acid) (50) are strongly bound to plasma proteins such that there is little free chemical in the circulation. Because only the free form can cross the placenta, be excreted by the kidney, enter the central nervous system, or be taken up by the liver, extensive protein binding will (i) tend to slow elimination processes that can occur at these sites and (ii) limit the amount of chemical that is free at any time to exert an adverse effect. Neonates have low protein binding capacity, both with regards to albumin and alpha-1-glycoprotein (2,6,8). Evidence that limitations in serum protein binding capacity may enhance toxicity due to greater percent of free chemicals has been suggested for lidocaine, cisplatin, and other drugs (51,52).

Placental transport and milk partitioning are also important. These phenomena do not take place within the neonate but affect the dose received in early life. Substance in the maternal circulation can be transferred across the placenta. The rate of placental transfer depends on molecular size, lipophilicity, and serum protein binding (53,54). Parameters for human and rodent PBPK pregnancy models have been presented (55), and several such models have been developed [e.g., Clewell et al. (56), Krishnan and Andersen (57)]. Similarly, the partitioning of chemicals into breast milk varies greatly (58) and can be described via modeling approaches in cases where empirical data are missing (59). A recent evaluation of the partitioning of lipophilic chemicals between human milk lipid and blood lipid indicates the importance of considering interindividual variability in this factor (58).

Immaturity of Hepatic Function in Early Life

The immaturity of hepatic function in early life has been demonstrated in two ways. In vitro data describe enzyme activity and protein levels in liver bank specimens covering a wide range of ages including in utero, early postnatal, older children, and adults (10,60–62). These studies involve immunological techniques to quantify levels of specific enzymes [e.g., cytochrome P450 (CYPs)], as well as functional assays using classical substrates for the enzymes being probed.

The second type of observation of hepatic immaturity has been in vivo data from pediatric pharmacokinetic studies involving drugs that are cleared primarily by hepatic extraction. Recent compilations of adult/child comparisons of classical pharmacokinetic parameters (half-life, clearance, Vd) (3,13–15) are available but several caveats should be noted: (i) much of the data are from children who are patients under medical treatment and so their health is compromised in some way; (ii) data for therapeutic drugs may be marginally relevant for environmental contaminants (e.g., slow clearance, poorly metabolized, highly fat-soluble agents); and (iii) the exposure levels in the clinical trials tend to be higher than exposure to environmental contaminants. However, comparative data do provide useful information on the development of pharmacokinetic systems in children.

Development of CYP Enzymes

While total CYP content of human hepatic microsomal protein at birth is roughly 30% of the adult level (62), there are a number of CYPs, including CYPs 1A2, 2C family, and 2E1, whose levels are very low at this time (2,4,10,62). These CYPs increase rapidly after birth but generally take six months to one year of life to approach adult levels, in general agreement with the time to reach adult half-life and clearance values for many drugs. Figure 1 shows that across 40 drugs for which data were compiled in a pediatric pharmacokinetic database, drug half-life was significantly longer in infants than in adults through two months of life, with this differential not significant in the two- to six-month group (3). The reason for the appearance of more rapid CYP maturation from the in vivo data may be

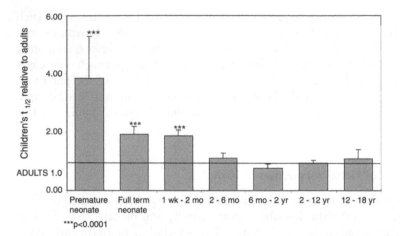

Figure 1 Analysis of children's pharmacokinetic database half-life results for full database—40 substrates. *Source*: Adapted from Ref. 3.

related to the larger size of the liver (relative to body weight) in neonates and increased hepatic blood flow relative to adults (46,47).

It is important to note that some CYPs, notably CYP3A7, are present at high levels in utero and at birth. Further, CYPs are inducible due to xenobiotic exposure both in the fetus and postnatally. As reviewed elsewhere, these factors need to be considered when assessing CYP function in early life (19).

DEVELOPMENT OF PHASE II CONJUGATION SYSTEMS

Phase II conjugation can represent a critical detoxification step. Pharmaco-kinetic data on compounds conjugated by UDP glucuronosyl transferase (UDPGT) suggest a progression from four-fold longer half-life in premature neonates to half-lives equal to adult by two to six months of age. These data are consistent with evidence that neonates are more susceptible to the side effects (anemia) of chloramphenicol, an antibiotic that requires glucuronida-tion for elimination (63,64). It should thus be expected that glucuronidation capacity in neonates through two months of life is limited compared to adults. However, the substrate specificity and developmental profiles for individual UDPGT forms should be given specific consideration, because different isozymes mature at different rates (65).

In contrast to glucuronidation, sulfation capacity appears to be adequate in neonates. For acetaminophen, sulfation predominates in early life until the onset of glucuronidation (66,67).

N-acetylation is a conjugation reaction of relevance to a number of car-cinogens, particularly aromatic amines. Most neonates (83%) demonstrate a slow rate of acetylation, while children by nine months of age had a slow/fast acetylator distribution similar to that seen in adults—approximately 50:50 (68,69).

Glutathione (GSH) conjugating ability exists in utero and at birth, due largely to a fetal glutathione transferase, glutathione-s-transferase (GST)-pi, the predominant form in early life. There are very limited data on the ontogeny of GSTs, but one report suggests that hepatic cytosolic levels of the GST-alpha and mu isoforms are somewhat reduced in the perinatal period relative to older children and adults (70). There is some degree of overlapping substrate specificity between GST-pi and other isozymes, but implications for child/adult differences are unclear.

The level of GSH, the nucleophilic tripeptide needed as a conjugation cofactor for this system, is also an important consideration. Premature infants who are deficient in GSH synthesis are more susceptible to retinopathy and possibly also to respiratory distress syndrome (71,72). Premature neonates tend to have greater levels of oxidized glutathione (GSSG) as compared to the reduced form (GSH) (71) in blood. Evidence of deficient GSH levels in early life comes from a study evaluating 44 healthy babies born to HIV-infected mothers. Only babies not exposed to anti-HIV drugs were studied (73). Plasma GSH levels were approximately 10% of adult levels at birth and rose through the first six months of life. Because the liver is the main source of plasma GSH, this may reflect an immaturity in tissue GSH as well.

Epoxide hydrolase (EH) metabolizes epoxides and can represent a bioactivation or deoxification process (74). EH protein levels in neonate liver microsomal protein were 42% of adult protein levels (75). Carbamazipine is metabolized to an epoxide and clearance of the epoxide via EH occurs more rapidly in adults (76).

DEVELOPMENT OF RENAL ELIMINATION

The kidney's ability to excrete endogenous and exogenous chemicals is immature at birth (3,7). Data suggest renal clearance rates in newborns that are three times less than in adults, although this immaturity can be considerably greater for certain drugs. Beyond two months of age, renal excretion can approach adult levels for some chemicals.

In early life, renal blood flow and glomerular filtration rate are substantially reduced, and transporter (secretory) systems in the proximal convoluted tubule are deficient (4,6,7). Plasma protein binding in neonates is substantially reduced; this factor combined with the immaturity in renal clearance can lead to substantially higher levels of free toxicant in plasma.

Disposition of Slowly Metabolized, Lipophilic Compounds

Much of the understanding communicated above comes from the study of readily metabolized and excreted toxicants. However, the fate of slowly metabolized, lipophilic toxicants is governed to a large extent by partitioning into lipid stores as opposed to other pathways described above. The smaller amount of lipids in neonates was mentioned earlier as a factor leading to the possibility of less retention of lipophilic toxicants. This agrees with the suggestion that the retention of lipophilic dioxins in the first year of life can be considerably less than in adults (77,78). Another factor that appears to contribute to faster dioxin clearance during infancy is an enhanced rate of fecal lipid excretion (77), which may augment excretion of toxicants that partition into lipids (78).

VARIABILITY IN PEDIATRIC PHARMACOKINETIC DATA

The trends for the ontogeny of pharmacokinetic systems are summarized in Table 2. These trends are generalizations that can be subject to a large amount of intersubject variability within an age group. This is because within any children's age grouping (e.g., one week to one month of age) children on either end of the age range are likely to diverge considerably from the average value for the parameter observed at the mean age, due to the rapid development that occurs during these early life stages. Even for children of exactly the same age, pharmacokinetic functions can vary due to interindividual differences in the rate at which hepatic and renal systems mature. This type of variability decreases toward maturity.

Analysis of the pediatric pharmacokinetic data set referred to above points out the high degree of variability that can occur in young infants (Fig. 2) (13). The individual child/adult half-life ratios span nearly two orders of magnitude at the earliest ages, with the bias in the data toward ratios above one (log 0 in Fig. 2), signifying longer half-life in neonates than adults. The figure shows a substantial number of cases in which the child/adult half-life ratio is greater than 3.2-fold, the standard risk assessment variability factor for pharmacokinetics, and a few cases in which there was more than a 10-fold difference. It is important to note that the variability in this data set comes from at least two sources: (i) variability across different pediatric populations of the same age undergoing pharmacokinetic testing (intersubject variability), and (ii) variability across the different drugs in the database (interchemical variability). However, beyond the first few months of life, variability in the data set is smaller and the data demonstrate that half-lives of these compounds are generally shorter in children than in adults.

DEVELOPMENT OF CHILDREN'S PBPK MODELS

As summarized above, many pathways that can affect toxicant activation, detoxification, and clearance are either immature (e.g., numerous CYPs, renal function) or overexpressed in early life (CYP3A7, CYP4A, GSTpi). One can speculate on the degree to which these child/adult differences may affect internal dosimetry for a particular toxicant in neonates and other

Table 2 Trends in Pharmacokinetic Function by Age Group and Pathway (Trends Reflect Likely Direction of Child/Adult Differences in Pathway Function)

PK pathway	Premature neonates	Neonate <1 mo	Early infant 1–2 mo	Mid-infant 3–5 mo	Late infant 6–11 mo	Toddler 1–2 yr	Older childhood
Oral absorption	↑[a]	↑[a]	↑[a]	↑[a]	↕	↕	↕
Dermal absorption	↑[b]	↕	↕	↕	↕	↕	↕
Lung absorption	↑[c]	↑[c]	↑[c]	↑[c]	↑[c]	↑[c]	↑[c]
Renal clearance	→	→	→	↕	↕	↕	↕
CYP1A2	→	→	→	↕	↑ Scale: $BW^{3/4}$	↑ Scale: $BW^{3/4}$	↑ Scale: $BW^{3/4}$
CYP2E1	→	→	→	↕	↑ Scale: $BW^{3/4}$	↑ Scale: $BW^{3/4}$	↑ Scale: $BW^{3/4}$
CYP3A family (except 3A7)	↓[d]	↓[d] →	↓[d] →	↕	↑ Scale: $BW^{3/4}$	↑ Scale: $BW^{3/4}$	↑ Scale: $BW^{3/4}$
CYP3A7	↑	↑	↕	↕	↓[e] Scale: $BW^{3/4}$	↓[e] Scale: $BW^{3/4}$	↓[e] Scale: $BW^{3/4}$
Other CYPs	→	→	→	↕	↑ Scale: $BW^{3/4}$	↑ Scale: $BW^{3/4}$	↑ Scale: $BW^{3/4}$

(Continued)

Table 2 Trends in Pharmacokinetic Function by Age Group and Pathway (Trends Reflect Likely Direction of Child/Adult Differences in Pathway Function) (*Continued*)

PK pathway	Premature neonates	Neonate <1 mo	Early infant 1–2 mo	Mid-infant 3–5 mo	Late infant 6–11 mo	Toddler 1–2 yr	Older childhood
Glucuronidation	↓[f]	↓[f]	→[f]	↑[f]	↑[f]	↔	↔
N-acetylation	↓[f]	↓[f]	↓[f]	↓[f]	↓[f]	↔	↔
GSTs	Uncertain[g]	Uncertain[g]	Uncertain[g]	Uncertain[g]	↔	↔	↔
Epoxide hydrolase	→	→	→	↔	↑	↑	↔
ADH[h]				↔	↔	↑	↑
					Scale: BW$^{3/4i}$	Scale: BW$^{3/4}$	Scale: BW$^{3/4}$

[a]Applies only when oral absorption is low (e.g., <50%).

[b]Increases in dermal absorption at 36 weeks gestation and earlier, with large increases possible prior to 32 weeks.

[c]Applies to deposition of particles and reactive gases in respiratory tract. Systemic uptake of nonreactive gases can also be increased from short-term acute exposure; however, under steady state conditions age group differentials for nonreactive gases cannot be predicted without PBPK modeling.

[d]CYP3A pathway activity low unless the chemical can also be a substrate for the fetal isozyme, CYP3A7, in which case metabolic rate can be higher in early life.

[e]CYP3A7 activity low at these ages but overlapping substrate specificity with CYP3A4 may keep metabolic activity at or slightly above adult levels.

[f]Immaturity of N-acetylation conjugating activity during first year of life causes slow acetylator phenotype regardless of genotype.

[g]GST neonate/adult ratio uncertain because preliminary data suggest that several isozymes are immature at birth while GSTpi is high at birth and other isozymes have not been evaluated. Given overlapping function between isozymes, the GST conjugating capacity for any given substrate is currently uncertain. Circulating GSH levels are low at birth and this may reflect a deficiency in tissue GSH.

[h]Suggestive evidence for immaturity of ADH system warrants using immaturity factors as for other metabolic systems (CYPs, glucuronidation).

Abbreviations: ADH, alcohol dehydrogenase; GSTs, glutathione s-transferase; CYP, cytochrome P450.

Source: Adapted From Ref. 86.

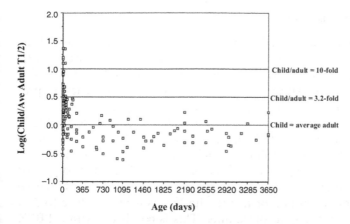

Figure 2 Child/adult half-life ratios across drugs and age groups. *Source*: Adapted from Ref. 13.

age groups. However, immaturity in a single pathway may influence internal dosimetry to a greater or lesser extent than expected based upon other pharmacokinetic factors such as (i) the presence of compensating metabolic or clearance pathways, (ii) blood flow limitations to the key metabolizing organ, and (iii) influence of differences in body composition or protein binding capacity (2,79). Therefore, it is important to evaluate child/adult differences in internal dosimetry with PBPK models. These models integrate many physiologic and functional parameters that affect chemical fate, yielding predictions of internal dose under a variety of exposure scenarios. Such predictive models are especially needed in assessing children's risks from environmental toxicants because obtaining pharmacokinetic data in children for such toxicants is typically not feasible for ethical reasons.

PBPK analyses have been useful in removing some of the uncertainty in extrapolation of exposure and risk between rodents and human adults (80–83). This approach should also be feasible for extrapolating dose from adults to children, with recent reviews pointing out how pharmacokinetic parameters may need adjustment for the development of PBPK models for children (2,42,84). However, to date there have been relatively few attempts at PBPK modeling in children (16–19,85). These efforts have adjusted adult PBPK models for known physiologic differences that occur in early life, and have also been adjusted for the ontogeny of metabolic systems in certain cases (16,19,85). However, for the most part, these modeling approaches have made predictions of toxicant dosimetry in children without the benefit of testing the model against actual pharmacokinetic data in children. This situation arises because, as stated above, there is a lack of pharmacokinetic data for environmental agents in children. However, these efforts demonstrate a reasonable modeling framework for assessing children's pharmacokinetics.

Children's PBPK Modeling Studies

Several case studies with environmental toxicants have recently been reported (16–18,85). An extensive analysis by Clewell et al. (16) used PBPK modeling to simulate internal exposures to parent compound and metabolite for a range of toxicants with differing physicochemical properties: isopropanol (water soluble, volatile), vinyl chloride and methylene chloride (lipophilic, volatile), perchlorethylene (lipophilic, volatile, poorly metabolized), tetrachlorodibenzo-p-dioxin (TCDD) (lipophilic, nonvolatile, poorly metabolized), and nicotine (water soluble, nonvolatile). Internal dosimetry was simulated across a wide range of ages from birth to 75. The results suggested that for most comparisons, infants less than six months of age had internal exposure of parent compound or metabolite that was within three-fold of the dose simulated for 25-year-old adults. This is consistent with a PBPK children's modeling effort for acrylamide and its active metabolite glycidamide (85). Due to the likely deficiency in GSH-mediated detoxification of both acrylamide and glycidamide, one may have anticipated considerably greater dosimetry in early life. However, the concomitant deficiency in the activating enzyme, CYP2E1, led to only modest differences in internal dose projections relative to adults. Other modeling case studies in children have involved inhaled solvents (17,18). In these simulations, PBPK models were adjusted for children's physiology, but not for the ontogeny of metabolic systems.

SUMMARY

Chemical dosimetry will likely differ across children's developmental stages and between children and adults. There are numerous differences in absorption, distribution, metabolism, and elimination, which could interplay to affect dosimetry in early life. Children's toxicokinetic responses may be more variable than in adults due to early life interindividual differences in rates of development of key physiologic and metabolizing systems. Risk assessments can begin to describe the implications of these child/adult differences by first appreciating how toxicokinetic mechanisms affect toxicant action and elimination, and then by considering the functional status of these mechanisms in early life. PBPK models are needed for quantitative evaluation of dosimetry differences across age groups. Initial models of toxicant dosimetry in children can be developed based upon general principles of physiological development and enzyme maturation. More specific models that utilize pharmacokinetic data for therapeutic drugs in children can help calibrate and refine model development. Ultimately, data from liver microsomal systems are needed to more fully explore the extent of children's metabolism of environmental toxicants. Application of such approaches will be instrumental in demonstrating whether the traditional interindividual uncertainty factor used in adult-based risk assessments (3.2-fold) is sufficient to cover sources of variability introduced by children's toxicokinetics.

REFERENCES

1. U.S. EPA. Summary Report of the Technical Workshop on Issues Associated with Considering Developmental Changes in Behavior and Anatomy when Assessing Exposure to Children, EPA/630/R-00/005, U.S. Environmental Protection Agency, Risk Assessment Forum, Washington, D.C., 2000. Available at http://nepis.epa.gov/pubtitleORD.htm.
2. Clewell HJ III, Teguarden J, McDonald T, et al. Review and evaluation of the potential impact of age and gender-specific pharmacokinetic differences on tissue dosimetry. Crit Rev Toxicol 2002; 32:329–389.
3. Ginsberg G, Hatts D, Sonawane B, et al. Evaluation of child/adult pharamacokinetic differences from a database derived from the therapeutic drug literature. Toxicol Sci 2002;66:185–200.
4. Alcorn J, McNamara PJ. Ontogeny of hepatic and renal systemic clearance pathways in infants. Part II. Clin Pharmacokinet 2002; 41:1077–1094.
5. Anderson BJ, McKee AD, Holford HG. Size, myths, and the clinical pharmacokinetics of analgesia in pediatric patients. Clin Pharmacokinet 1997; 33:313–327.
6. Kearns GL, Reed MD. Clinical pharmacokinetics in infants and children. A reappraisal. Clin Pharmacokinet 1989; 17(suppl 1):29–67.
7. Morselli PL. Clinical pharmacology of the perinatal period and early infancy. Clin Pharmacokinet 1989; 17(suppl 1):13–28.
8. Besunder JB, Reed MD, Blumer JL. Principles of drug biodisposition in the neonate. A critical evaluation of the pharmacokinetic-pharmacodynamic interface. Part I. Clin Pharmacokinet 1988; 14:189–216.
9. Hines RN, McCarver DG. The ontogeny of human drug-metabolizing enzymes: phase I oxidative enzymes. J Pharmacol Exp Ther 2002; 300:355 360.
10. Cresteil T. Onset of xenobiotic metabolism in children: toxicological implications. Food Addit Contam 1998; 15(suppl):45–51.
11. Vieira I, Sonnier M, Cresteil T. Development expression of CYP2E1 in the human liver. Hypermethylation control of gene expression during the neonatal period. Eur J Biochem 1996; 238:476–483.
12. Johnsrud EK, Koukouritaki SB, Divakaran K. Human hepatic CYP2E1 expression during development. J Pharmacol Exp Ther 2003; 307(1):402–407.
13. Hattis D, Ginsberg G, Sonawane B, et al. Differences in pharmacokinetics between children and adults—II. Children's variability in drug elimination half-lives and in some parameters needed for physiologically-based pharmacokinetic modeling. Risk Anal 2003; 23:117–142.
14. Renwick AG, Dorne JL, Walton K. An analysis of the need for an additional uncertainty factor for infants and children. Regul Toxicol Pharmacol 2000; 31:286 296.
15. Dorne JL, Walton K, Renwick AG. Uncertainty factors for chemical risk assessment: human variability in the pharmacokinetics of CYP1A2 probe substrates. Food Chem Toxicol 2001; 39:681–696.
16. Clewell HJ III, Gentry PR, Covington TR, et al. Evaluation of the potential impact of age and gender-specific pharmacokinetic differences on tissue dosimetry. Toxicol Sci 2004; 79:381–393.

17. Price K, Haddad S, Krishnan K. Physiological modeling of age-specific changes in the pharmacokinetics of organic chemicals in children. J Toxicol Environ Health A 2003; 66:417–433.
18. Pelekis M, Gephart LA, Lerman SE. Physiological-model-based derivation of the adult and child pharmacokinetic intraspecies uncertainty factors for volatile organic compounds. Regul Toxicol Pharmacol 2001; 33:12–20.
19. Ginsberg G, Hattis D, Sonawane B. Incorporating pharmacokinetic differences between children and adults in assessing children's risks to environmental toxicants. Toxicol Appl Pharmacol 2004; 198:164–183.
20. Gow PJ, Ghabrial H, Smallwood RA, et al. Neonatal hepatic drug elimination. Pharmacol Toxicol 2001; 88:3–15.
21. McCarver DG, Hines RN. The ontogeny of human drug-metabolizing enzymes: phase II conjugation enzymes and regulatory mechanisms. J Pharmacol Exp Ther 2002; 300:361–366.
22. Bruckner JV. Differences in sensitivity of children and adults in chemical toxicity. Regul Toxicol Pharmacol 2000; 31:280–285.
23. U.S. EPA. Guidance Manual for the Integrated Exposure Uptake Biokinetic Model for Lead in Children, EPA/540/R-93/081. PB93–963510, U.S. Environmental Protection Agency, Washington, D.C., 1994.
24. NRC (National Research Council), Pesticides in the Diets of Infants and Children, National Academy Press, Washington, D.C., 1993.
25. Bowers TS, Cohen JT. Blood lead slope factor models for adults: comparisons of observations and predictions. Environ Health Perspect 1998; 106(suppl 6): 1569–1576.
26. Teichberg S, Isolauri E, Wapnir RA, et al. Development of the neonatal rat small intestinal barrier to nonspecific macromolecular absorption: effect of early weaning to artificial diets. Pediatr Res 1990; 28:31–37.
27. Lecce JG. Selective absorption of macromolecules into intestinal epithelium and blood by neonatal mice. J Nutr 1972; 102:69–75.
28. Catassi C, Bouncci A, Coppa GV, et al. Intestinal permeability changes during the first month: effect of natural versus artificial feeding. J Pediatr Gastroenterol Nutr 1995; 21:383–486.
29. Vukavic T. Timing of the gut closure. Pediatr Gastroenterol Nutr 1984; 3: 700–703.
30. Boucher FD, Modlin JF, Weller S, et al. Phase I evaluation of zidovudine administered to infants exposed at birth to the human immunodeficiency virus. J Pediat 1993; 122:137–144.
31. Walsh CT. The influence of age on the gastrointestinal absorption of mercuric chloride and methyl mercury chloride in the rat. Environ Res 1982; 27: 412–420.
32. Kostial K, Kello D, Jugo S, et al. Influence of age on metal metabolism and toxicity. Environ Health Perspect 1978; 25:81–86.
33. ATSDR (Agency for Toxic Substances and Disease Registry), Toxicological Profile for Lead, U.S. Department of Health and Human Services, Public Health Service, Atlanta, GA, 1999.
34. Heubi JE, Balistreri WF, Suchy FJ. Bile salt metabolism in the first year of life. J Lab Clin Med 1982; 100:127–136.

35. U.S. EPA. Dermal Exposure Assessment: Principles and Applications, EPA/600/8-91/011B, U.S. Environmental Protection Agency, Washington, D.C., 1992. Available at http://nepis.epa.gov/pubtitleORD.htm.
36. Wester RC, Maibach HF, Surinchak J, et al. Predictability of in vitro diffusion systems. Effect of skin types and ages on percutaneous absorption of trichlorban. In: Bronaugh RL, Maibach HI eds. Percutaneous Absorption. New York, NY: Marcel Dekker 1985:223–226.
37. Bonina FP, Montenegro L, Micali G, et al. In vitro percutaneous absorption evaluation of phenobarbital through hairless mouse, adult and premature human skin. Int J Pharmacol 1993; 98:93–99.
38. Barrett DA, Rutter N. Percutaneous lignocaine absorption in newborn infants. Arch Dis Child 1994; 71:F122–F124.
39. Harpin VA, Rutter N. Barrier properties of the newborn infant's skin. J Pediat 1983; 102:419–425.
40. Barker N, Hadgraft J, Rutter N. Skin permeability in the newborn. J Invest Dermatol 1987; 88:409–411.
41. Sarangapani R, Gentry PR, Covington TR, et al. Evaluation of the potential impact of age- and gender-specific lung morphology and ventilation rate on the dosimetry of vapors. Inhal Toxicol 2003; 15:987–1016.
42. Ginsberg G, Foos BP, Firestone MP. Review and analysis of inhalation dosimetry methods for application to children's risk assessment. J Toxicol Environ Health A 2005; 68:573–615.
43. Lerman J, Gregory GA, Wills MM, Eger EI. Age and solubility of volatile anesthetics in blood. Anesthesiology 1984; 61:139–143.
44. Kimbell JS, Subramaniam RP. Use of computational fluid dynamic models for dosimetry of inhaled gases in the nasal passages. Inhal Toxicol 2001; 13:325–334.
45. Morselli PL, Franco-Morselli R, Bossi L. Clinical pharmacokinetics in newborns and infants. Age-related differences and therapeutic implications. Clin Pharmacokinet 1980; 5:485–527.
46. Gibbs JP, Murray G, Risler L, et al. Age-dependent tetrahydrothiophenium ion formation in young children and adults receiving high dose busulfan. Cancer Res 1997; 57:5509–5516.
47. Murry DJ, Crom WR, Reddick WE, et al. Liver volume as a determinant of drug clearance in children and adolescents. Drug Metab Dispos 1995; 23:1110–1116.
48. Blanco JG, Harrison, PL, Evans WE, Relling MV. Human cytochrome P450 maximal activities in pediatric versus adult liver. Drug Metab Dispos 2000; 28:379–382.
49. Saunders NR, Knott GW, Dziegielewska KM. Barriers in the immature brain. Cell Mol Neurobiol 2000; 20:29–40.
50. Templin MV, Stevens DK, Stenner RD, et al. Factors affecting species differences in the kinetics of metabolites of trichloroethylene. J Toxicol Environ Health A 1995; 44:435–447.
51. Zemlickis D, Klein J, Moselliy G, Koren G. Cisplatin protein binding in pregnancy and the neonatal period. Med Pediatr Oncol 1994; 23:476–479.
52. Kakiuchi Y, Kohda Y, Miyabe M, Momose Y. Effect of plasma alpha1-acid glycoprotein concentration on the accumulation of lidocaine metabolites during continuous epidural anaesthesia in infants and children. Int J Clin Pharm Th 1999; 37:493–498.

53. Illsley NP, Hall S, Penfold P, Stacey TE. Diffusional permeability of the human placenta. Contrib Gynecol Obst 1985; 13:92–97.
54. Waddell WJ, Marlowe C. Transfer of drugs across the placenta. Pharmacol Ther 1981; 14:375–390.
55. Young JF, Branham WS, Sheehan DM, et al. Physiological "constants" for PBPK models for pregnancy. J Toxicol Environ Health A 1997; 52:385–401.
56. Clewell HJ III, Gearhart JM, Gentry PR, et al. Evaluation of the uncertainty in an oral reference dose for methylmercury due to interindividual variability in pharmacokinetics. Risk Anal 1999; 19:547–558.
57. Krishnan K, Andersen ME. Physiological pharmacokinetic models in the risk assessment of developmental neurotoxicants. In: Slikker W, Chang L, eds. Handbook of Developmental Neurotoxicity. New York, NY: Academic Press, 1998:709–725.
58. Aylward LL, Hays SM, Lakind JS, Ryan JJ. Rapid communication: partitioning of persistent lipophilic compounds, including dioxins, between human milk lipid and blood lipid: an initial assessment. J Toxicol Environ Health A 2003; 66:1–5.
59. Byczkowski JZ, Kinkead ER, Leahy HF, et al. Computer simulation of the lactational transfer of tetrachloroethylene in rats using a physiologically-based model. Toxicol Appl Pharmacol 1994; 125:228–236.
60. Hakkola J, Tanaka E, Pelkonen O. Developmental expression of cytochrome P-450 enzymes in human liver. Pharmacol Toxicol 1998; 82:209–217.
61. Tanaka E. Clinically important pharmacokinetic drug-drug interactions: role of cytochrome P450 enzymes. J Clin Pharmacol Ther 1998; 23:403–416.
62. Treyluyer JM, Cheron G, Sonnier M, Cresteil T. Cytochrome P450 expression in sudden infant death syndrome. Biochem Pharmacol 1996; 52:497–504.
63. Vest MF. The development of conjugation mechanisms and drug toxicity in the newborn. Biol Neonate 1965; 8:258–266.
64. Mulhall A, deLouvois J, Hurley R. Chloramphenicol toxicity in neonates: its incidence and prevention. Br Med J 1983; 287:1424–1427.
65. Gregus Z, Klaassen CD. Hepatic disposition of xenobiotics during prenatal and postnatal development. In: Polin RA, Fox F, eds. Fetal and Neonatal Physiology. Philadelphia, PA: Saunders, 1998:1492–1493.
66. Miller RP, Roberts RJ, Fischer LJ. Acetaminophen elimination kinetics in neonates, children, and adults. Clin Pharmacol Ther 1976; 19:284–294.
67. Levy G, Khanna NN, Soda DM, et al. Pharmacokinetics of acetaminophen in the human neonate: formation of acetaminophen glucuronide and sulfate in relation to plasma bilirubin concentration and D-glucaric acid excretion. Pediatrics 1975; 55:818–825.
68. Szorady I, Santa A, Veress I. Drug acetylator phenotypes in newborn infants. Biol Res Preg 1987; 8:23–25.
69. Pariente-Khayat A, Pons G, Rey E, Richard M-O. Caffeine acetylator phenotyping during maturation in infants. Pediatr Res 1991; 29:492–495.
70. Strange RC, Howie AF, Hume R, Matharoo B. The developmental expression of alpha-, mu- and pi-class glutathione S-transferases in human liver. Biochim Biophys Acta 1989; 993:186–190.
71. Nemeth I, Boda D. Blood gluathione redox ratio as a parameter of oxidative stress in premature infants with IRDS. Free Radical Bio Med 1994; 16:347–353.

72. Papp A, Nemeth I, Karg E, Papp E. Glutathione status in retinopathy of prematurity. Free Radical Bio Med 1999; 27:738–743.
73. Chantry CJ, Rodriguez JL, Febo I, et al. Plasma glutathione concentrations in non-infected infants born from HIV-infected mothers: developmental profile. Puerto Rican Health Sci J 1999; 18:267–272.
74. Jerina DM. Metabolism of aromatic hydrocarbons by the cytochrome P-450 system and epoxide hydrolase. Drug Metab Dispos 1983; 11:1–4.
75. Ratanasavanh D, Beaune P, Morel F, et al. Intralobular distribution and quantitation of cytochrome P-450 enzymes in human liver as a function of age. Hepatology 1991; 13:1142–1151.
76. Eichelbaum M, Tomson T, Tybring G, Bertilsson L. Carbamazepine metabolism in man induction and pharmacogenetic aspects. Clin Pharmacokinet 1985; 10:80–90.
77. Kreuzer P, Csanady GyA, Baur C, Kessler W. 2,3,7,8-Tetrachlorodibenzo-p-dioxin (TCDD) and congeners in infants. A pharmacokinetic model of human lifetime body burden by TCDD with special emphasis on its uptake by nutrition. Arch Toxicol 1997; 71:382–400.
78. Lorber M, Phillips L. Infant exposure to dioxin-like compounds in breast milk. Environ Health Perspect 2002; 110:A325–332.
79. Kedderis GL. Extrapolation of in vitro enzyme induction data to humans in vivo. Chem Biol Interact 1997; 107:109–121.
80. Andersen ME, Clewell HJ III, Gargas ML, et al. Physiologically-based pharmacokinetics and risk assessment process for methylene chloride. Toxicol Appl Pharmacol 1987; 87:185–205.
81. Bois FY, Zeise L, Tozer TN. Precision and sensitivity of pharmacokinetic models of cancer risk assessment: tetrachloroethylene in mice, rats, and humans. Toxicol Appl Pharmacol 1990; 102:300–315.
82. Rao HV, Ginsberg GL. A physiologically-based pharmacokinetic model assessment of methyl t-butyl ether in groundwater for a bathing and showering determination. Risk Anal 1997; 17:583–597.
83. Hattis D, White P, Koch P. Uncertainties in pharmacokinetic modeling for perchloroethylene: II. Comparison of model predictions with data for a variety of different parameters. Risk Anal 1993; 13:599–610.
84. Price P, Chaisson C, Conolly RB, Young JS. A Project to Model Inter-and Intra-Individual Variation in Physiological Factors Used in PBPK Models of Humans, LINEA, Inc., Portland, ME, 2002.
85. Walker K, Hattis D, Russ A, Ginsberg G. Physiologically-Based Pharmacokinetic Modeling for Acrylamide—Risk Implications of Polymorphisms and Developmental Changes in Selected Metabolic Enzymes, Report from the George Perkins Marsh Institute, Clark University, and the Connecticut Department of Public Health to the U.S. Environmental Protection Agency under Cooperative Agreement #827195-0, 2004.
86. Ginsberg G, Hattis D, Miller R, et al. Pediatric Pharmacokinetic Data: Implications for Environmental Risk Assessment for Children. Pediatrics 2004; 113: 973–983.

12

Sensitive Populations and Risk Assessment

Michael Dourson and Daniel Drinan
Toxicology Excellence for Risk Assessment (TERA), Cincinnati, Ohio, U.S.A.

INTRODUCTION

Health Canada (1) and the U.S. Environmental Protection Agency (EPA) [e.g., U.S. EPA (2–7)] and others have developed guidance, which provides some limited advise on identifying the sensitive subpopulation for human health risk assessment. The International Programme on Chemical Safety (IPCS) (8) has also provided some limited guidance. Adoption of these latter guidelines in particular might lead to more logically constructed arguments and documentation of the nature of sensitivity (or susceptibility), as well as the fraction of the population identified as sensitive. This chapter introduces both general principles to protecting the sensitive individual, including the use of two specific uncertainty factors (UFs), and mechanistic toxicology and epidemiology information as the basis for identifying a sensitive subpopulation.

GENERAL PRINCIPLES (9)

"Safe" or subthreshold doses are defined by a number of health and regulatory agencies worldwide, each of which considers effects in sensitive populations. A variety of different names are used for these subthreshold doses, such as: Health Canada's tolerable daily intake or concentration (1); IPCS's tolerable intake (TI) (8,10); U.S. Agency for Toxic Substances and Disease Registry's minimum risk level (11); U.S. EPA's reference dose (RfD) (5,12) or reference concentration (RfC) (3,5,13,14); or the World Health Organization's acceptable daily intake (15,16). In general, many of the underlying assumptions, judgments of critical effect,[a] and choices of UFs (or safety factors) are similar among health and regulatory agencies in estimating these subthreshold doses.

The basic approach involves the application of "safety" or UFs to the no-observed-adverse-effect level (NOAEL) or lowest-observed-adverse-effect level (LOAEL) for the critical effect as discussed in the hazard identification. Several good reviews of this area are available [e.g., Dourson et al. (17), Kalberlah and Schneider (18), and U.S. EPA (5)].

The basic equation to determine the subthreshold doses is:

$$\text{subthreshold doses} = \text{critical effect level}/\text{UF} \tag{1}$$

A brief description of commonly used UFs and their bases is shown in Table 1. These factors are considered to be necessary reductions in the exposure level, based on scientific judgments of available toxicity, toxicodynamic and toxicokinetic data, and inherent uncertainty. Although not all health organizations apply these factors as discrete divisors, most groups consider uncertainties in the following areas: inter-human, experimental animal to human, and less than ideal data, such as extrapolation from less than chronic results to lifetime, extrapolation from a LOAEL rather than a NOAEL, and estimation of a subthreshold exposure based on an incomplete set of studies, including those that test experimental animals of different ages. Moreover, two of these areas, inter-human variability and interspecies extrapolation (UF_H and UF_A), are the current focus of intense study using toxicokinetics and toxicodynamic data as replacements of default UFs. These data-informed choices are referred to as chemical-specific adjustment factors (CSAFs) (8,10,22).

Two of these other areas, that of inter-human variability and an incomplete set of studies (i.e., UF_H and UF_D), specifically account for uncertainty regarding sensitive populations. For example, the area of inter-human variability generally uses a default UF of 10 to account for the variability in response between the population mean and sensitive subgroups [e.g., IPCS (8,10)]. This assumes that there is variability in response from one human to the next and that this variability may not have been detected in the study. This factor also assumes that subpopulations of humans exist which are more sensitive to the toxicity of the chemical than the average population. Individual susceptibility depends on both toxicokinetic and toxicodynamic mechanisms, and these mechanisms may be classified into

---

[a] The critical effect is the first adverse effect, or its known (and immediate) precursor, that occurs as dose rate increases (5,9).

Table 1 Description of Typical UF and MF in Deriving RfDs or RfCs

Standard UF	*General guidelines*[a]
H (inter-human)	Use a 10-fold factor as a default when extrapolating from valid experimental results from studies using prolonged exposure to average healthy humans. This factor is intended to account for the variation in sensitivity among the members of the human population
A (laboratory animal to human)	For RfDs, use a 10-fold factor as a default when extrapolating from valid results of long-term studies on experimental animals when results of studies of human exposure are not available or are inadequate. For RfCs, this factor is reduced to three-fold when a NOAEL (HEC) is used as the basis of the estimate. In either case, this factor is intended to account for the uncertainty in extrapolating animal data to humans
S (subchronic to chronic)	Use a 10-fold factor as a default when extrapolating from less than chronic results on experimental animals or humans. This factor is intended to account for the uncertainty in extrapolating from less than chronic NOAELs to chronic NOAELs
L (LOAEL to NOAEL)	Use a 10-fold factor as a default when deriving an RfD or RfC from a LOAEL, instead of a NOAEL. This factor is intended to account for the uncertainty in extrapolating from LOAELs to NOAELs.
D (incomplete database to complete)	Use a 10-fold factor as a default when extrapolating from valid results in experimental animals when the data is "incomplete." This factor is intended to account for the inability of any single study to adequately address all possible adverse outcomes
MF	Use professional judgment to determine an additional UF termed as a MF that is greater than zero and less than or equal to 10. The magnitude of the MF depends upon the professional assessment of scientific uncertainties of the study and database not explicitly treated above (e.g., the number of animals tested). The default value for the MF is 1. Please note that the MF is being phased out by the U.S. EPA (5).

[a]Professional judgment is required to determine the appropriate value to be used for any given UF; (see text). The values listed in this table are default values that are frequently used by the U.S. EPA in the absence of data.
Note: The maximum UF used with the minimum confidence database for an RfD is 10,000; for an RfC it is 3000.
Abbreviations: NOAEL, no-observed-adverse-effect level; MF, modifying factor; UFs, uncertainty factors; RfDs, reference doses; RfCs, reference concentrations; HEC, high equivalent concentration; LOAEL, lowest-observed-adverse-effect level.
Source: From Refs. 12–14 and 19–21.

three types: factors that increase the concentration of active substance; factors that augment the reaction of the active chemical with the target tissue; and factors that promote the sequence of events between the initial reaction and final adverse effect (23). While this factor seeks to provide protection for sensitive members of the population, the IPCS (10) specifically states that "idiosyncratic hypersusceptibility (excessive reaction following exposure to a given dose of a substance compared with the large majority of those exposed to the same dose) in a few individuals would not be the basis for the derivation of the TI..."

UNCERTAINTY FACTORS THAT ACCOUNT FOR SENSITIVITY[b]

As more fully described by the U.S. EPA (5), dose–response analysis for potentially susceptible subpopulations is often done as part of the overall dose–response analysis for health effects for any particular chemical. Life stages may include the developing individual before and after birth up to maturity (e.g., embryo, fetus, young child, and adolescent), adults, or aging individuals. Other susceptible subpopulations may include people with specific genetic polymorphisms that render them more vulnerable to a specific agent or people with specific diseases or preexisting conditions (e.g., asthmatics). The term may also refer to gender differences, lifestyle choices, or nutritional state.

It is important to recognize that little basis currently exists for a priori identification of susceptible subpopulations for many chemicals. Although the evaluation of effects in various segments of the experimental animals or human population, such as those mentioned above, can identify susceptible subpopulations for a particular chemical, consideration of the chemical's effects and mode of action may suggest specific susceptible populations. For example, people with preexisting kidney or lung disease would have less functional reserve, and would be expected to be more susceptible to chemicals that target the kidney or lung, respectively. The use of mechanistic toxicology to identify susceptible populations is discussed later.

As can be seen in Table 2, several issues must be considered in assessing the potential for some subpopulations, including different life stages, to have greater susceptibility than others. These include the timing (life stage)–response relationship, indicating greater susceptibility to exposure at some life stages than at others; whether effects are of a different type in identifiable subgroups of the population; and the dose–response relationship; that is, whether effects are observed at different levels of exposure in different subpopulations. Another important consideration is whether effects are observed at the same dose but with a shorter latency in different subpopulations. Additionally, differences among groups in terms of the seriousness and reversibility of effects must be considered.

For example, an agent may produce relatively mild and reversible neurological effects in adults, but produce permanent behavioral impairment following in utero exposure at the same tissue dose. It is also important to keep

[b] Please note that some of this text is taken from the U.S. EPA (5) and Dourson et al. (24).

Table 2 Factors for Evaluating Evidence Regarding Identification and Characterization of Susceptible Subpopulations[a]

Factor	Increased weight	Decreased weight
Timing (life stage)–response relationship	Effects occur at greater magnitude at one or more life stage(s)	No difference in effects at different life stage(s)
Type of effect	Different types of effects in specific subpopulations	Same effect(s) across all potential subpopulations
Dose–response relationship	Effect occurs at lower exposures in one or more subpopulation(s)	No evidence for differential dose–response across different subpopulations
Latency of effect	Latency to observed effect different in specific subpopulations	No difference between subpopulations in latency to effect
Seriousness/reversibility of effects	Effects different in seriousness or degree of reversibility in specific subpopulations and/or differences in later consequence of an initially reversible effect	No differences between subpopulations in seriousness and/or reversibility of effects, or in later consequences of an initially reversible effect

[a]From the U.S. EPA. Subpopulations may be defined by gender, individuals at different life stages (fetus, child, adult, and elderly), differences in genetic polymorphisms, and/or preexisting diseases or conditions that may result in differential sensitivity to adverse effects from exposure to a specific toxic agent.
Source: From Ref. 5.

in mind that effects that may initially appear to be reversible may reappear later or be predictive of later adverse outcomes. This is probably best exemplified by certain outcomes following a developmental exposure; for example, an initial depression in birth weight or weight gain or subtle developmental retardation may be indicators of more serious abnormalities later in life.

Because of this a priori lack of identification of susceptible subpopulations for many chemicals, a number of scientists have investigated whether a 10-fold factor accurately accounts for the variability between the average and sensitive human in response to chemicals [e.g., Dourson and Stara (19), Hattis et al. (25), Kaplan et al. (26), Sheehan and Gaylor (27), Calabrese et al. (28), Calabrese and Gilbert (29), Hattis and Silver (30), and Renwick and Lazarus (31)]. The results of these studies differ somewhat, in part due to the incorrect supposition in some of these analyses that the 10-fold factor was to account for the total range of human variability; rather, this UF is intended to account for the range in variability from the population average to the sensitive human, as mentioned above. In fact, analyses by Renwick and Lazarus (31) and Dourson et al. (24) show that the default value of 10 for inter-human variability appears to be protective for most chemicals when starting from a median response, or by inference, from a NOAEL assumed to be from an average group of humans. The former authors demonstrate this from the perspective of toxicodynamics and toxicokinetic uncertainties. The latter authors show this by way of general principles. The original analysis by Hattis and Lynch (this volume) evaluated variability and uncertainty relative to the adequacy of the default UF value of 10 to address the distribution of data on human interindividual variability from therapeutic compounds. However, in this instance, additional constraints were placed on the "protective" nature of the UF.

For example, Dourson et al. (24) show that UF_H applied to the NOAEL or benchmark dose (BMD) does not reflect the complete distribution of human sensitivities, but rather reflects the range of sensitivities expected between the lower range of a normal distribution in the overall population and the sensitive subgroup. This is because UF_H is commonly applied to the NOAEL or BMD estimated for humans. Human NOAELs or BMDs are usually projected from those observed in laboratory animals by dividing by another UF, UF_A (not further discussed here), which is meant to account for differences in sensitivity between species (Fig. 1A). This projected NOAEL or BMD for humans is expected to reflect the rate of response in the lower range of a normal distribution of human responses, because this is what the NOAEL or BMD reflect in the animal study (Fig. 1B). In some cases, the NOAEL and BMD can be measured or estimated from human studies. If so, some assurance is needed that the NOAEL and BMD are not derived from a subpopulation of resistant individuals. In that case, the NOAEL and BMD might not reflect the rate of response in the lower range of a normal distribution of human responses, by definition.

Figure 2A shows a trimodal distribution composed of sensitive, average, and resistant humans.[c] When interpreted properly and starting from a NOAEL or BMD of the average group of humans, UF_H accounts for overall variability in the human population of much greater than 10-fold, perhaps between 100- and 1000-fold or more. For example, see Figure 2B where variability spans approximately three orders of magnitude.

Such appropriate interpretation also allows modification of UF_H when NOAELs are available for a known sensitive or resistant human subgroup or if human toxicokinetics or toxicodynamics are known with some certainty. In such cases, this default value of 10 for UF_H should be adjusted (either increased or decreased) or replaced accordingly. Health groups around the world do this on occasion.

---

[c] Note here that the NOAEL or BMD appears to be less than the 10% mark shown more clearly in Figure 1B. However, the NOAEL in Figure 2A is the same as Figure 1B. In the case of Figure 2A, two populations must be added to obtain the 10% response, that of the sensitive and average humans.

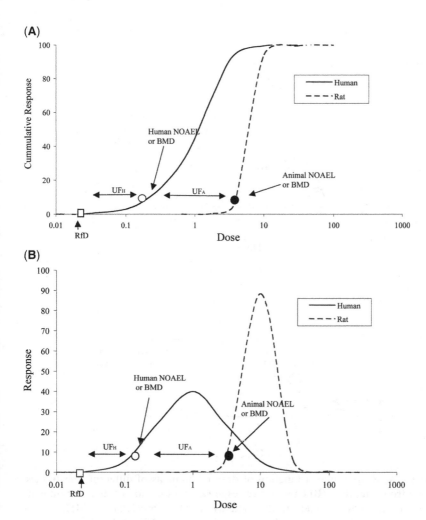

Figure 1 (**A**) Cumulative response as a function of dose for humans and rats. Data are hypothetical, but approximate real situations. (**B**) Response as a function of dose for humans and rats. Hypothetical data are the same as in (**A**). *Abbreviations*: BMD, benchmark dose; NOAEL, no-observed-adverse-effects level; RfD, reference dose, UF, uncertainty factor.

As discussed more extensively by Dourson et al. (24), another of the commonly used UFs also addresses sensitive subgroups, that of an incomplete set of studies, or UF$_D$. Often risk assessors conduct a dose–response assessment in the absence of data from multiple studies. This creates the question as to whether data from another species, or data from different types of bioassays, including those conducted in differently aged experimental animals

(A)

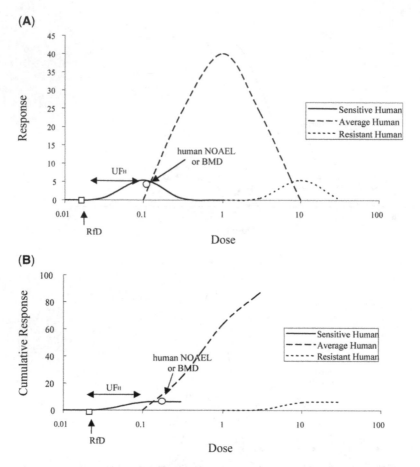

Figure 2 (**A**) Response as a function of dose for humans of different sensitivities. Data are from Figure 1. (**B**) Cumulative response as a function of dose for humans of different sensitivities. Data are the same as in (**A**). *Abbreviations*: BMD, benchmark dose; NOAEL, no-observed-adverse-effects level; RfD, reference dose, UF, uncertainty factor.

(such as reproductive or developmental toxicity studies) would yield lower NOAELs. A common way to address this uncertainty is by dividing by an additional UF, based on the assumption that the critical effect can be discovered in a reasonably small selection of toxicity studies.

Dourson et al. (32) analyzed this assumption through a study of frequency histograms of NOAEL ratios for chronic dog, mouse, and rat studies, and for reproductive and developmental toxicity studies in rats. On average, chronic rat and dog studies yielded similar NOAELs; reproductive and developmental toxicity studies were less likely to have the lowest

NOAELs, although the lowest values were sometimes from these studies. Chronic mouse studies occasionally resulted in the determination of a critical effect. These authors concluded that several bioassays are needed in order to develop a high confidence estimate of a subthreshold dose, and that if one or more bioassays are missing, including bioassays on younger experimental animals, then a factor should be used to address this scientific uncertainty.

Although, the value of this factor was not specified, the U.S. EPA later used the results of this study to place values of 3- or 10-fold for this UF depending on the availability of studies (33). Evans and Baird (34) later addressed this factor more quantitatively through the use of regression and nonparametric analyses. These investigators concluded that the value of UF_D depended on the definition of a complete database. This factor was also the focus of a workshop held by Toxicology Excellence for Risk Assessment in September 2005 (35), where the broader implications of missing reproductive or developmental toxicity studies were analyzed by comparing various types of in vivo studies for more than 150 diverse chemicals. The distribution of various effect level ratios (e.g., NOAEL) was shown and used to calculate the probability that current default database UF values adequately account for missing reproduction or developmental toxicity studies.

UF_D also has relevance for issues surrounding the Food Quality Protection Act (FQPA) safety factor, or at least its toxicity component. The U.S. EPA (36) discusses this issue at length, showing that UF_D and the toxicity component of the FQPA overlap completely. The U.S. EPA (36) concludes that when the UF_D has been used properly, the need for the FQPA safety factor is reduced. Fenner-Crisp (37) and Dourson et al. (24) also discuss this comparison and make similar conclusions.

MECHANISTIC INFORMATION AND SENSITIVE POPULATIONS[d]

In risk assessment, high-quality human toxicology and epidemiology data that focus on NOAEL information are optimum and can be used to directly calculate an RfD or RfC, with animal studies providing supporting evidence. High-quality human studies remove the necessity of interspecies extrapolation, which must be performed when only experimental animal studies exist. However, if less than ideal human NOAEL data are all that can be found, such data can still be used to inform the choice of a UF based in part on an understanding of sensitive populations and the extrapolation of data from experimental animals. These factors would be more accurate

---

[d] Please note that portions of this text are taken from Dourson et al. (38)

than the usual 10-fold defaults, using guidelines of the IPCS (8) for CSAFs, for example. Human data can also be used that do not directly focus on NOAEL information. These data, often gathered after accidental exposures, can provide insight into mechanisms of toxicity from acute exposures to chemicals tested only on laboratory animals. In addition, this type of information can identify human-specific effects not seen in experimental studies using different species.

Thus far, the types of human data used in risk assessment have ranged from case studies to well-defined epidemiological studies. Epidemiological information is collected from multiple sources, including surveillance, public health statistics, geographic correlation studies, and the cohort and case–control studies designed to identify associations and sometimes to support inferences about cause and effect. For example, much epidemiological information is gathered from case studies of accidental exposures at work, "unusual natural contaminants, or as the aftermath of an accident such as an explosion, or industrial release." That being said, deliberate human studies also take place. Two examples of this are in the testing of environmental and pharmaceutical chemicals, both through a well-established ethical process, where volunteers are exposed to small amounts of a substance. When humans are exposed to contaminants in their everyday environment, the focus of risk research is to identify exposures that do not evoke "biochemical and/or physiological changes." Pharmaceutical testing, on the other hand, is designed to identify doses of compounds that provide the most benefit with the fewest side effects.

Sensitive populations can sometimes be identified from epidemiology studies and understanding of toxicokinetics and toxicodynamics as described briefly above, or by an understanding of the underlying mechanism(s) of toxicity. Often these concepts are intertwined. As examples of how risk assessment scientists have used these phenomena in their determination of dose–response, Table 3 describes why reduced UFs were appropriate for several chemicals found in the U.S. EPA's IRIS (33) and elsewhere.

For example, a one-fold UF was suggested for fluorine, nitrate,[e] and nitrite when relevant studies provided a NOAEL for the critical effect in a sensitive population, such as infants or children. A one-fold factor was also used for benzoic acid and manganese because large populations have been exposed and homeostatic processes might be involved. A three-fold UF was suggested for arsenic, barium, perchlorate, and selenium because

(Text continues on page 264.)

---

[e] Haber et al. (41) further described the use of mechanistic data within the sensitive infant population by the U.S. EPA to derive, in essence, a CSAF and reference dose (RfD) for nitrate. In this case, a full understanding of the mechanism of nitrate toxicity in the infants led to a high confidence RfD that was equivalent to the NOAEL of the critical effect in the sensitive subgroup (33). Thus, an overall uncertainty factor of 1 was sufficient.

Table 3 Reduced UF_H Based on Understanding of Mechanism, Toxicokinetics, and/or Toxicodynamics

Chemical	UF_H	Reason for reduced UF_H
Acrylonitrile 107-13-1	5.8	The 5.8-fold CSAF for within human variability was reduced from the usual 10-fold due to specific data on toxicokinetics. The overall UF was 180 (3.2 for interspecies extrapolation, 5.8 for intraspecies variation, and 10 to extrapolate from subchronic to chronic duration) (39).
Arsenic, inorganic 7440-38-2	3	The UF of 3 was to account for both the lack of data to preclude reproductive toxicity as a critical effect and to account for some uncertainty in whether the NOAEL of the critical study accounts for all sensitive individuals. A value of 10 was not used because the epidemiology study was sufficiently large and varied so that some sensitive individuals judged were included
Barium 7440-39-3	3	A UF 3 was used to account for differences between adults and children and for lack of an adequate developmental toxicity study. A value of 10 was not used because the epidemiology study was judged to include sensitive individuals
Benzoic acid 65-85-0	1	A UF of 10 for the protection of sensitive subgroups was considered unnecessary; although reactions to benzoate and structurally related compounds do occur, a UF of 10 would be of little value to the sensitive individuals, because the RfD is based on estimated intake of benzoic acid and benzoate as food preservatives
Boron 7440-42-8	6.3	This CSAF was based on known variability in clearance of 2.0 between pregnant women and their nonpregnant counterparts, and a default value of 3.16 for toxicodynamics
Fluorine (soluble fluoride) 7782-41-4	1	UFs were not deemed necessary because the NOAEL is that of the critical effect

(Continued)

Table 3 Reduced UF_H Based on Understanding of Mechanism, Toxicokinetics, and/or Toxicodynamics (*Continued*)

Chemical	UF_H	Reason for reduced UF_H
		(i.e., dental fluorosis) in a sensitive population of humans (i.e., children) for a length of exposure that encompasses both the critical effect and the sensitive population
Manganese 7439-96-5	1	The information used to determine the RfD for manganese was taken from many large populations consuming normal diets over an extended period of time with no adverse health effects. As long as physiologic systems are not overwhelmed, humans exert an efficient homeostatic control over manganese such that body burdens are kept constant with variation in the manganese content of the diet. The information providing a chronic NOAEL in many cross-sections of human populations, taken in conjunction with the essentiality of manganese, warrants a UF of 1
Nitrate 14797-55-8	1	A UF of 1 was employed because available data define the no-observed-adverse-effect level for the critical toxic effect (methemoglobinemia) in the most sensitive human subpopulation (infants)
Nitrite (1) 14797-65-0	1	No UF was used in the derivation of the RfD because the NOAEL was of the critical toxic effect (i.e., methemoglobinemia) in the sensitive human population (i.e., infants). The length of exposure encompassed both the critical effect and the sensitive population
Perchlorate	3	Data were sufficient to estimate an overall UF of 3-fold with this NOAEL based on expected differences in toxicokinetics and toxicodynamics between children a known sensitive subgroup, and pregnant women and their fetuses, a

(Continued)

Table 3 Reduced UF$_H$ Based on Understanding of Mechanism, Toxicokinetics, and/or Toxicodynamics (*Continued*)

Chemical	UF$_H$	Reason for reduced UF$_H$
		second identified sensitive subgroup for perchlorate, and concerns about the overiodination of this population. From ref. 40.
Selenium and compounds 7782-49-2	3	A UF of 3 was applied to the NOAEL to account for sensitive individuals. A full factor of 10 was not deemed necessary because similar NOAELs were identified in two moderately sized human populations exposed to selenium levels in excess of the RDA throughout a lifetime so that some sensitive individuals judged were included
Silver 7440-22-4	3	A UF of 3 was applied to account for minimal effects in a sensitive subpopulation that has exhibited an increased propensity for the development of argyria. The critical effect observed is a cosmetic effect, with no associated adverse health effects. Also, the critical study reports on only one individual who developed argyria following an intravenous dose of 1 g silver (4 g silver arsphenamine). Other individuals did not respond until levels five times higher were administered. No additional UF for less than chronic to chronic duration is needed because the dose has been apportioned over a lifetime of 70 yrs
Zinc and compounds 7440-66-6	3	A UF of 3 was used, based on a minimal LOAEL from a moderate-duration study of the most sensitive humans and consideration of a substance that is an essential dietary nutrient

Note: Data are from the U.S. EPA's IRIS unless otherwise stated.
Abbreviations: CSAFs, chemical specific adjustment factors; UF, uncertainty factor, NOAEL, no-observed-adverse-effects level; RFD, reference dose; RDA, recommended daily allowance; LOAEL, lowest-observed-adverse-effect level.
Source: From Refs. 33, 39 and 40.

populations were studied that likely included sensitive individuals, and/or relevant mechanistic data were missing a piece of pertinent information, yet enough mechanistic data, and/or toxicokinetic and toxicodynamic knowledge was available to warrant a reduction from the usual default 10-fold UF. In cases such as acrylonitrile, boron, silver, or zinc, specific data on toxicokinetics and toxicodynamics or severity of critical effect allowed the development of factors other than default, sometimes through the use of CSAFs.

SUMMARY

Risk assessment is a scientific endeavor that is designed to protect the human population, including sensitive subgroups, from adverse effects of chemical exposures. In general, sensitive human subgroups are different from their normal counterparts by variation in toxicokinetics and toxicodynamics. While some of these variances may predispose individuals to risk from one chemical, they may also be protective of risk from other chemicals. A UF is commonly used to address this within human variability (UF_H); another UF, database incompleteness (UF_D), is designed to address the inability of any single study to adequately test for all outcomes in differently aged groups.

Variation in the critical effect among human populations may be due to differences in a chemical's life-stage impact, dose–response relationship, latency, and/or seriousness and reversibility. Toxicokinetics and toxicodynamics vary during human growth and development, and toxicokinetics is variable in humans of different ages and health status. Mechanisms and dynamics of toxic effect are sometimes different among human groups, leading some of these groups to be more sensitive than others.

Distinctions among variations in critical effect, toxicokinetics and toxicodynamics, and/or mechanism of toxic effect are often difficult to make, and in fact, based on future better understandings of the underlying biology, such distinctions may not really exist. A better understanding of this area will lead to the replacement of default values for UFs with actual data, and the resulting risk assessments will better protect public health, while at the same time be more precise.

REFERENCES

1. Meek ME, Newhook R, Liteplo RG, et al. Approach to assessment of risk to human health for priority substances under the Canadian Environmental Protection Act. Environ Carcinogen Ecotoxicol Rev 1994; C12:105–134.
2. U.S. EPA, Guidelines for Developmental Toxicity Risk Assessment. EPA/600/FR-91/001. Fed Reg 1991; 56:63798.
3. U.S. EPA, Methods for Derivation of Inhalation Reference Concentrations and Application of Inhalation Dosimetry, EPA/600/8-90/066F, U.S. Environmental

Protection Agency, Office of Research and Development, Washington, DC, 1994. Available at http://nepis.epa.gov/pubtitleORD.htm.

4. U.S. EPA, Guidelines for Reproductive Toxicity Risk Assessment. Review Draft, EPA/600/AP-94/001, U.S. Environmental Protection Agency, Office of Research and Development, Washington, DC, 1994.

5. U.S. EPA, A Review of the Reference Dose and Reference Concentration Processes, EPA/630/P-02/002F, U.S. Environmental Protection Agency, Risk Assessment Forum, Washington, DC, 2002. Available at http://cfpub.epa.gov/ncea/cfm/recordisplay.cfm?deid=55365.

6. U.S. EPA, Guidelines for Carcinogen Risk Assessment, EPA/630/P-03/001B, U.S. Environmental Protection Agency, Washington, DC, 2005. Available at http://www.epa.gov/iris/cancer032505.pdf.

7. U.S. EPA, Supplemental Guidance for Assessing Cancer Susceptibility from Early-life Exposure to Carcinogens, EPA/630/R-03/003F, U.S. Environmental Protection Agency, Risk Assessment Forum, Washington, DC, 2005. Available at http://www.epa.gov/iris/children032505.pdf.

8. IPCS (International Programme on Chemical Safety), Guidance Document for the Use of Data in Development of Chemical-specific Adjustment Factors (CSAF) for Interspecies Differences and Human Variability in Dose/concentration Response Assessment, World Health Organization, Geneva, 2001. Available at www.who.int/entity/ipcs/publications/methods/harmonization/en/csafs_guidance_doc.pdf.

9. Haber LT, et al. Noncancer risk assessment: principles and practice in environmental and occupational settings. In: Bingham E, Cohrssen B, Powell CH, eds. Patty's Toxicology. Vol. 1. 5th ed. New York, NY: John Wiley & Sons, Inc., 2001.

10. IPCS (International Programme on Chemical Safety), Assessing Human Health Risks of Chemicals: Derivation of Guidance Values for Health-based Exposure Limits, Environmental Health Criteria 170, International Programme on Chemical Safety, World Health Organization, Geneva, Switzerland, 1994. Available at www.inchem.org/documents/ehc/ehc/ehc170.htm.

11. Pohl HR, Abadin HG. Utilizing uncertainty factors in minimal risk levels derivation. Regul Toxicol Pharmacol 1995; 22:180–188.

12. Barnes DG, Dourson ML. Reference dose (RfD): description and use in health risk assessments. Regul Toxicol Pharmacol 1988; 8:471–486.

13. Jarabek AM. The application of dosimetry models to identify key processes and parameters for default dose-response assessment approaches. Toxicol Lett 1995; 79:171–184.

14. Jarabek AM. Interspecies extrapolation based on mechanistic determinants of chemical disposition. J Human Ecol Risk Assess 1995; 1:641–642.

15. Lu F. Safety assessments of chemicals with thresholded effects. Regul Toxicol Pharmacol 1985; 5:460–464.

16. Lu F. Acceptable daily intake: inception, evolution, and application. Regul Toxicol Pharmacol 1988; 8:45–60.

17. Dourson ML, Felter SP, Robinson D. Evolution of science-based uncertainty factors in noncancer risk assessment. Regul Toxicol Pharmacol 1996; 24:108–120.

18. Kalberlah F, Schneider K. Quantification of Extrapolation Factors, Fb 797, Dortmund, Berlin, 1998.

19. Dourson ML, Stara JF. Regulatory history and experimental support of uncertainty (safety) factors. Regul Toxicol Pharmacol 1983; 3:224–238.
20. Jarabek AM. Inhalation RfC methodology: dosimetric adjustments and dose-response estimation of non-cancer toxicity in the upper respiratory tract. Inhal Toxicol 1994; 6:301–325.
21. Dourson ML. Methods for establishing oral reference doses (RfDs). In: Mertz W, Abernathy CO, Olin SS, eds. Risk Assessment of Essential Elements. Washington, DC: ILSI Press, 1994:51–61.
22. Meek ME, RenwickA, Chanian E, et al. Guidelines for application of compound specific adjustment factors (CSAF) in dose/concentration response assessment. Comm Toxicol 2001; 7:575–590.
23. Grandjean P. Individual susceptibility to toxicity. Toxicol Lett 1992; 64/65:43–51.
24. Dourson M, Charnley G, Scheuplein R. Differential sensitivity of children and adults to chemical toxicity: II. Risk and regulation. Regul Toxicol Pharmacol 2002; 35:448–467.
25. Hattis D, Erdreich L, Ballew M. Human variability in susceptibility to toxic chemicals–a preliminary analysis of pharmacokinetic data from normal volunteers. Risk Anal 1987; 7:415–426.
26. Kaplan N, Hoel D, Portier C, et al. An evaluation of the safety factor approach in risk assessment. In: Branbury Report 26: Developmental Toxicology: Mechanisms and Risk, Branbury Reports, Cold Spring Harbor, 1987.
27. Sheehan DM, Gaylor DW. Analysis of the adequacy of safety factors. Teratology 1990; 41:590–591.
28. Calabrese EJ, Beck BD, Chappell WR. Does the animal to human uncertainty factor incorporate interspecies differences in surface area? Regul Toxicol Pharmacol 1992; 15:172–179.
29. Calabrese EJ, Gilbert CE. Lack of total independence of uncertainty factors (UFs): implications for the size of the total uncertainty factor. Regul Toxicol Pharmacol 1993; 17:44–51.
30. Hattis D, Silver K. Human interindividual variability—a major source of uncertainty in assessing risks for noncancer health effects. Risk Anal 1994; 14:421–431.
31. Renwick AG, Lazarus NR. Human variability and noncancer risk assessment—an analysis of the default uncertainty factor. Regul Toxicol Pharmacol 1998; 27:3–20.
32. Dourson ML, Knauf LA, Swartout JC. On reference dose (RfD) and its underlying toxicity data base. Toxicol Ind Health 1992; 8:171–189.
33. U.S. EPA, Integrated Risk Information System (IRIS), U.S. Environmental Protection Agency, National Center for Environmental Assessment, Washington, DC, 2005. Available at http://www.epa.gov/iris.
34. Evans JS, Baird SJS. Accounting for missing data in noncancer risk assessment. Human Ecol Risk Assess 1998; 4:291–317.
35. TERA (Toxicology Excellence for Risk Assessment), Peer Consultation on the Scientific Rationale for Approaches to Derive Database and Toxicodynamic Uncertainty Factors to Protect Children's Health, 2005. Available at http://www.tera.org/peer/UFD/UFDWelcome.htm.
36. U.S. EPA, Determination of the Appropriate FQPA Safety Factor(s) in Tolerance Assessment, U.S. Environmental Protection Agency, Office of Pesticide

Programs, Washington, DC, 2002. Available at http://www.epa.gov/pesticides/trac/science/determ.pdf.

37. Fenner-Crisp P. The FQPA 10x safety factor—How much is science? How much is sociology? Human Ecol Risk Assess 2001; 7:107–116.
38. Dourson ML, Andersen ME, Erdreich LS, et al. Using human data to protect the public's health. Regul Toxicol Pharmacol 2001; 33:234–256.
39. The Sapphire Group I. Toxicological Review of Acrylonitrile (CAS No. 107–13-1). Final, Prepared for The Acrylonitrile Group, Washington, DC, 2004. Available at http://www.tera.org/peer/AN/ANToxicologicalReviewDocument-revised.pdf.
40. Strawson J, Zhao Q, Dourson M. Reference dose for perchlorate based on thyroid hormone change in pregnant women as the critical effect. Regul Toxicol Pharmacol 2004; 39:44.
41. Haber LT, Maier A, Zhao Q, et al. Applications of mechanistic data in risk assessment–the past, present, and future. Toxicol Sci 2001; 61:32–39.

13

Statistical Issues in Physiologically Based Pharmacokinetic Modeling

Weihsueh A. Chiu

National Center for Environmental Assessment, Office of Research and Development, U.S. Environmental Protection Agency, Washington, D.C., U.S.A.

INTRODUCTION

Physiologically based pharmacokinetic (PBPK) models, like all computational models, are simplified representations of a more complex system. As reviewed elsewhere in this text (e.g., Chapters 6, 7, and 10), their utility lies in their ability to make predictions concerning situations or populations for which there are no direct in vivo data on internal dose. In particular, PBPK models are designed to determine the relationship between external exposure to one or more chemicals and internal measures of biologically relevant dose, and may be used to extrapolate across species, routes, levels, or patterns of exposure, individuals, and life stages. As PBPK modeling has matured from primarily a research endeavor to one focused on application to risk assessment, it has become increasingly important to characterize the uncertainty and variability in their predictions. However, rigorous statistical characterization of uncertainty and variability has not yet been routinely applied, as PBPK models and the data that underlie them present a number of challenges to traditional statistical inference. This chapter presents a brief introduction to application of statistical methods to PBPK models, with a particular focus on the concepts of identifiability and sensitivity as they relate to data, PBPK models, and their predictions.

SENSITIVITY, IDENTIFIABILITY, AND THE ROLE OF STATISTICS

As depicted in Figure 1, a **PBPK** model (M), with its parameters θ, is a quantitative means of integrating physiological and pharmacokinetic (e.g., absorption, distribution, metabolism, excretion) data (D), collected under experimental conditions C_D, so as to predict measurements of internal dose (Y) under exposure conditions that are of interest to risk assessment (C_Y). The data D typically include both chemical-independent data (e.g., physiological tissue volumes and blood flows) and chemical-specific data (in vitro partition coefficients, metabolic rates, in vivo time courses of blood, or tissue concentrations). These data provide the constraints as to what model(s) and values for their parameters are appropriate, given the current state of scientific knowledge. The risk assessment-relevant predictions Y, and the exposure conditions C_Y, are determined both by toxicological information about sites and modes of action of toxicity and by human exposure patterns. Ideally, **PBPK** model-based analyses should not only make "point estimate" predictions based on the data, but also characterize the likely *range* of predictions that are consistent with available data.

The relationships among the data, model, and predictions can be decomposed into two general properties: sensitivity and identifiability, as shown in Figure 1. Sensitivity is the relationship between the model and risk assessment-relevant predictions. This has often been quantified through a "local" sensitivity coefficient, in which model parameters are individually varied (e.g., by 1%) from their baseline values and compared with the resulting variations in predictions (1–3). However, such local sensitivity analysis may not be indicative of "global" sensitivity—so that sensitivity may be different with different baseline parameters or with different parameters varied simultaneously. As an example, consider a typical case in which a compound is metabolized in the liver under Michaelis–Menten kinetics. At high doses, metabolism depends strongly on the maximum rate of metabolism V_{max} and only weakly on the affinity parameter K_m, whereas, at low doses, metabolism depends only on their ratio: V_{max}/K_m. Thus, predictions for

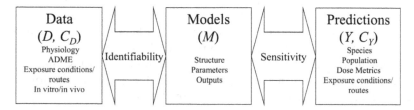

Figure 1 Illustration of the components of developing and using a physiologically based pharmacokinetic model, and their relationships with respect to identifiability and sensitivity.

metabolism at low and high doses have different sensitivities, so sensitivity analyses are often performed at different doses (1–3). In addition, considering only low doses, if V_{max}/K_m is very small, then metabolism will only be weakly dependent on hepatic blood flow, whereas if V_{max}/K_m is very large, then metabolism will be strongly dependent on hepatic blood flow. Thus, depending on the value of V_{max}/K_m, the sensitivity of metabolism to hepatic blood flow will be quite different. While these are relatively simple examples that can be easily checked, the aggregate changes in sensitivities across the entire parameter space may be substantial and are difficult to analyze without a more global approach (4,5).

Identifiability, on the other hand, is the relationship between the data and the model, the key question being: are there sufficient data to *identify* the structure of the model and all its parameters? Ideally, one wants a model that is neither so simple that it leaves out important biological information nor so complex that there are insufficient data to constrain all its parameters. In the simple example of Michaelis–Menten metabolism in the liver, if all the in vivo data are at low doses so that metabolism is linear, then the maximum rate of metabolism V_{max} and the affinity parameter K_m are not *separately* identifiable—only the ratio of V_{max}/K_m ("intrinsic clearance") can be estimated. In addition, if V_{max}/K_m is in actuality very large, then the in vivo data may be able to identify only a lower bound on even their ratio, because metabolism depends most strongly in that case only on hepatic blood flow.

In order to characterize the uncertainty in model predictions, both sensitivity and identifiability need to be considered together. For instance, a high sensitivity to a well-identified parameter may be less important for characterizing uncertainty than a low or moderate sensitivity to a parameter that is relatively poorly identified. Because of the nonlinearity of PBPK models, their relatively large number of parameters, and the complex array of data necessary to consider together, a rigorous, transparent process is needed to ensure that the model, its parameters, the resulting predictions, and their uncertainties are accurately estimated. However, typical approaches to PBPK modeling, which have been less than satisfying in this regard for characterizing uncertainty, may take the following steps:

1. Fix physiological parameters at values found in the literature.
2. Fix partition coefficients at in vitro values.
3. Optimize metabolic parameters (by least squares) based on in vivo data (e.g., gas uptake studies).
4. Occasionally, adjust parameters, such as alveolar ventilation rate, cardiac output, and partition coefficients, to achieve better model fit.
5. Use same parameter values for making model predictions, perhaps with some local sensitivity analysis.

Such an approach leads to some discomfort for a number of reasons:

- The physiological, in vitro, and in vivo data are typically all taken from different individuals. The above approach assumes that all these individuals are identical, thus ignoring interindividual variability, and that the measurements have no error.
- Inaccurate selection of the "fixed" parameters (steps 1 and 2) will propagate to the optimized ones (step 3).
- The selection of what parameters to fix and what to optimize is somewhat ad hoc.
- No rigorous presentation of the uncertainty in model predictions can be derived.

In reality, all the parameters of the model have some degree of uncertainty and variability, and all the data have some degree of error, and the practice of statistical inference offers, in principle, a rigorous theoretical framework for accounting for them (6). With respect to model parameters, these can all be summarized in a "likelihood function:"

$$P(D|\theta M), \tag{1}$$

the probability of observing the data D given the model M with parameters θ. This likelihood function is the basis for both classical (frequentist) and Bayesian approaches to statistical inference. In particular, under a frequentist approach, maximum likelihood may be used to derive the "best-fit" parameters and their confidence limits (either approximate or exact). With Bayesian approaches, prior distributions for model parameters are defined based on previously collected data or knowledge, and these distributions are "updated" with the data at hand to yield posterior distributions for model parameters (7,8). In either case, population variability can be incorporated through a "hierarchical" structure in which parameters for an individual are drawn from a sampling distribution that defines the population characteristics, such as a normal distribution with a population mean and standard deviation.

The main advantage of a statistical framework is the requirement to be completely explicit about the assumptions underlying the analysis. As an example, the "nonstatistical" approach above can be understood as a set of particular assumptions within the statistical framework, as follows. First, partition the data D and parameters θ each into three groups: physiological data D_{ph} and parameters θ_{ph}, partition coefficient data D_{pc} and parameters θ_{pc}, and in vivo time-course data D_{iv} with data points $y_{iv}(t)$ and metabolic parameters θ_{met}. The "nonstatistical" approach above amounts to assuming the following:

1. Physiological parameters fixed at $\theta_{ph}^* = D_{ph}$.
2. Partition coefficients fixed at $\theta_{pc}^* = D_{pc}$.

3. Find the "optimized" values of parameters θ_{met} as those values θ^*_{met} by maximizing the log likelihood, assumed to be the sum over in vivo time courses of the squared deviations between the in vivo data y_{iv} and the model predictions $y_{M,\theta}$: $\ln[P(D_{iv}|\theta_{met}M)] = \Sigma_{y,t}$ $[y_{iv}(t) - y_{M,\theta^*,\theta_{met}}(t)]^2$.
4. If the fit is not adequate, manually adjust one or more of θ^*_{ph} or θ^*_{pc}.
5. Use fixed parameters for prediction, perhaps with sensitivity analysis using one-by-one variation of each parameter.

A statistical approach with more realistic assumptions may be that all individual parameters θ have a lognormal (LN) sampling distribution,[a] with population mean μ and standard deviation σ. The measured physiological and partition coefficients have independent data errors with standard deviations ε_{ph} and ε_{pc}, respectively, and the in vivo data are assumed to be lognormally distributed around the model prediction, with error standard deviation ε_{iv}. The approach would be to specify and use the likelihood function as follows:

1. Physiological parameters lognormally distributed $D_{ph} \sim LN\{\mu_{ph}, (\sigma^2_{ph} + \varepsilon^2_{ph})^{\frac{1}{2}}\}$.
2. Partition coefficients lognormally distributed $D_{pc} \sim LN\{\mu_{pc},(\sigma^2_{pc} + \varepsilon^2_{pc})^{\frac{1}{2}}\}$.
3. In vivo data lognormally distributed $y_{iv}(t) \sim LN\{y_{M,\theta}(t), \varepsilon_{iv}\}$, with each θ drawn from its sampling distribution $\theta \sim LN(\mu, \sigma)$.
4. Joint likelihood of model parameters can be used to assess "global" sensitivity of model predictions.

This approach is computationally intensive due to the large number of parameters and data, although a number of established methods, such as nonlinear mixed effects models and Markov chain Monte Carlo integration, can be used to perform these analyses (6,7,12). Moreover, while the statistical approach outlined above certainly depicts a more realistic picture system as a whole, this does not *necessarily* mean that the large number of existing PBPK modeling studies using ad hoc approaches are completely invalid. The extent to which the use of a point estimate versus the full distribution matters will necessarily be case-specific, depending on the data at hand, the structure of the model, and the desired model outputs. The examples below provide some illustrative case studies.

---

[a] Lognormal distributions are often used because variability distributions of many biological and pharmacokinetic parameters appear positively skewed (9,10), but other distributions could be used where appropriate (11).

274   *Chiu*

CASE STUDIES

Example 1: Human Route-to-Route Extrapolation for Vinyl Chloride

Route-to-route extrapolation provides a simple illustrative case of when the results may not be sensitive to statistical estimation methods. Chiu and White (13) show that in a prototypical PBPK model for a volatile organic chemical (VOC), the relationship between oral "dose" (mg/kg-day) and inhalation exposure concentration "C" (mg/L) giving the same internal dose can be summarized as in Table 1. The equivalence substantially depends only on four parameters: the alveolar ventilation rate Q_P (L/day), the hepatic blood flow Q_l (L/day), the blood–air partition coefficient P_B, and the first-order metabolism coefficient k_{met} (note that route-to-route extrapolation is typically done at low exposures where metabolism is first order).

In the case of vinyl chloride, the U.S. EPA's health risk assessment made extensive use of the Clewell et al. (14,15) PBPK model. The model parameter estimation method for this model was typical of the "fix some, estimate others" methodology described above [U.S. EPA (16), Appendix B].

The physiological parameters are the current EPA reference values (U.S. EPA, 1988), except for alveolar ventilation in the human, which was calculated from the standard EPA value for the ventilation rate in the human, 20 m^3/day, assuming a 33% pulmonary dead space. The partition coefficients for Fischer-344 (F344) rats were taken from Gargas et al. (1989), and those for Sprague-Dawley rats were taken from Barton et al. (1995). The Sprague-Dawley values were also used for modeling of Wistar rats. Blood/air partition coefficients for the other species were obtained from Gargas et al. (1989), and the corresponding tissue/blood partition coefficients were estimated by dividing the Sprague-Dawley rat tissue/air partition coefficients by the appropriate blood/air value.

The affinity for the 2E1 pathway (KM1) in the rat, mouse, and hamster was set to 0.1 on the basis of studies of the competitive interactions between CYP2E1 substrates in the rat (Barton et al., 1995; Andersen et al., 1987b). The affinity used for the non-2E1 pathway (KM2) in the mouse and rat was set during the iterative fitting of the rat total metabolism, glutathione depletion, and rate of metabolism data, described below. The capacity parameters for the two oxidative pathways (VMAX1C and VMAX2C) in the mouse, rat, and hamster were estimated by fitting the model to data from closed-chamber exposures with each of the species and strains of interest (Barton et al., 1995; Bolt et al., 1977; Clement, 1990; Gargas et al., 1990). After the other parameters were scaled from animal weights obtained from individual studies, the model was exercised for

Table 1 Steady State Route-to-Route Relationship for Different Dose Metrics Derived from a Typical Volatile Chemical PBPK Model Metabolized in the Liver

Case	Route-to-route relationship	Dose metric equivalence	
I	$C = \text{Dose} \times \text{BW} \times Q_P^{-1} \times \{1 + Q_P/(Q_l P_B)\}$	Rate of metabolism, parent concentration in liver, any metabolite concentration[a]	
II	$C = \text{Dose} \times \text{BW} \times Q_P^{-1} \times 1$	Parent concentration in arterial blood, nonhepatic tissues[b]	Fully saturated metabolism
III	$C = \text{Dose} \times \text{BW} \times Q_P^{-1} \times (1 + k_{met}/Q_l)^{-1}$	Parent concentration in arterial blood, nonhepatic tissues[b]	Linear metabolism

[a]Note that paradoxically, route-to-route equivalence for these dose metric are independent of metabolism parameters. This is because at steady state they depend only on equivalent delivery to the liver between routes—if the routes provide the same input to the liver, then the same amount of metabolism, etc. will occur. In particular, for the oral route, there is 100% delivery to the liver, whereas for the inhalation route, a fraction of that which is inhaled is cleared by exhalation before systemic circulation. Also, the actual concentrations of any metabolite will depend on its own pharmacokinetic properties (and PBPK submodel, if available), but these are independent of the exposure route of the parent compound, so do not impact the route-to-route equivalence.
[b]Note that the actual concentrations in nonmetabolizing tissues depend also on the tissue:blood partition coefficients, but these are independent of the exposure route of the parent compound, so do not impact the route-to-route equivalence.

Abbreviations: PBPK, physiologically base pharmacokinetic; C, concentration; Dose, mass per unit time per unit body weight; BW, body weight; Q_P, alveolar ventilation rate; Q_l, hepatic blood flow rate; P_B, blood–air partition coefficient; k_{met}, effective first-order metabolic clearance rate.
Source: From Ref. 13.

optimization to a single pair of values, VMAX1C and VMAX2C, to be used for all of the data on a given sex/strain/species.

The application of the PBPK model of interest here was the route-to-route extrapolation between the human reference dose and reference concentration (the PBPK model was also used for species extrapolation, but this application is not discussed here). In particular, an oral human equivalent dose (HED) and an inhalation human equivalent concentration (HEC) were both derived from an oral rat study no-observed-adverse-effect level (NOAEL) using the PBPK model. The HEC was therefore essentially a human route extrapolation of the HED. Because the dose metric of interest was the total metabolites produced per day per unit liver volume, the appropriate route-to-route equivalence corresponds to "Case I" in Table 1:

$$HEC = HED \times BW \times \{Q_P^{-1} + (Q_l P_B)^{-1}\} \qquad (2)$$

Chiu and White (13) showed that this formula reproduces the exact value derived from the full PBPK model. Importantly, for the purposes of risk assessment predictions (rather than fitting to in vivo data), values for the physiological parameters are generally taken directly from the literature rather than derived from optimized (e.g., study-specific) values. Moreover, because the RfD and RfC have uncertainty factors for accounting for human variability, so the route-to-route extrapolation need only account for uncertainty.

It is instructive to compare the results of Clewell et al. (14,15) with updated data on the parameters of interest. As shown in Table 2, these values put together imply an HEC at the point of departure of 3.1 mg/m^3, with an uncertainty geometric standard deviation of 1.24 (calculated by Monte Carlo simulation, in which 1000 values of each parameter were sampled according to their assumed distribution). This is coincidentally exactly 1.24-fold greater than the value from the Clewell et al. (14,15) model of 2.5 mg/m^3. Therefore, the point estimate value reported in the vinyl chloride assessment, using the Clewell et al. (14,15) PBPK model was consistent with a value derived using a more comprehensive characterization of uncertainty. Importantly, two of the parameters contributing the most to the uncertainty—body weight and cardiac output—are completely *independent* of the chemical itself. In addition, these are somewhat a matter of "scenario definition," i.e., the reference values are calculated based on a specific "default" or "reference" individual with a particular body weight, ventilation rate, etc.

More generally, this example shows a case of limited sensitivity—only a few of the model parameters had *any* impact on the route-to-route prediction—and high identifiability—the parameters of interest were all fully identified by the data. In such cases, a statistical approach to characterizing uncertainty may not be necessary to achieve accurate results.

Table 2 Parameters and Predictions for Route-to-Route Extrapolation Using the Vinyl Chloride Physiologically Based Pharmacokinetic Model

Parameter/ prediction	Central estimate for males (geometric mean)	Uncertainty in central estimate (geometric standard deviation)	Values based on Clewell et al. (14,15)
BW	83 kg[a]	–	70 kg
Q_P	7.8 L/min[b]	1.06	9.7 L/min
Q_C	6.6 L/min[c]	1.21	6.7
Q_l/Q_C	0.26[d]	1.05	0.26
P_B	1.16[e]	1.2	1.16
HEC/HED (calculated)	53 kg/L-min	1.24	43 kg/L-min
HEC at point of departure[f]	3.1 mg/m^3	1.24	2.5 mg/m^3

[a]For body weight, Price et al. (2003) reports that for males, National Health and Nutrition Examination Survey (NHANES) III showed a mean body weight of 83 kg.
[b]The EPA Exposure Factors Handbook (U.S. EPA 1997) contains an extensive review of inhalation rates (which, assuming 30% pulmonary dead space, can be converted to alveolar ventilation rate by multiplying by 0.7), and recommend a typical long-term value of 15.2 m^3/day for males. Price et al. used that review in combination with NHANES III data to estimate a mean of 17 m^3/day for males. Values used are geometric mean and geometric standard deviation of these two values converted to alveolar ventilation in L/min.
[c]Average cardiac output itself is not well-estimated. Price et al. (2003) used two different approaches and derived values of 5.45 and 8.0 L/min for males. Values used are geometric mean and geometric standard deviation of these two values.
[d]U.S. EPA (2006) recently collected from the published literature 59 individual measurements of hepatic blood flow at rest as a fraction of cardiac output. Values used are geometric mean and geometric standard error of the mean.
[e]For the blood–air partition coefficient, in vitro measurements (in this case 1.16 from Gargas et al. 1989) are generally thought to be reliable, with measurement error and interindividual variability thought to be no more than around 20%.
[f]HED at the point of departure (the NOAEL in a rat feeding study) was 0.085 mg/kg-day, but value reported in U.S. EPA (2000) was rounded to 0.09 mg/kg-day.
Abbreviations: Q_P, alveolar ventilation rate; Q_C, cardiac output; Q_l/Q_C, fraction of cardiac output as hepatic blood flow; P_B, blood–air partition coefficient; HEC, human equivalent air concentration; HED, human equivalent dose.
Source: From Refs. 16–20.

However, conducting a statistically rigorous analysis in such cases is likely to be fairly straightforward and pose a limited computational burden.

Example 2: Uncertainty in Amount of Perchloroethylene Metabolized at Low Exposures

Perchloroethylene (PERC) is a dry-cleaning solvent and a common environmental contaminant. Because it is presumed that metabolites, rather than the parent compound, are responsible for PERC toxicity, a number of groups have developed human PBPK models to estimate the amount of PERC metabolized by the body. Most of these are point estimates developed using the somewhat "ad hoc" procedures described above, and these have produced widely varying estimates for the percent of inhaled PERC that is metabolized at low doses, ranging from 1.7% to 23% (bars without error bars in Fig. 2). Because of the ad hoc nature of the different estimation procedures, and the fact that all the human exposure data analyzed were >10 ppm (predominantly > 50 ppm), it is difficult to discern whether one should prefer one or another estimate. From the discussion above, it would appear that rigorous statistical analysis could help to resolve these discrepancies, and determine the extent to which these differences are artifacts of the ad hoc estimation process or a reflection of true uncertainty in the data.

Two analyses, those by Bois et al. (12) and Chiu and Bois (21) (bars with error bars in Fig. 2), performed such formal statistical analyses, in particular using Bayesian population statistical modeling. The following explicit assumptions were made in these analyses [Chiu and Bois (21) is actually an update to Bois et al. (12)]:

- All model parameters have truncated lognormal population distributions.
- All population means had uncertainties described by truncated lognormal population distributions.

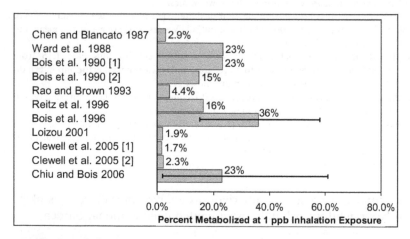

Figure 2 Predictions for the amount metabolized at 1 ppb perchloroethylene inhalation exposure from eight published analyses. All estimates are point estimates except for Bois et al. and Chiu and Bois (actually an update of the former), which include uncertainty and population variability (error bars are 95% confidence intervals). *Source*: From Refs. 12, 21–28.

- All population variances had scaled inverse-chi-square distributions.
- Parent compound in exhaled breath and blood from one study (29), comprising six individuals, was analyzed.

The results of the original analysis (12) were somewhat paradoxical in that the predictions seemed somewhat high compared to previous analyses—thus, instead of leading to a better understanding, they just added one more to the set of apparently mutually inconsistent analyses. However, the reanalysis by Chiu and Bois (21) concluded that these earlier results had not reached an appropriate level of convergence. Moreover, Chiu and Bois (21) show that when the full range of uncertainty and variability is taken into account, their analysis implies 2% to 61% metabolized, a range consistent with *all* previous point estimates. Their conclusion, then, is that the amount metabolized at low doses is not strongly constrained by the data at hand.

This conclusion is actually quite easy to understand intuitively. The only data available on PERC in human are PERC in exhaled (or alveolar) air, PERC in venous blood, and trichloroacetic acid (TCA) in blood and urine. Where available, data on PERC in exhaled air indicates recovery of 80% to 100%, depending on assumptions about the ventilation rate, implying a percent metabolized of 0% to 20% (29). Therefore, small differences in exhaled breath predictions have a large impact on the inferred metabolized dose. In fact, Chiu and Bois (21) reported the estimated percent metabolized at 50 ppm to be 3%, with 95% confidence interval from 0.45% to 26%, very similar to the range based on the empirically reported recovery. In addition, postexposure PERC venous blood concentrations reported by Monster et al. (29) increased slightly more than two-fold with a two-fold increase in exposure concentration from 72 to 144 ppm, suggestive of slightly less clearance and thus consistent with partial saturation of metabolism. The uncertainties in the observed range are therefore magnified by possible non-linearity in extrapolating to low doses. This is exemplified by the much wider confidence interval of 2% to 61% reported by Chiu and Bois (21) for the percent metabolized at 1 ppb exposures. With respect to TCA, while some TCA data are available at as low as 10 ppm (30), ameliorating somewhat the low-dose extrapolation issue, the fraction of total metabolism to TCA is unknown and likely to be dose-dependent. Therefore, TCA recovery (typically on the order of a percent or two) only provides a lower bound on the amount metabolized. Thus, it is not surprising that a wide range of different PBPK model assumptions, leading to very different predictions as to the amount metabolized at low dose, can yield similarly acceptable consistency with the in vivo data available.

In this case, some of the key parameters to which the predictions are sensitive—namely the metabolism parameters V_{max} and K_m—are precisely those that are poorly identified by the data.

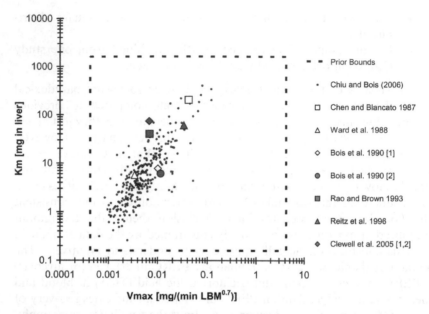

Figure 3 V_{max} and K_m values from eight published analyses of perchloroethylene pharmacokinetics. All parameters were converted to the same units as those in Chiu and Bois [note in particular that the unit for K_m (mg in liver) used by Chiu and Bois is not the same as that typically used in physiologically based pharmacokinetic model models (mg/L in venous blood leaving liver)]. All analyses are point estimates except for Chiu and Bois, which included a Bayesian analysis of uncertainty and variability. Points shown for Chiu and Bois are 300 random samples of the population means for V_{max} and K_m (i.e., reflecting uncertainty in the population means); the scatter would be greater if population variability were also included (see Fig. 2). Also included are the bounds on the prior distributions used in that analysis. *Source*: From Refs. 21–28.

Such a case is one in which PBPK model-based point estimates based on typical ad hoc methods for parameter estimation may be *particularly* misleading because they do not capture the range of uncertainty in a prediction of interest to risk assessment. These conclusions are illustrated in Figure 3, which shows 300 random samples from the Chiu and Bois (21) analysis depicting the uncertainty in the population mean values of V_{max} and K_m, along with the published V_{max} and K_m values for the point estimates in Figure 2 [units converted to match those of Chiu and Bois (21)]. Thus, the answer to the question posed above as to whether differences in reported point estimates are artifacts or a reflection of true uncertainty in the data appears clear: all the point estimates are within the envelope of the uncertainty derived from Chiu and Bois (21). The utility of using a rigorous statistical methodology, then, is to quantify these uncertainties in a transparent and reproducible manner. The analysis of Chiu and Bois (21) is one step toward that goal, but is limited in that it considers only a single in vivo data set. In addition, the apparent lack of convergence of the original Bois et al. (12) results points to the need for careful implementation and interpretation of these results. For more complex models and data sets, current computing algorithms and processing power may actually be insufficient to perform a "complete" analysis of all data and parameters in a realistic time-frame. In such cases, it may be preferable to limit the scope of the analysis, for instance, by analyzing a tractable subset of parameters or data sets.

CONCLUSIONS

With the maturation of PBPK modeling from primarily a research tool to one focused on risk assessment applications, characterization of the sensitivity, uncertainty, and variability in these models has become increasingly important. However, while the application of rigorous statistical methods has as yet been limited, as reliance on PBPK models increases, so too will scrutiny of the assumptions and data underlying their development. In some cases, such as the route-to-route extrapolation example above, the use of "ad hoc" methods may not be highly inaccurate. However, in other cases, only application of statistical methods can adequately characterize the range of PBPK model predictions consistent with available data. Furthermore, a statistical framework provides a means by which the critical assumptions and analyses in the development of a PBPK model can be made explicit and transparent, a key requirement for regulatory decisions based on model predictions. Finally, statistical analyses can help to direct research toward critical data that will have the greatest impact on reducing remaining uncertainties.

REFERENCES

1. Clewell HJ III, Lee TS, Carpenter RL. Sensitivity of physiologically based pharmacokinetic models to variation in model parameters—methylene chloride. Risk Anal 1994; 14:521–531.
2. Evans MV, Crank WD, Yang HM, et al. Applications of sensitivity analysis to a physiologically based pharmacokinetic model for carbon tetrachloride in rats. Toxicol Appl Pharmacol 1994; 128:36–44.
3. Simmons JE, Boyes WK, Bushnell PJ, et al. A physiologically based pharmacokinetic model for trichloroethylene in the male long-evans rat. Toxicol Sci 2002; 69:3–15.
4. Bois FY, Woodruff TJ, Spear RC. Comparison of three physiologically based pharmacokinetic models of benzene disposition. Toxicol Appl Pharmacol 1991; 110:79–88.

5. Spear RC, Bois FY, Woodruff T, et al. Modeling benzene pharmacokinetics across three sets of animal data: parametric sensitivity and risk implications. Risk Anal 1991; 11:641–654.

6. Bernillon P, Bois FY. Statistical issues in toxicokinetic modeling: a bayesian perspective. Environ Health Perspect 2000; 108:883–893.

7. Jonsson F, Bois FY, Johanson G. Assessing the reliability of PBPK models using data from methyl chloride-exposed, non-conjugating human subjects. Arch Toxicol 2001; 75:189–199.

8. Smith TJ, Lin YS, Mezzetti M, et al. Genetic and dietary factors affecting human metabolism of 1,3-butadiene. Chem Biol Interact 2001; 135–136:407–428.

9. Hattis D, Erdreich L, Ballew M. Human variability in susceptibility to toxic chemicals—a preliminary analysis of pharmacokinetic data from normal volunteers. Risk Anal 1987; 7:415–426.

10. Hattis D, Banati P, Goble R, et al. Human interindividual variability in parameters related to health risks. Risk Anal 1999; 19:711–720.

11. Hattis D, Ginsberg G, Sonawane B, et al. Differences in pharmacokinetics between children and adults—II. Children's variability in drug elimination half-lives and in some parameters needed for physiologically-based pharmacokinetic modeling. Risk Anal 2003; 23:117–142.

12. Bois FY, Gelman A, Jiang J, et al. Population toxicokinetics of tetrachloroethylene. Arch Toxicol 1996; 70:347–355.

13. Chiu WA, White P. Steady-state solutions to PBPK models and their applications to risk assessment I: route-to-route extrapolation of volatile chemicals. Risk Anal 2006; 26:1.

14. Clewell HJ III, Covington TR, Crump KS, et al. The Application of a Physiologically Based Pharmacokinetic Model for Vinyl Chloride in a Noncancer Risk Assessment, Prepared by ICF Kaiser/Clement Associates under EPA contract number 68-D2-0129 for the U.S. Environmental Protection Agency, National Center for Environmental Assessment, Washington, DC, 1995.

15. Clewell HJ III, Gentry PR, Gearhart JM, et al. The Development and Validation of a Physiologically Based Pharmacokinetic Model for Vinyl Chloride and its Application in a Carcinogenic Risk Assessment for Vinyl Chloride, Prepared by ICF Kaiser for the U.S. Environmental Protection Agency, Office of Health and Environmental Assessment, and the Occupational Safety and Health Administration, Directorate of Health and Standards Programs, Washington, DC, 1995.

16. U.S. EPA, Toxicological Review of Vinyl Chloride, EPA/635/R-00/004, U.S. Environmental Protection Agency, Integrated Risk Information System, Washington, DC, 2000.

17. Price PS, Conolly RB, Chaisson CF, et al. Modeling interindividual variation in physiological factors used in PBPK models of humans. Crit Rev Toxicol 2003; 33:469–503.

18. U.S. EPA, Exposure Factors Handbook. Volume I: General Factors, EPA/600/P-95/002FA, U.S. Environmental Protection Agency, Office of Research and Development, Washington, DC, 1997.

19. U.S. EPA, Use of Physiologically Based Pharmacokinetic Models to Quantify the Impact of Human Age and Interindividual Differences in Physiology and Biochemistry Pertinent to Risk (Final Report), EPA/600/R-06/014A, U.S.

Environmental Protection Agency, Office of Research and Development, Washington, DC, 2006.

20. Gargas ML, Burgess RJ, Voisard DE, et al. Partition coefficients of low-molecular-weight volatile chemicals in various liquids and tissues. Toxicol Appl Pharmacol 1989; 98:87–99.

21. Chiu WA, Bois FY. Revisiting the population toxicokinetics of tetrachloroethylene. Arch Toxicol 2006; 80:382.

22. Bois FY, Zeise L, Tozer TN. Precision and sensitivity of pharmacokinetic models of cancer risk assessment: tetrachloroethylene in mice, rats, and humans. Toxicol Appl Pharmacol 1990; 102:300–315.

23. Chen CW, Blancato JN. Role of pharmacokinetic modeling in risk assessment: perchloroethylene as an example. In: Pharmacokinetics in Risk Assessment, Drinking Water, and Health. Vol. 8. Washington, DC: National Academy of Sciences, National Research Council, 1987:369–390.

24. Clewell HJ, Gentry PR, Kester JE, et al. Evaluation of physiologically based pharmacokinetic models in risk assessment: an example with perchloroethylene. Crit Rev Toxicol 2005; 35:413–433.

25. Loizou GD. The application of physiologically based pharmacokinetic modelling in the analysis of occupational exposure to perchloroethylene. Toxicol Lett 2001; 124:59–69.

26. Rao HV, Brown DR. A physiologically based pharmacokinetic assessment of tetrachloroethylene in groundwater for a bathing and showering determination. Risk Anal 1993; 13:37–49.

27. Reitz RH, Gargas ML, Mendrala AL, et al. In vivo and in vitro studies of perchloroethylene metabolism for physiologically based pharmacokinetic modeling in rats, mice, and humans. Toxicol Appl Pharmacol 1996; 136:289–306.

28. Ward RC, Travis CC, Hetrick DM, et al. Pharmacokinetics of tetrachloroethylene. Toxicol Appl Pharmacol 1988; 93:108–117.

29. Monster AC, Boersma G, Steenweg H. Kinetics of tetrachloroethylene in volunteers; influence of exposure concentration and work load. Int Arch Occup Environ Health 1979; 42:303–309.

30. Volkel W, Friedewald M, Lederer E, et al. Biotransformation of perchloroethene: dose-dependent excretion of trichloroacetic acid, dichloroacetic acid, and N-acetyl-S-(trichlorovinyl)-L-cysteine in rats and humans after inhalation. Toxicol Appl Pharmacol 1998; 153:20–27.

14

Drug Development and the Use of Pharmacokinetics/Toxicokinetics in Selecting the First Dose of Systemically Administered Drugs in Humans—A Nonclinical Perspective

Abigail Jacobs

Center for Drug Evaluation and Research, Food and Drug Administration, Silver Spring, Maryland, U.S.A.

THE OVERALL DRUG DEVELOPMENT/APPROVAL PROCESS

Drug development is a long and complicated process, involving the contributions and interactions of many persons over a considerable period of time (Fig. 1).

Lead Drug Candidate Selection

During the drug discovery phase of drug development, a drug sponsor may evaluate drug candidates in a variety of in silico, in vitro, or in vivo screens, to decide if there is biologic plausibility that a drug may work, whether it may have unacceptable toxicity (e.g., liver, kidney, or genotoxicity), and whether the oral bioavailability or tissue distribution will be reasonable. Some companies may use biomarkers or toxicogenomics/proteomics in a few animals to explore this.

The views expressed in this document are those of the author, not necessarily of the U.S. FDA.

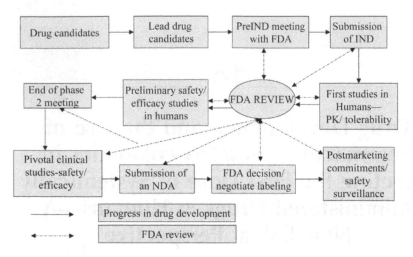

Figure 1 Major steps in the drug development process. *Abbreviations*: IND, investigational new drug; FDA, Food and Drug Administration; NDA, new drug application; PK, pharmacokinetic.

Some companies may choose to use an exploratory investigational new drug (IND) application (1) to help in drug candidate selection. One of the options of the exploratory INDs is the use of a microdose for humans (defined as less than 1/100th of the dose calculated to yield a pharmacological effect of a test substance and a maximum dose of $\leq$100 µg) to explore bioavailability and biodistribution. Advances in bioanalytical techniques of positron emission tomography (2) and accelerator mass spectrometry (3) have enabled the conduct of these microdosing studies. The hope is that the pharmacokinetic (PK) properties at a microdose will scale up to the therapeutic dose. However, findings from microdosing studies may not always scale up to the clinical dose. Examples would include when first-pass metabolism, plasma protein binding, or gastrointestinal transporter mechanisms are saturable at oral therapeutic doses and if dose-dependent kinetics occur within the normal therapeutic range.

Other options for an exploratory IND besides microdosing include study of pharmacologic effects or mode of action studies. The dose and duration allowed in exploratory IND studies is more limited than in traditional first studies in humans and thus the amount of nonclinical (non-human) data needed is also more limited than in traditional first IND studies in humans. However, the amount of PK data needed may be substantial for the microdosing scenario (4). After a lead compound is selected, with or without an exploratory IND, further development of the drug product needs to follow the traditional IND route.

Further Evaluation of a Lead Drug Candidate

After a lead candidate is selected, the studies agreed to and suggested by the international conference on harmonization (ICH) guidances will be planned and/or conducted. These would include local tolerance, genotoxicity, repeat dose toxicity, developmental/reproductive toxicity, chronic toxicity, carcinogenicity, and safety pharmacology [generally adverse pharmacologic effects on the cardiovascular, central nervous system, and respiratory systems, per the U.S. Food and Drug Administration (FDA) (5,6)]. When these studies need to be submitted to the FDA is dependent on the stage of drug development and duration of the proposed clinical protocol per the FDA (7). Although chronic toxicology or carcinogenicity studies may not be needed at the earliest stages of drug development, they may be planned and initiated early in drug development, if the therapeutic indication for the drug is non-life-threatening and chronic, so as not to delay drug approval if clinical studies are not of long duration.

PK/toxicokinetic (TK) studies are often conducted and are desirable for all pivotal studies conducted in animals. One incentive for capturing high-quality quantitative PK data in animals is that these data may help in interpreting toxicology studies and in selection of the amount of the first dose to be given to humans—not so high as to cause irreversible toxicity nor so low that time is wasted in finding tolerable doses. Another incentive is that several of the possible ICH criteria for selection of a high dose for carcinogenicity studies in rodents of nongenotoxic drugs may be based on plasma area under the curve (AUC) values in animals and humans. Among other options, the high dose selected in the rodent species may be a dose that corresponds to a plasma AUC that is at least 25 times (taking into account metabolites and protein binding) the human AUC at the maximum recommended dose (8) or may be a dose at which there is saturation of absorption, rather than the dose being a maximum tolerated dose (MTD). This means that for relatively nontoxic drugs, an enormous amount of drug is not expended for such studies.

Preinvestigational New Drug Meetings

When the drug sponsor has put together its tentative drug development plan, it would generally have a preIND (preIND application) meeting with the center for drug evaluation and research (CDER) FDA to discuss the pharmacology/toxicology profile of the drug, the proposed indication, the general development plan, and any other issues that may have arisen.

Investigational New Drug Submission and Review

The sponsor may choose to submit an IND, which includes a detailed clinical protocol with proposed initial and highest doses to be used, together with supporting data from in vitro studies and from animals, chemistry and manufacturing information, and data on the stability of the drug. The general requirements for an IND submission are described in 21CFR 312.20–312.23. The FDA reviews the proposed clinical protocol with regard to the appropriate number of subjects, whether they should be healthy or patients, the initial dose to be used, the dose escalation plan, the highest dose to be used, the amount and frequency of monitoring, and the criteria for removal of humans from the study. The review time for the initial submission by a sponsor requesting to conduct clinical studies is often as short as two weeks and is always less than 30 days. These earliest clinical studies in humans often involve measurement of PK properties of the drug, studies of metabolic interactions, and dose ranging for tolerability. Guidance for the selection of the first dose for healthy humans is described in a CDER guidance (9). This guidance encapsulates the CDER practice of many years, whereby the first dose for the first time a drug is given to healthy humans is selected. PK and TK data are often very helpful in this endeavor. Before there are PK data from humans, the extrapolation by CDER of drug effects in animals to humans is generally done using interspecies scaling based on body surface area (BSA), and this process is described later in more detail, under the section of "Selection of the First Dose in Humans."

Investigational Studies and Informed Consent

Investigational drug studies are not officially approved by the FDA, but proposed studies may proceed if the FDA does not put the studies on hold. For all studies conducted in humans, persons who will receive the drug (or placebo) sign an informed consent form (21CFR50.20–50.27). The informed consent form describes the potential risks of the drug that are known at the time: initially risks anticipated only from in vitro or animals studies, and later also risks uncovered from studies in humans (clinical studies). An investigational review board reviews the proposed protocol and the informed consent form.

Preliminary Safety and Efficacy Studies in Humans (Phase 2)

If a sponsor decides after the initial studies in humans to proceed with development of a drug product, additional nonclinical (non-human) studies to support new clinical (human) protocols will be submitted to the U.S. FDA, generally according to the timing relative to stage of drug development, as described in the ICH-M3 document (7). Studies in animals of the same duration as the duration proposed for the clinical studies will be needed. Based on what is learned in the first clinical studies and the additional studies in animals, a new protocol is then submitted to the FDA to evaluate the preliminary safety and efficacy of the drug in humans. These studies attempt to determine if there is any hint of efficacy, the usefulness of biomarkers, and information on dose selection, therapeutic end points, and the number of patients that will be needed for the pivotal studies, together with an assessment of the adverse effects seen, generally after multiple dosing.

End of Phase 2

When "Phase 2" studies have been completed and evaluated, an end of phase 2 meeting is held between the drug sponsor and the FDA. At this meeting, agreement is reached on the remainder of the development program regarding what is needed to support a therapeutic claim and to evaluate safety. The reproductive/developmental toxicology studies in animals need to be completed before the pivotal clinical studies are conducted, so that it can be determined if women intending to become pregnant or already pregnant can participate or if women participating may need to use contraceptive methods. Discussion between the FDA and the drug sponsor would include details of the proposed pivotal clinical studies in humans, including the number of study groups, the number of patients and the patient population to be studied, therapeutic end points or biomarkers to be evaluated, and statistical issues. Discussion would also include whether further pharmacology/toxicology studies other than those already planned were needed, as well as required chemistry manufacturing/stability data. The population of humans participating in the pivotal clinical trials needs to be representative of persons who will use the drug if it is approved for marketing. Proposed protocols for the definitive PK evaluations of drugs in patients with the proposed therapeutic indication at the maximum recommended human dose, with the to-be-marketed formulation of the drug, would also be presented. Often, it is important to evaluate the PKs of the drug in persons who have impaired renal or liver function, as well as in persons with normal liver and kidney function.

Pivotal Clinical Studies for Safety and Efficacy (Phase 3)

At some point, according to the plans of the drug sponsor, the detailed clinical protocols for the pivotal studies in humans are then submitted to the FDA for review. At this stage, as well as at each stage of drug development, all information that is known about a drug from in vitro, animal, and human studies is integrated and used to inform on the design and nature of studies that are to be conducted next. In the United States, assessment of male fertility, female fertility, and embryo–fetal development in animals should be completed before women of childbearing potential using birth control are enrolled in Phase 3 trials.

NDA Submission

After the pivotal clinical studies for safety and efficacy have been conducted and evaluated by the drug sponsor, the sponsor may decide to submit a new drug application (NDA) for approval of the drug product for a particular indication and patient population at particular doses and dosing regimens. The general requirements for an NDA submission are given in 21CFR part 314. The submission will include all information known about a drug, including all nonclinical (non-human) and clinical (human) studies conducted (per ICH or requested), as well as all chemistry and manufacturing information. Long-term studies of six months in rats and six-month (for a life-saving therapy), nine-month (for a new molecular entity that is a member of an already existing pharmacologic class), or 12-month studies (for the first drug in its pharmacologic class) in nonrodents will need to be in the NDA if these studies have not already been conducted and reported upon and the condition being treated is chronic. Carcinogenicity studies in two species will generally be needed for drugs for chronic conditions. Proposed product labeling for patients and health care professionals will also be included in the NDA.

The FDA review team (clinicians, statisticians, pharmacologists/ toxicologists, chemists, statisticians, regulatory project managers, drug advertising reviewers) first decides by day 74 after submission of an NDA that the appropriate studies have been conducted and are included in the submission for evaluation and whether it will accept the filing of the NDA. If the NDA submission is considered adequate for review, a review is initiated by all the various disciplines involved in the drug review. The review team generally meets midway in the review cycle (the entire review cycle being six months for drugs designated as priority, and nine months for nonpriority drugs). The various reviewers generally have other review assignments as well for other drug products during this period.

Approval, Phase 4 Studies and Postmarketing Surveillance

Each reviewer and the review team make one of three recommendations: approval, approvable (with a list of information or studies needed for approval), or not approvable. Safety signals to watch postmarketing are identified and the need for restricted distribution or a medication guide is considered. Conduct of postapproval clinical or nonclinical studies may be a condition of approval. If the overall recommendation is for approval, the FDA and the drug sponsor negotiate the wordings of the proposed product labeling/package insert, which describes safe and effective use of the product.

More details on selection of the first dose to be given to humans are described in the following sections.

SELECTION OF THE FIRST DOSE IN HUMANS

When humans have never before been administered a particular drug, the doses causing no toxicity and those causing toxicity need to be estimated from non-human studies. It is no easy task to determine the maximum safe starting dose of drugs in initial clinical studies in humans. The reviewing division at the CDER/FDA must decide whether to concur with proposed doses and monitoring within a few weeks after a drug sponsor has submitted a package of nonclinical studies (in silico, in vitro, or in vivo in non-human animals) and a proposal for clinical studies to be conducted in humans. The selection of the first dose in humans is based on an integrated assessment of the pharmacology, toxicology, and PK/TK data across several species. Human experience with other drugs in the same pharmacologic or chemical class may also be informative. Sometimes the FDA may need to propose lower doses and additional monitoring. In doing so, PK and TK data are of critical importance. (*Note*: In discussing the use of these data, this discussion will not address drugs for oncology or other life-threatening condition indications, which may use a different paradigm, nor will it discuss the dosing of children with drugs approved for adults.)

PHARMACOLOGY AND TOXICOLOGY DATA

Data or information on the drug's proposed pharmacologic and toxicologic activity are provided by the drug sponsor to support the dose and duration of the proposed first study. Such data include clinical signs, macroscopic and microscopic lesions, clinical chemistry and hematology values, and exaggerated pharmacodynamic effects in a rodent and nonrodent species, generally rats and dogs. An ICH core battery of safety pharmacology data (generally adverse pharmacologic effects on physiologic function after a single intravenous dose) is used to assess the cardiovascular, central nervous system, and respiratory systems (5,6). Toxicology and safety pharmacology data are often accompanied by nonclinical PK/TK data, and some detailed absorption, distribution, metabolism, and elimination (ADME) data, but not always. Collection of such data is strongly recommended and generally received, but such data in animals are not strictly required (8,10), and the quality and quantity of such data from animals may vary at the earliest stage of drug development. Collection of PK data in humans is very important. Such data are used throughout drug development to help relate effects in animals to effects in humans, in evaluating possible drug interactions, and in assessing the length of time needed to eliminate the drug from the body. The clinical PK data are results that are described in the final labeling and the packet insert for the product, when it is approved by the FDA for marketing. The drug in plasma AUC values in animals is used to calculate multiples of human exposure when carcinogenicity and reproductive/developmental toxicology studies are described in the final labeling and the packet insert for the product.

The early PK/TK data for a drug might include the following:

1. Human, rat, and dog metabolism in liver slices, in vitro
2. In vivo CYP P450 and other metabolism in the non-human species of interest

3. Absolute bioavailability in animals, if an intravenous formulation of the drug product can be made
4. Serum protein binding for humans and the species of interest
5. Tissue distribution, preferably at two time points
6. The PK parameters of elimination rate, AUC for plasma concentration, maximum concentration (C_{max}), volume of distribution, all after single and multiple doses in the non-human test species of interest

PK data describe the ADME of the drug, whereas TK data describe various measures of systemic exposure in a study whose purpose is to evaluate toxicity.

WITHIN SPECIES EVALUATIONS

As part of the review of a proposed first dose in humans, one must first decide whether the species for which there are toxicology data is an appropriate species and then determine at which doses in the studies adverse effects occur or do not occur [no-observable-adverse-effects levels (NOAELs)], together with an evaluation of reversibility and monitorability of the adverse effects. These decisions can be made in part after assessing whether a plasma steady state has been reached in the study and whether the dose-limiting toxicity correlates with a particular PK/TK parameter. The reviewer initially determines what the dose-limiting toxicity is and its reversibility, whether the C_{max} and AUC values are linear with dose, and then inspects the PK/TK data on other points. These include whether the increase in C_{max} or AUC values is more or less than dose proportional and the volume of distribution and its relationship to plasma volume. If the increase in C_{max} or AUC values is more than dose proportional, saturation of metabolism might be indicated. If this occurs at BSA equivalents of the proposed human doses, one might escalate the doses in humans more slowly. If the increase in C_{max} or AUC values is less than dose proportional, then saturation of absorption is indicated. If the volume of distribution is large, and not limited to the plasma volume, then the drug may be tightly bound somewhere and accumulation may be indicated. The steepness of a response for serious or irreversible/nonmonitorable effects would be noted. This is important because a small change could result in massive toxicity. Tissue distribution studies might indicate accumulation of drug or metabolite in particular tissues.

Of importance is the determination of whether dose-limiting toxicity correlates with time above a threshold, C_{max}, steady state concentration (Css), AUC, or total dose values. All such possible scenarios have consequences when selecting the first dose in humans. The true extent of a drug's toxicity can be missed if the elimination rate in the animals is so rapid that there is essentially no drug in the system after a few hours, if dosing was only once per day. In this case, the animal may be able to recover from the effect completely before receiving the next dose the following day. When the half-life is short, multiple dosing per day (if originally by gavage) may be indicated or a feed study may be indicated so that the animals are continually exposed, just as the humans may be. In the case of one drug, the sponsor decided to split the dose over the course of the day, since the half-life was short. Surprisingly, the toxicity was more severe for the same total dose and occurred earlier when the dose was split into two portions, and even more toxic when the dose was administered four times per day, rather than once per day. In this case, the dose-limiting toxicity correlated with time above a threshold. The agreed-to first dose in humans was a fraction of the dose giving toxicity. Based on measurements of PK in humans after the first dose, subsequent doses in humans were limited by plasma concentrations (which needed to be below a certain value).

When toxicity correlates with total dose, as with irreversible inhibition of essential enzymes or poisoning of mitochondria, a clinical study may need to be limited by total dose (or total plasma AUC) rather than by dose per day. Sometimes such effects may be predicted from the pharmacologic activity of the drug, but not always. A short-term nonclinical or clinical study may not indicate that such a life-threatening toxicity is lurking. Rather, a study in animals of longer duration than the initial study may be needed. If a cardiovascular effect in a safety pharmacology study is associated with a C_{max} value, it may make sense to limit the corresponding first dose in humans to a dose corresponding to a fraction of the C_{max}, rather than the plasma AUC.

CROSS SPECIES COMPARISONS

Comparisons across species allow one to better assess what the first dose should be. Allometric scaling has long been used in estimating corresponding doses of drugs across species (11,12). The human equivalent dose (HED) has conventionally been based on body weights and an allometric exponent and has been calculated as $(HED = \text{animal dose} \times W_{animal}/W_{human})^{(1-b)}$. For mg/m^2 normalization, b would be 0.67, but several studies indicated that the MTD in carcinogenicity studies in rodents scaled best when $b = 0.75$, and $b = 0.75$ was used by the U.S. EPA for extrapolation of doses in carcinogenicity studies (13). The use of $b = 0.75$ has a large effect on the conversion factor for smaller species such as mice and rats. Mice, however, are not commonly used to support selection of the first dose in healthy human studies, and the nonrodents used are generally large. When the CDER guidance for the first dose in humans was developed, it was decided to stay with BSA comparisons that correlated with $b = 0.67$. This was because CDER/FDA had considerable experience with BSA comparisons, the drug development community used BSA conversions, and the BSA comparisons were more conservative than some other approaches. Further, plasma AUC values in rats and humans of pharmaceuticals correlated fairly well when doses were normalized to mg/m^2 (14). Data on BSA of various species are found in the literature [e.g., Spector (15)]. The CDER guidance on the start dose in humans gives a table (Table 1) for conversion of animal doses to HEDs, based on BSA (9). For example, if a NOAEL for dogs in a study is 18 mg/kg then the HED in mg/kg is 18 mg/kg divided by 1.8 = 10 mg/kg.

If similar effects occur across species at similar plasma AUC values and the AUC and clearance values are those that would be predicted based on mg/m^2 scaling, then it is more likely that the effects will occur in humans at doses scaled to mg/m^2. If dose-limiting toxicities and the doses at which they occur do not correlate with plasma AUC or clearance across species, then it is probably not advisable to use mg/m^2 BSA comparisons to select the first dose in humans without lowering the dose further. Lack of correlation could be due to differences in oral bioavailability. Scaling across species based on BSA assumes similar bioavailability across species. The PKa (the negative log of the acid ionization constant) of a substituent on a drug molecule is the pH at which the ionizing group is 50% unionized. In solutions at pH values well below the ionization constant, the substituent is unionized, and unionized drugs may have much greater bioavailability than ionized drugs. Thus, if oral bioavailability is low in rats (e.g., 4%), fairly low in fasted dogs, considerably higher in fed dogs (30%), and the drug has a PKa in the range of three to four, then lower stomach pH of fed dogs may be responsible for the higher bioavailability in dogs. This suggests that the bioavailability in humans (with lower stomach pH than rats) will be at least as high as the dogs and any scaling of effects in rats should take this into account. Furthermore, the potential for large variability in oral bioavailability in humans also exists, generally considered to be about 10-fold, but could be more or less than 10-fold. Metabolism may differ across species and it is recognized that in vitro metabolism does not always correlate to in vivo metabolism with regard to similarities or differences. Sometimes a drug may be more extensively metabolized in a non-human species than in humans or vice versa, or metabolized to different pharmacologically or toxicologically active chemicals across species.

Even when the active pharmaceutical ingredient has previously been studied or approved for use in humans, PKs is very important and may be used in bridging from one formulation to another and in selection of a first dose in humans for new formulations, since the formulation is key in determination of bioavailability. One drug had been approved at a particular dose, and it was noted in the drug labeling that increased bioavailability occurred after a very fatty meal in animals and led to QT prolongation and ventricular arrhythmias (occasionally seen in humans), which corresponded to a plasma concentration above a certain Css. When the active

Table 1 Calculation and Conversion Factors to Convert Animal Dose to HED Based on BSA

Species	Reference body weight (kg)	Working weight range[a] (kg)	BSA (m²)	To convert data in mg/kg to dose in mg/m² multiply by k_m	To convert animal dose in mg/kg to HED[b] in mg/kg, either	
					Divide animal dose by	Multiply animal dose by
Human	60	—	1.62	37	—	—
Child[c]	20	—	0.80	25	—	—
Mouse	0.020	0.011–0.034	0.007	3	12.3	0.081
Hamster	0.080	0.047–0.157	0.016	5	7.4	0.135
Rat	0.150	0.080–0.270	0.025	6	6.2	0.162
Ferret	0.300	0.160–0.540	0.043	7	5.3	0.189
Guinea pig	0.400	0.208–0.700	0.05	8	4.6	0.216
Rabbit	1.8	0.9–3.0	0.15	12	3.1	0.324
Dog	10	5–17	0.50	20	1.8	0.541
Primates						
Monkeys[d]	3	1.4–4.9	0.25	12	3.1	0.324
Marmoset	0.350	0.140–0.720	0.06	6	6.2	0.162
Squirrel monkey	0.600	0.290–0.970	0.09	7	5.3	0.189
Baboon	12	7–23	0.60	20	1.8	0.541
Micro-pig	20	10–33	0.74	27	1.4	0.730
Mini-pig	40	25–64	1.14	35	1.1	0.946

[a]For animal weights within the specified ranges, the HED for a 60 kg human calculated using the standard k_m value will not vary more than ±20% from the HED calculated using a k_m value based on the exact animal weight.

[b]Assumes 60 kg human. For species not listed or for weights outside the standard ranges, HED can be calculated from the formula: HED = animal dose in mg/kg × (animal weight in kg/human weight in kg)$^{0.33}$.

[c]The k_m value is provided for reference only since healthy children will rarely be volunteers for phase 1 trials.

[d]For example, cynomolgus, rhesus, and stumptail.

Abbreviations: HED, human equivalent dose; BSA, body surface area.

pharmaceutical ingredient was reformulated in a lipid vehicle from a tablet, the drug sponsor wanted to dose the humans at the same approved dose. However, studies in monkeys indicated that the new formulation was 10 times more bioavailable than the original tablet. This greater bioavailability was confirmed in humans after the first dose, which was 1/10th that originally proposed by the drug sponsor and which corresponded to a fraction of no-effect dose/Css concentration in animals.

The elimination rate of the drug in one species may be less than expected from scaling to BSA from another species. If protein binding is very different among the non-humans and humans, then scaling cannot be by mg/m^2 alone and plasma AUC values need to be corrected for free drug, since the toxicity would correlate with the concentration of free drug, rather than the total drug concentration.

USE OF PK IN PICKING THE FIRST DOSE IN HUMANS AND UNCERTAINTIES

The selection of the first dose in humans depends on whether the population is considered healthy or whether it is a patient population with moderate or serious disease. More toxicity in patients is acceptable for more serious conditions. For normal volunteers, the first dose is generally determined by extrapolation from a NOAEL for a dose-limiting effect in the most sensitive species, followed by conversion to a HED with allometric scaling based on BSA. The recommended safety factor is 10, which may be increased or decreased under certain conditions (9). However, sometimes C_{max} or AUC values in all the species tested suggest that correlation is better with dose in mg/kg or something else. The safety factor might be increased if there is a steep dose response for serious toxicity, especially if the steep dose response is seen in multiple species, or if there are severe nonmonitorable toxicities (such as cardiac necrosis, brain vacuolization, or nonimmune vasculitis) or toxicities without premonitory signs (sudden death with no warning that the patient is about to die). The safety factor might be reduced if the drug is a member of a well-characterized class with similar metabolic profile, bioavailability, and toxicity across all species, including in humans for other members of the class, and the dose response for toxicity is moderate to shallow and is consistent across tested species with regard to appropriately scaled dose and exposure. Nonclinical data on bioavailability, the metabolite profile, and plasma drug levels associated with toxicity may influence the selection of the NOAEL. Other types of scaling could be based on what is known about a class of drugs.

As part of an evaluation of the multiples of human exposure at which the MTD for pharmaceuticals occurred in rats after oral administration for 13 weeks (14), a comparison of the actual AUC values in humans was made to the AUC values predicted by scaling by BSA. BSA scaling was clearly a better predictor than dose in mg/kg in general, but it sometimes either vastly overpredicted or underpredicted what the human AUC value would be. For 36/57 (63%) of the drugs, scaling by using BSA predicted the human AUC value within two-fold (0.5–2) (Table 2). For 49/57 (86%) of the drugs, the BSA predicted value was within four-fold (0.25–4) of the AUC value. In contrast, the mg/kg ratio clearly overestimated the human exposure. For 51/57 (89%) of the drugs, the mg/kg ratio of human to animal doses predicted an exposure ratio that was greater than four times the actual exposure ratio, as estimated from PK data. The results confirm the importance of measuring PK in toxicity studies for use in any extrapolations to humans and the importance in recognizing when even scaling by BSA may not be appropriate.

The first doses in humans may be selected based on PK factors when appropriate. Reigner and Blesch (16) described the use of a pharmacokinetically guided approach using systemic exposure to select starting doses for humans, as well as three other methods for estimating the starting dose for humans for noncytotoxic compounds referred to as dose by factor, dose by similar drug, and comparative. For the pharmacokinetically guided approach, they suggested selecting the starting dose as the AUC obtained at the NOAEL in the nonclinical studies times the clearance in humans predicted by allometric scaling from three species, and acknowledged the usual caveats for allometric scaling. They then applied a safety factor of three. Mahmood et al. (17) demonstrated in a regulatory research project that interspecies clearance data in animals could be used to estimate a suitable starting dose for humans, provided that the data were obtained from a dose that produced no adverse effects. The scaling of clearance from the animal data that were in the literature was performed using three different methods for 10 drugs that had been studied in at least three species and for which there were data in the literature. Observed clearance in humans was then compared to predicted clearance, and recommended first-time dose in humans was compared by different approaches: dose versus clearance; HED, calculated as the animal dose in mg/kg × (animal weight in kg/human weight in kg)$^{0.33}$ versus clearance; NOAEL and no safety factor; and use

Table 2 Prediction of AUC by Body Surface Area Scaling

AUC expected/AUC measured	Number of drugs
0.01–0.24	6
0.25–0.49	9
0.50–2.0	36
2.1–5.0	4
5.1–9	2

Abbreviation: AUC, area under the curve.

of a correction factor applied to the clearance in the species whose clearance in mg/kg was closest to predicted human clearance. The correction factor was obtained by dividing the clearance in the chosen species by the predicted human clearance. For two of the drugs, predicted clearance was much greater than observed clearance in humans and was attributed to clearance being greater than the respective blood flows. The evaluation of the authors was that of the four approaches; HED versus clearance and use of a correction factor were most suitable for selecting the first dose in humans. Unfortunately, CDER rarely has data from three species. The CDER guidance for first dose in normal volunteers (9) generally recommends scaling by normalizing doses to BSA equivalents. However, it indicates that with sufficient data and, in some instances, it may be possible to develop a PK model to predict human doses and concentrations, rather than using the model as an ancillary tool. The caveats of using PK data to model what happens to the drug are that (i) human bioavailability and metabolism may differ significantly from the animals, (ii) the mechanism of toxicity may not be known and may be due to toxic accumulation in a peripheral compartment, or (iii) toxicity may be due to an unidentified metabolite. For example, a drug was administered intravenously a single time, disappeared from the plasma in 15 minutes, yet accumulated in the spleen and liver such that operating room lights six months later caused damage to the liver, when the abdomen was opened. In this case, evaluation of plasma concentrations was not helpful in assessing the correlations with toxicity. However, studies of tissue distribution or mass balance would have indicated the accumulation in a peripheral compartment.

USE OF PK IN PICKING THE MAXIMUM DOSE TO BE USED IN HUMAN VOLUNTEERS

How much toxicity is reasonable for humans who are healthy? In some cases, PKs might be used to help select the maximum dose to be used in human volunteers. If animal studies did not achieve an MTD or no serious toxicity was observed in animals, the initial study might be allowed to proceed up to the maximum plasma levels achieved in the animal studies, provided no unacceptable adverse reactions occurred in the humans. However, in the case of a serious, unmonitorable, adverse effect, some fraction of the NOAEL dose or the Css in animals could be sufficient to result in cessation of dose escalation.

CONCLUSIONS

PKs and TKs data are useful adjuncts to the pharmacology and toxicology data in selecting appropriate doses of drugs for study in humans. Throughout drug development, these PK/TK/ADME data help bridge effects seen in animals to effects seen in humans, help select the appropriate non-human species for evaluation of toxicity, and help predict drug interactions.

No drug is without risk, and persons receiving a drug early in drug development may receive no benefit. By the time a drug has been evaluated in controlled clinical studies, an initial assessment can be made of its risks and benefits. Sometimes adverse effects affecting only a small percentage of the persons taking a product may not show up until many more persons than were in clinical trials are exposed to the drug. The approval of a drug means that for the intended population and the intended therapeutic indication, at the recommended dose, the risks of taking a drug are reasonable considering the potential benefit. In the end, each health care provider and the patient make such a decision, based on individual circumstances.

REFERENCES

1. U.S. Food and Drug Administration (FDA), Guidance for Industry, Investigators, and Reviewers Exploratory IND Studies. Draft guidance, 2005.
2. Bergstrom M, Grahnen A, Langstorm B, et al. Positron emission tomography microdosing: a new concept with application in tracer and early clinical drug development. Eur J Clin Pharmacol 2003; 59(5–6):357–366.
3. Lappin G, Garner RC. Current perspectives of 14C-isotope measurement in biomedical accelerator mass spectrometry. Anal Bioanal Chem 2004; 378(2): 356–364.
4. Sandhu P et al. Evaluation of microdosing strategies for studies in preclinical drug development: demonstration of linear pharmacokinetics in dogs of a nucleoside analog over a 50-fold dose range. Drug Metab Dispos 2004; 31(11): 1254–1259.
5. U.S. Food and Drug Administration (FDA), Guidance for Industry. Safety Pharmacology Studies for Human Pharmaceuticals. International Conference on Harmonization, ICH-S7A, 2001.
6. U.S. Food and Drug Administration (FDA), Guidance for Industry. Nonclinical Evaluation of the Potential for Delayed Ventricular Repolarization (QT interval prolongation) by Human pharmaceuticals. International Conference on Harmonization, ICH-S7B, 2005.
7. U.S. Food and Drug Administration (FDA), Guidance for Industry. Nonclinical Safety Studies for the Conduct of Human Clinical Trials for Pharmaceuticals. International Conference on Harmonization, ICH-M3, 1997.
8. U.S. Food and Drug Administration (FDA), Guideline for Industry. Dose Selection for Carcinogenicity Studies of Pharmaceuticals. International Conference on Harmonization, ICH-S1C, 1995.
9. U.S. Food and Drug Administration (FDA). Guidance to Industry Estimating the Maximum Safe Starting Dose in Initial Clinical Trials for Therapeutics in Adult Health Volunteers, 2005.
10. U.S. Food and Drug Administration (FDA), Guideline to Industry. Toxicokinetics: The Assessment of Systemic Exposure in Toxicity Studies. International Conference on Harmonization, ICH-S3A, 1999.

11. Freireich EJ et al. Quantitative comparison of toxicity of anticancer agents in mouse, rat, hamster, dog, monkey, and man. Cancer Chemother Rep 1966; 50(4):219–244.
12. Schein PS et al. The evaluation of anticancer drugs in dogs and monkeys for the prediction of qualitative toxicities in man. Clin Pharmacol Ther 1970; 11:3–40.
13. U.S. EPA. A cross-species scaling factor for carcinogen risk assessment based on equivalence of mg/kg 0.75/day. Fed Reg 1992; 57:24152–24173.
14. Contrera J et al. A systemic exposure based alternative to the MTD for Carcinogenicity studies of human therapeutics. J Am Coll Toxicol 1995; 14:1–10.
15. Spector WS, ed. Handbook of Biological Data. Philadelphia: W.B. Saunders, 1956:175.
16. Reigner BG, Blesch KS. Estimating the starting dose for entry into humans: principles and practice. Eur J Clin Pharmacol 2002; 57(12):835–845.
17. Mahmood I et al. Selection of the first-time dose in humans: comparison of different approaches based on interspecies scaling of clearance. J Clin Pharmacol 2003; 43(7):692–697.

15

Pharmacokinetic/Physiologically Based Pharmacokinetic Models in Integrated Risk Information System Assessments

Robert S. DeWoskin

National Center for Environmental Assessment, Office of Research and Development, U.S. Environmental Protection Agency, Research Triangle Park, North Carolina, U.S.A.

John C. Lipscomb

National Center for Environmental Assessment, Office of Research and Development, U.S. Environmental Protection Agency, Cincinnati, Ohio, U.S.A.

Chadwick Thompson, Weihsueh A. Chiu, and Paul Schlosser

National Center for Environmental Assessment, Office of Research and Development, U.S. Environmental Protection Agency, Washington, D.C., U.S.A.

Carolyn Smallwood, Jeff Swartout, and Linda Teuschler

National Center for Environmental Assessment, Office of Research and Development, U.S. Environmental Protection Agency, Cincinnati, Ohio, U.S.A.

Allan Marcus

National Center for Environmental Assessment, Office of Research and Development, U.S. Environmental Protection Agency, Research Triangle Park, North Carolina, U.S.A.

The views expressed in this article are those of the author(s) and do not necessarily reflect the views or policies of the U.S. EPA.

INTRODUCTION—INCREASING USE OF PK/PBPK MODELS IN IRIS

This chapter addresses the increasing use of pharmacokinetic (PK) and physiologically based PK (PBPK) models in the derivation of reference values for the U.S. Environmental Protection Agency's (EPA's) Integrated Risk Information System (IRIS) Program. IRIS is an electronic database containing information on human health effects that may result from exposure to various chemicals in the environment (1). Initially developed in the mid-1980s, the U.S. EPA created IRIS to provide agency risk assessors with consistent high-quality information on the hazardous effects of chemical substances for use in decision-making and regulatory activities. The IRIS database has subsequently become available to risk assessors worldwide, and is a major resource supporting efforts to protect the public health.

There are currently over 500 chemical assessments available in IRIS. Not all contain a complete set of reference values; however, this is not from a lack of trying to develop values whenever possible. Generally, the reason a reference value is not developed is inadequate or nonexistent data (bioassay or human) to support the value through the extensive peer and public reviews to which each assessment is subjected. Advances in PBPK modeling now offer a scientifically sound means to extract the most value from the available data, and to bridge data gaps. PBPK models integrate our current understanding of the physiological and biochemical determinants of chemical disposition in the body, and provide a means to predict internal doses from hypothetical exposures.

This chapter addresses (i) the use of PK and PBPK models in IRIS assessments to derive reference values, (ii) the main quality criteria in the evaluation of a PBPK model intended for use in an assessment, and (iii) examples of PK or PBPK models that have been used in IRIS assessments or other U.S. EPA programs.

THE USE OF PK AND PBPK MODELS IN THE IRIS PROGRAM

The reader is referred to the detailed discussion in previous chapters on deriving reference values (Chapters 1,2) and the role of PBPK models in risk assessment (Chapter 6). This chapter specifically addresses how PK and PBPK models are used to derive reference values in the IRIS Program.

Derivation of Reference Values in an IRIS Assessment

Reference values are developed in an IRIS assessment when data are available and of sufficient quality to support a quantitative characterization of a toxin's dose–response relationship. Currently, reference values are only derived to protect against an increased incidence of noncancer effects or cancer following a chronic (i.e., lifetime) oral or inhalation exposure.

For noncancer effects, the reference dose (RfD) for an oral exposure is usually expressed in mg (or μg)/kg body weight/day, and the reference concentration (RfC) for an inhalation exposure is expressed in mg/m^3 in air for a continuous (24 hours) exposure. The reference values for carcinogenic effects are expressed either as a slope factor, in units of proportion of a population affected per mg/kg-day of exposure, or as a unit risk, which is the excess lifetime cancer risk estimated to result from continuous exposure to an agent at a concentration of 1 μg/L in water, or 1 $μg/m^3$ in air.[a]

As discussed in previous chapters, the default procedure for deriving a reference value is to identify a point of departure (POD) based on a no-observed-adverse-effect level (NOAEL), lowest-observed-adverse-effect level (LOAEL), or benchmark analysis for the critical effect, and then to adjust the POD with uncertainty factors (UF) to account for data gaps (i.e., uncertainties) in the estimate of the human dose–response. The five UFs that are available for use in an IRIS assessment are (i) the UF_A for interspecies extrapolation (default value $UF_A = 10$); and (ii) UF_H for intraspecies (i.e., human) variability (default value $UF_H = 10$), the UF_L for LOAEL-to-NOAEL extrapolation (default value $UF_L = 10$), the UF_S for subchronic-to-chronic study extrapolation (default value from 1 to 10), and the database UF (UF_D) for lack of a complete database (default value from 1 to 10). Of these, the UF_A and UF_H can be further broken down into a fraction that addresses the PK differences (between species or among individuals in a population) or the pharmacodynamic (PD) differences.[b] In IRIS, these toxicokinetics (TK) and toxicodynamics (TD) fractions are equally weighted for the UF_A and UF_H.[c] Thus, one can talk about the $UF_{A\text{-PK}}$ of $10^{1/2}$ (3.16), a $UF_{A\text{-PD}}$ of $10^{1/2}$, a $UF_{H\ PK}$ of $10^{1/2}$, and a $UF_{H\text{-PD}}$ of $10^{1/2}$.

When there are no data or models to improve upon the default values for PK and PD differences, the two equally weighted parts are combined to reconstitute the original factor and the default value of 10. When data, empirically fitted models, or biologically based models are available to provide a better estimate of PK or PD differences, the improved estimates replace that part of the default factor. When only the PK or the PD part is replaced, the remaining part, i.e., the $10^{1/2\ (3.16)}$ value, is rounded to 3. For example, if a PBPK model analysis yielded an estimated fivefold increase in the human internal dose at a target site compared with the test

---

[a] The unit risk is interpreted as follows: if unit risk $= 2 \times 10^{-6}$ per μg/L, then two excess cancer cases (upper bound estimate) are expected to develop per 1,000,000 people if exposed daily for a lifetime to 1 μg of the chemical in 1 L of drinking water.

[b] The terms toxicokinetic (TK) or toxicodynamic (TD) are also valid and of comparable meaning. PK and PD are used in this chapter for consistency with the use of the PK and PBPK acronyms for the models.

[c] There is currently no formal guidance for weighting the $UF_{H\text{-PK}}$ and $UF_{H\text{-PD}}$, and the available data can be used to support a weighting other than 3.16 and 3.16.

animal internal dose based on PK differences (e.g., different metabolic profiles for a parent compound), this value of 5 would replace the default value of 3.16, and the total adjustment to develop a human equivalent dose (HED) would be $15 = 5$ (for PK differences based on the model results) $\times$ 3 (UF$_{A-PD}$).

PK and PBPK Model Uses in IRIS

Among biologically based models, PBPK models address the PK differences, and biologically based dose–response (BBDR) models address the PD differences. Either model, if adequately developed and tested, provides a more chemical-specific and scientifically sound approach to bridging data gaps than do the UFs. Figure 1 illustrates the PK events (PBPK models) and PD events (BBDR models) that models simulate between an administered dose and an observed response. Figure 1 also lists basic physiological and chemical-specific data needed to develop the models. Because of the rising costs of animal testing and ethical considerations in obtaining human data, PBPK model development and kinetic studies are increasingly being seen as a cost-effective alternative.

PBPK model development continues to outpace BBDR models primarily due to the relatively straightforward processes involved in PKs compared to PDs and the availability of the data needed to develop the models. Only recently, however, have even PBPK models been available

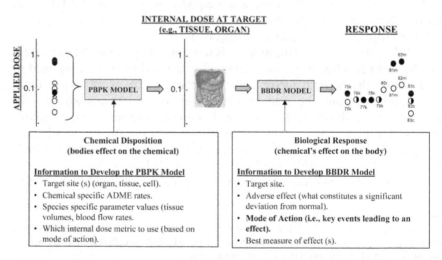

Figure 1 Role of biologically based models in simulating pharmacokinetic events (PBPK models) and pharmacodynamic events (BBDR models) that occur between the administered dose and the observed response. *Abbreviations*: PBPK, physiologically based pharmacokinetic; BBDR, biologically based dose–response; ADME, absorption, distribution, metabolism, excretion.

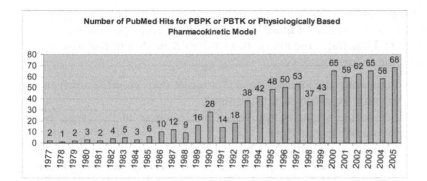

Figure 2 Increasing development and use of physiologically based pharmacokinetic models in the past 15 years (tabulated as of December 2005). *Abbreviations*: PBPK, physiologically based pharmacokinetic; PBTK, physiologically based toxicokinetics. Roughly 80–90% of the PBPK models are for organic chemicals, 10% for inorganics.

for use in risk assessment. Figure 2 presents the results of a MEDLINE search from 1977 to the present of all indexed fields using keywords for PBPK or PBTK (TK) models. Although a crude barometer of the progress that has been made in developing PBPK models for risk assessment, the numbers

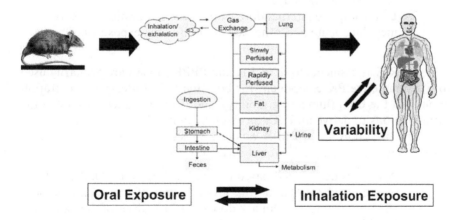

- **Aggregate Exposures – One Chemical, Multiple Exposure Routes**
- **Cumulative Exposure – Multiple Chemicals, Same or Different Routes**
- **Duration Extrapolation – Single Dose to Continuous Exposure**

Figure 3 Physiologically based pharmacokinetic model applications to improve the scientific support in deriving a human reference value when data are limited.

suggest that most models have been published within the last 10 to 15 years, and that a steady stream of models are now being developed for use.

Figure 3 illustrates the many uses for PBPK models in risk assessment including the following:

1. Estimating the internal dose at a target that would result from an external exposure in a test animal.
2. Estimating a human equivalent internal dose at the target based on animal kinetic data.
3. Estimating an animal or human internal dose following exposure from one route of exposure (e.g., inhalation) based upon dose–response and kinetic data from a different route of exposure (e.g., oral exposure).
4. Estimating variability in the human internal dose within a population (to identify sensitive subpopulations) due to individual differences that affect the pharmacokinetics (e.g., tissue volumes, blood flows, enzyme profiles).
5. Estimating an internal dose of a chemical following exposure from several different routes (i.e., an aggregate exposure from simultaneous or sequential dermal, oral, and/or inhalation exposures).
6. Estimating the internal dose of single or multiple chemicals from exposure to a mixture via the same or multiple routes of exposure (cumulative exposure).
7. Estimating the internal dose from different durations of exposures including acute and other less-than-life time exposures.

In IRIS assessments to-date, PK and PBPK models are primarily used to (i) estimate the PK component for both the UF_A (interspecies extrapolation) and the UF_H (human variability), and (ii) to conduct route-to-route extrapolations to derive an RfC, an RfD, or unit risk.

The procedure for a PBPK model analysis to replace the $UF_{A\text{-}PK}$ in deriving an RfD or an RfC^d would be as follows:

1. A POD is determined based on the NOAEL, the LOAEL if there was not a NOAEL, or the lower bound of a benchmark dose (BMDL) from a benchmark dose (BMD) analysis of the "administered" dose–response relationship.
2. An internal dose metric is chosen as the most closely correlated with the response from an analysis of the dose–response relationships for

---

[d] See the RfC methodology for the non-PBPK model approach to deriving the human equivalent concentration.

different metrics and an understanding of the chemical's mode of action (MOA).[e]

3. The test animal PBPK model is run to predict the value of the internal dose metric that results from exposure to "administered dose POD." This value of the internal dose is then the "internal dose POD." As an alternate approach to deriving the internal dose POD, a BMD analysis can be conducted on the model-derived internal doses for each of the administered doses in the study to develop a BMDL for an internal dose.

4. The human PBPK model is then iteratively run until a human equivalent administered oral dose (e.g., as a drinking water exposure) or inhalation dose (as a continuous exposure) is determined that would yield an internal dose value equal to the "internal dose POD" from step (3) above.

5. The resulting HED is divided by the appropriate UFs to establish the RfD or RfC.

A variant of the above approach involves dividing the dose metric associated with the administered dose POD by the UFs before exercising the human model to determine equivalent doses (3). The outcome is essentially the same if the PK processes in the test species and human are in the linear range at the administered doses. If uncertain about whether or not the dose might saturate any of the nonlinear processes, both procedures should be tried and the decision as to which to use should be based upon what is known about the MOA and which would be best supported by the science.

When a PBPK model is used to derive an RfD or an RfC, the magnitude of the combined UFs applied to the POD is identical to that of the nonmodeling (i.e., default) approach except, as discussed above, the fact that the kinetic component of the interspecies UF is replaced by the results of the PBPK model analysis (i.e., $UF_{A-PK} = 1$ instead of 3.16). The remaining PD component UF_{A-PD} is rounded to 3, and the total UF_A for a model-based derivation is equal to the model-derived estimate for PK differences times 3 (UF_{A-PD}).

A PBPK model can also be used to conduct a route-to-route extrapolation if absorption from different routes of interest (e.g., oral and inhalation) is represented. As discussed in Chapter 6 by Thompson et al., the PBPK model can be run to derive an administered dose from any represented route of exposure that would be needed to produce a given internal dose. So, once an internal dose amount from one route of exposure is

---

[e] The mode of action (MOA) is a phrase used within the U.S. EPA to refer to a description of the key events (as in critical path) at the cell (or putative target) level of organization that leads to the adverse effects, in contrast to a description of all of the molecular and biochemical events in the overall process.

available (either directly from experimental data or estimated by the PBPK model), then the model can estimate the administered dose for the other route of exposure.

There are a number of a priori conditions that must be met for a route-to-route extrapolation, including (i) there are no "significant" portal of entry effects, (ii) the chemical is systemically absorbed and distributed, (iii) first pass effects are well-represented in the model, and (iv) sufficient kinetic data are available to calibrate and test the model for the routes of interest. It is, of course, best to have a human PBPK model so that the extrapolations are based directly on an estimate of the human internal dose. If only an animal model is available, however, the route-to-route extrapolation can be conducted based on animal data, and the default uncertainties factors (or default procedure for an RfC) used to extrapolate from the test animal dose to the HED.

A PBPK model can also be used to estimate the PK component (UF_{H-PK}) of the interindividual variability factor using a Monte Carlo approach (see Chapter 6 by Thompson et al.). The accuracy of the variability estimate using the Monte Carlo analysis depends upon how well-characterized the distributions are for the most sensitive parameters. As a caveat, Monte Carlo sampling can lead to results that are not physiologically realistic if the parameter distributions are poorly characterized or some parameters are closely correlated to other parameters (e.g., respiratory rate and cardiac output). Improved sampling methods and ways to characterize human variability are areas of active research (4–7).

EVALUATION OF PBPK MODELS USED IN IRIS ASSESSMENTS

Over the past few years, the U.S. EPA has increased its staff and research capability in computer simulation modeling to both advance the science and evaluate the increasing numbers of PBPK models submitted for review. In early 2005, the U.S. EPA's National Center for Environmental Assessment (NCEA) formed an internal working group known as the PK Working Group (PKWG), whose primary goal is to provide technical consult to the IRIS program chemical managers (CMs) in their evaluation and use of PBPK models in IRIS assessments. In IRIS, the CM is the point person who develops the assessment and shepherds it through the multiple layers of review. The PKWG consists primarily of toxicologists and modelers within NCEA, but benefits from participating scientists throughout the U.S. EPA.

Biological models are presented to the PKWG by the IRIS CMs in an initial scoping session to discuss the available kinetic data and proposed applications. A subgroup is then formed within the PKWG to provide consult to the CM throughout the remaining assessment development. Scoping sessions have now become a routine occurrence early in the assessment development cycle to allow time for model testing or further development for specific applications.

The PKWG advice reflects the combined expert judgment of the group members. The PKWG, however, has no authority over the direction of the assessment or its conclusions. Those decisions remain within the purview of the CM. Consensus is not required within the PKWG, as there are many areas of ongoing research in model development and evaluation. The PKWG consult does, however, strive to provide the rationale for all determinations and is available, as needed, for continuing discussions.

At present, there are no formal guidelines for evaluating PBPK models. An excellent general approach that covers many of the key areas has been developed by Clark et al. (8) and is briefly summarized in Table 1. The PKWG does place a strong emphasis on model transparency including the availability of model code (preferably annotated), command files

Table 1 Model Evaluation—Highlights of a Six-Step Process (with Sample Subheadings) for Evaluating Physiologically Based Pharmacokinetic Models

Assessment of the model purpose
  Risk assessment needs (e.g., dose–response analysis, cumulative risks, route-to-route)
  Generic issues (e.g., target organ, age, exposure route, species)
Assessment of the biological characterizations and model structure
  Consideration of alternate model structures
  Appropriate tissue compartments
  Relevant physiological parameters and biochemical parameters
  Presence of cross-chemical analogies and chemical-specific characterizations
  Life-stage considerations
  Appropriate exposure scenarios
Assessment of the mathematical descriptions
  Check model equations (unit accuracy, mass balance, and blood flow balance)
  Are mathematical descriptions of the biology reasonable
Assessment of the computer implementation
  Review model code for proper syntax and mathematical structure
  If resources allow, independently reimplement model in different computer language (or alternatively, transform model into a simpler system of differential equations and solve)
Parameter analysis and assessment of the model fit
  Review parameter documentation (model derived, literature derived, nonidentifiable)
  Uncertainty in parameters perform
  Sensitivity analysis
  Evaluate goodness of fit (model predictions versus data) and statistical analysis
Assessment of any specialized analyses
  Monte Carlo methods to quantify uncertainty and human variability
  Parameter Covariance to reflect physiological interdependencies

Source: Adapted from Ref. 8.

(i.e., run conditions), data (with citations and preferably characterized for quality) used to calibrate or test the model, and supporting documentation (e.g., optimization routines and rationale for choice of parameters that were optimized). This information allows independent verification of model output and model behavior under different run conditions, including replication of any submitted model predictions and quality control checks for coding errors or structural flaws (e.g., lack of mass balance).

The overall context within which a model is evaluated should also consider a common mantra among modelers that "all models are wrong, but some are useful." This simple phrase reminds us that a model represents a theory, and that like a theory

1. All models are simplified representations of reality,
2. Every model will most likely evolve and improve as more is learned about the real system,
3. For many practical applications a "perfect" model is not needed, only one that further reduces the uncertainty and provides a more scientifically supportable reference value than would otherwise be obtained using the default adjustment factors.

SPECIFIC EXAMPLES OF USE OF PK/PBPK MODELS IN IRIS SYSTEM

The availability of a PBPK model for a given chemical does not necessarily mean that the model will be used in an assessment. There are many considerations, including:

*   The adequacy of the model structure,
*   The quality of the data needed to calibrate and test the model,
*   Whether the specie(s) represented in the model are the same as those in the critical dose–response studies,
*   How much is known about the mode of action to determine an appropriate dose metric, and
*   The ever-present time and resource constraints if the model needs to be peer evaluated or further elaborated.

This section presents some representative examples of PK and PBPK models that have or have not been used in IRIS assessments and how their use affected the resulting reference values compared with derivations based on the default UFs. The examples begin with assessments where models were available but not used (and why not)—chloroform, methyl ethyl ketone (MEK), and acetone. These are followed by examples where PBPK models were used for interspecies extrapolations for noncancer reference values [ethylene glycol monobutyl ether (EGBE), vinyl chloride (VC), dichloromethane (DCM), xylene] or cancer slope factors (VC), or for route-to-route extrapolation (VC to extrapolate from the oral-to-inhalation route). Examples are then presented where PK data alone suffice to replace or modify an interspecies and/or variability factor (boron). Finally, an example of a combined exposure–kinetic model is presented—the Integrated Exposure, Uptake, and Biokinetic[f] (IEUBK) model—used to estimate blood lead (PbB) levels in children from exposure to lead in the environment.

These examples were prepared by individual members of the NCEA PKWG, whose names are listed at the start of the text they authored as an acknowledgement of their important contribution to this chapter.

PBPK Models Available But Not Used to Derive a Reference Value

Robert S. DeWoskin and Chad Thompson

Chloroform

The 2001 IRIS assessment for chloroform (9) provides an example of a PBPK model that is not used directly to derive a reference value, rather as a support for the assessment conclusion that the mode of action for the carcinogenic effects of chloroforms leads to a nonlinear dose–response relationship.

The chloroform assessment developed an RfD based on a 7.5-year oral exposure study in dogs (10). An oral slope factor was not derived because the weight of evidence indicated that the observed increased incidence of liver and kidney tumors in several species by several exposure routes only occurred at high dose levels, which produced cytotoxicity, and that the carcinogenic response was associated with regenerative hyperplasia in response to cytolethality. Thus, the RfD was sufficient to protect against noncancer effects (including cytolethality and regenerative hyperplasia) as well as against increased risk of cancer.

A PBPK model developed by Corley et al. (11) was available at the time, and was parameterized for rats, mice, and humans. The Corley model was further elaborated by Reitz et al. (12) to include a dynamic component to simulate the effects of chloroform metabolism on cell killing in the liver, and to provide insight into possible modes of action for chloroform liver carcinogenicity in humans. The PBPK model was not used in the assessment to derive reference values because the model was not parameterized for the dog, which was the test species for the critical effect and POD used to derive the RfD. The modeling results by Reitz et al. (12), however, did support the assessment conclusions that the hepatotoxicity of chloroform (and hence the potential for carcinogenicity) was strongly dependent on the rate of metabolism, which in turn depended upon dose rate.

In summary, the available chloroform PBPK model was not used because it was not parameterized for the test species (dog) for the designated

---

[f] A biokinetic model is another name (albeit infrequently used today) for a PBPK model.

critical effect used to derive the RfD. Modeling results from the PBPK model in conjunction with a dynamic model were used, however, to support the assessment conclusion that chloroform's mode of action for carcinogenic effects would lead to a nonlinear dose–response relationship.

Methyl Ethyl Ketone

Methyl Ethyl Ketone (MEK) (13) PBPK models were available for humans (14,15) and rats (16,17) at the time of the 2003 IRIS assessment for MEK. PBPK models were unavailable for other species. The available models, however, were not used because they were developed from limited data and had not been adequately tested.

The MEK RfD was based on developmental effects in rats following a drinking water exposure to 2-butanol (an MEK metabolite). PK and toxicologic data in rats supported the use of 2-butanol as an appropriate surrogate for MEK. The data indicated that (i) orally administered 2-butanol was almost completely converted to MEK and its metabolites within 16 hours, (ii) peak MEK blood concentrations occurred at similar times after administration of 1776 mg/kg 2-butanol (7 to 8 hours) or 1690 mg/kg MEK (4 to 5 hours), and (iii) common metabolites (3-hydroxy-2-butanone and 2,3-butanediol) were formed and eliminated with similar kinetics after the administration of 2-butanol or MEK (16,18). Comparable PK data for 2-butanol and MEK in humans were not available; however, evidence for metabolic conversion of MEK to 2-butanol in humans supported the assumption that rats and humans metabolize 2-butanol similarly.

The POD of 657 mg 2-butanol/kg-day was derived from a BMD analysis of the dose–response for decrease in mean postnatal day 21 body weight. Because 2-butanol was used as a surrogate for MEK, a molar adjustment was needed to account for the different molecular weights of the two chemicals. A total UF of 1000 was applied to the adjusted POD of 639 mg of MEK/kg-day: $UF_A = 10$ interspecies extrapolation from animals to humans, $UF_H = 10$ to protect the sensitive subgroups; and $UF_D = 10$ for database deficiencies. The resulting RfD was 0.6 mg/kg-day.

A human PBPK model (14,15) was developed to describe the dose-dependent elimination kinetics of MEK in humans following inhalation exposure to low concentrations of MEK. The model has eight compartments—lungs, gastrointestinal (GI) tract, liver, richly perfused tissue, poorly perfused tissue, fat, muscle, and blood. Liira et al. (14) developed kinetic data for two men who received separate exposures of 25, 200, or 400 ppm MEK in an inhalation chamber for 4 hours. Venous blood samples were taken during each exposure, and for 8 hours thereafter. The metabolism of MEK was assumed to occur only in the liver, and was described by Michaelis–Menten kinetics. Liira et al. (14) reported that the model predicted concentrations similar to observed blood concentrations of MEK in 17 male volunteers exposed to 200 ppm. The authors noted that the kinetic constants represented healthy male subjects.

Research utilizing Sprague–Dawley rats (16,19) identified the pathways of MEK metabolism and permitted a calculation of rate constants for the elimination of MEK and its metabolites from the blood, as well as for the metabolic transformations. The data were used as the basis for a perfusion-limited PBPK model for MEK (16) to predict blood concentrations of 2-butanol and its metabolites. The differential equations of the model accounted for (i) the elimination of 2-butanol and its metabolites from the blood at rates linearly proportional to blood concentrations, (ii) transport between the blood and the liver compartments, and (iii) metabolic conversions in the liver. Metabolic conversions were described with Michaelis–Menten saturation kinetics and included rates for bidirectional conversions between 2-butanol and MEK, unidirectional conversion of MEK to 3-hydroxy-2-butanone, and bidirectional conversions between 3-hydroxy-2-butanone and 2,3-butanediol. Kinetic constants in the model were optimized to fit the in vivo blood concentration data.

Thrall et al. (17) developed a PBPK model for MEK in F344 rats, from experimentally determined partition coefficients using in vitro vial equilibration technique and in vivo measurements of MEK uptake in rats exposed to 100 to 2000 ppm MEK in a closed, recirculating gas uptake system. The model included both a saturable metabolic pathway (described with Michaelis–Menten kinetic constants) and a nonsaturable first-order pathway. The model provided adequate predictions (based on visual inspection) of exhaled MEK concentrations following inhalation, intravenous, intraperitoneal, or oral administration of MEK to rats.

One notable difference between Thrall et al. (17) and Dietz et al. (16) models was the peak exhaled breath concentrations following oral gavage. Dietz et al. (16) found peak MEK concentrations in blood 4 hours after oral gavage (1690 mg/kg MEK), whereas Thrall et al. (17) reported peak MEK concentrations in exhaled air 1 hour after oral gavage (50 mg/kg MEK).

The assessment noted that the Thrall et al. (17) model could be used to simulate kinetics if human values for model parameters were used. Such a model would need to be tested against relevant human data, and that had not occurred at the time of the assessment.

In summary, three PBPK models for MEK were available at the time of the assessment, but had been developed based on a limited number of data sets in rats and humans. The predictive capabilities of the models had not been adequately tested, and none of the models were parameterized for rats and humans sufficient to support replacement of the interspecies UF. Data to support the use of the MEK PBPK models for route-to-route extrapolation were also limited or not available.

Acetone

Acetone (20) is a major metabolite of isopropanol. PBPK models for acetone have been developed for rats and humans using TK data for both isopropanol and acetone from different routes of exposure (3,21,22). The models represent uptake (lungs and skin), metabolism (liver), and fat storage with slowly and rapidly perfused compartments. The models were adequately tested for predicting the kinetics of human exposure from the inhalation pathway, but not the oral pathway. There is no mention in the assessment of evaluation results for the rat PBPK model.

The reference values were derived using the NOAEL/LOAEL approach. Based on studies by the National Toxicology Program (NTP) (23) and Dietz et al. (24), a LOAEL of 1700 mg/kg-day (20,000 ppm dose group) and a NOAEL of 900 mg/kg-day (10,000 ppm dose group) were identified for increased incidence of mild nephropathy. These values were calculated as time-weighted average doses based on body weights, water consumption, and nominal drinking water concentrations. The dose–response relationship was sharply J shape with 1 of 10 control exhibiting mild nephropathy, 0 of 10 for the intermediate groups (200, 400, 900 mg/kg-day for male rats), and then 9 of 10 at both the high doses (1700 and 3400 mg/kg-day). This non-graded response precluded a BMD analysis, and may have influenced the decision about using a PBPK model.

The assessment did not derive an RfC because of insufficient human and animal data. Overall, the most pronounced effect of acetone reported in human inhalation studies was irritation of the eyes and respiratory tract.

Although the rat and human PBPK model for acetone were available at the time of the assessment, they were not used to support an interspecies extrapolation, variability estimate, or route-to-route extrapolation (RfD to an RfC). The human PBPK model for oral exposure was not used for the derivation of the HED because the human model had not been tested against human data. The PBPK models were also not used for route-to-route extrapolation because the oral exposure models had not been adequately tested.

Since the publication of the IRIS assessment for acetone, additional modeling results have been published for the rat and human PBPK model, including proposed use in deriving both an RfD and an RfC (25). This work will be considered in future revisions of the IRIS assessment for acetone.

PBPK MODELS OR PK DATA USED IN THE DERIVATION OF REFERENCE VALUES

The following examples illustrate use of a PBPK model or PK data to replace or modify the kinetic component of the interspecies UF (UF_{A-PK}) or the human variability UF (UF_{H-PK}) or to conduct a route-to-route extrapolation.

Ethylene Glycol Monobutyl Ether

Chad Thompson

A PBPK model for EGBE was used in the 1999 IRIS assessment, in conjunction with a BMD analysis to derive an RfD and an RfC (26). The PBPK model results were used to replace the default UF_{A-PK}. The combined BMD/PBPK model approach was considered the most appropriate for the RfD derivation because it was considered to be the method that incorporated the most mechanistic information and had the best characterization of the dose–response relationship. Compared to the default approach, the BMD/PBPK model derivations were slightly (17%) more conservative for the RfD, and about 2.6-fold less conservative for the RfC. This result highlights the fact that sophisticated modeling approaches have the potential for improving risk assessment without bias.

Model Structure, Focus, and History

The EGBE PBPK model was developed for both rats and humans and simulated absorption from multiple routes of exposure including oral (gavage), drinking water, intravenous infusion, and dermal (liquids and vapor) (27). The objective of the model was to estimate the concentration of 2-butoxyacetic acid (BAA; the primary metabolite of EGBE) in the target tissue (blood) of rats and humans. The model used allometrically scalable physiological and biochemical parameters (e.g., blood flows, tissue volumes, and metabolic capacity) in place of standard values for a 70 kg human. This scaling normalized the values of these parameters to the actual body weights, a feature that was used to develop specific values for the subjects in several human kinetic studies. The model also accounted for changes in the values of physiological parameters for humans when exercising. Model parameter values were developed for the rat to further evaluate the model and to support interspecies comparisons of BAA kinetics in the blood compartment. BAA formation was assumed to occur only in the liver, and the rat liver perfusion data of Johanson et al. (28) were scaled to the human.

A second model was linked to the EGBE PBPK model specifically to track the disposition of BAA following its formation in the liver. The kidney was added to the BAA model as the organ of elimination for BAA. All other metabolic routes for EGBE (formation of ethylene glycol and glucuronide conjugate) were combined, because they were used only to account for the total disposition of EGBE in the rat metabolism studies and not for cross-species extrapolations. Contrary to observations in rats, Corley et al. (29) found no evidence of metabolites in urine that would indicate that humans form conjugates of EGBE or ethylene glycol. Thus, these pathways, which were lumped together in the Corley et al. (27) rat model to simulate rat kinetic data, were not active in the human model simulations. Additional detail on values for other important model parameters (e.g., blood:air and tissue:blood partition coefficients, protein binding) is discussed in the assessment (26).

MOA, Choice of Dose Metric, and Specific Application

Hematologic and hepatocellular changes were noted in both sexes of rats. In females, both hematologic and hepatocellular changes were noted at the low-dose level (58.6 mg/kg-day), using water consumption rates and body weights measured during the last week of exposure. Only hepatocellular cytoplasmic changes (HCCs) were observed in male rats.

Dose–response information on the hematologic effects in female rats was selected as the basis for the oral RfD. The proposed MOA for EGBE's hemolytic effects begins with interactions between EGBE metabolites and cellular molecules in erythrocytes, followed by erythrocyte swelling, then lysis. Critical measures for this hemolysis MOA include increased mean corpuscular volume (MCV) and decreased red blood cell (RBC) count. Erythrocyte swelling (i.e., MCV) is considered a precursor to lysis; however, some studies indicate that this may not always be a consistent sequence, depending upon the exposure regimens. Therefore, the assessment chose the more sensitive of the two end points, increased MCV, to use in the derivation of the RfD.

In rats, the primary toxic metabolite responsible for hemolysis is BAA. Blood BAA level (or C_{max} BAA) has been shown to be a more appropriate dose metric for hemolysis than area under the curve (AUC). In the model simulations, the dosing to rats and humans was represented as an exposure of the entire dose of EGBE from drinking water over a 12-hour period (i.e., waking state) each day, 7 days/wk; and confirmation was obtained that steady-state levels were reached.

Comparison of Model Application to Default Value

Four approaches were compared in deriving the RfD, and each approach resulted in the application of a different array of UFs. The approaches used included a standard default method, a BMD analysis, a PBPK model, and a combined BMD/PBPK approach. Before summarizing the results of each approach, the selection of certain UFs is described.

For interspecies extrapolation (UF_A), in vivo and in vitro studies indicate that, pharmacodynamically, humans are less sensitive than rats to the hematologic effects of EGBE. For this reason, a fractional component of the UF_{A-PD} was considered. However, the in vivo relative insensitivity of humans cannot be quantified at this time, so a value of 1 was used to account for PD differences between rats and humans ($UF_{A-PD} = 1$). This value was used in all of the approaches. In the standard default and BMD approaches, where PKs was not explicitly considered, an overall UF_A of 3 (1 for PDs × 3 for PKs) was used. When the PBPK model was employed, the overall UF_A was set to 1.

For all RfD calculation approaches, a value of 1 was selected for extrapolating the results from a subchronic study to chronic exposures (UF_S). Although no chronic oral studies are currently available for EGBE, there does not appear to be a significant increase in the severity of hemolytic effects beyond one to three weeks of oral (30) or inhalation (31) EGBE exposures.

For the default and PBPK approaches, a value of 3 was selected for extrapolating a LOAEL to a NOAEL (UF_L). A value of less than 10 is justifiable because there was information that indicated that the LOAEL was very near the threshold level for the hematologic effects of concern. For the two approaches employing BMD analysis, a value of 1 was used because of the minimal and precursive nature of the critical lesion (cell swelling as measured by increased MCV) and the fact that a BMD_{05} for a minimally adverse effect is typically deemed to be equivalent to a NOAEL for continuous data sets.

For the UF_D, a value of 1 was used for all approaches. While no chronic oral studies or adequate human data are available for EGBE, oral and inhalation dose–response data indicate that there would be little if any increase in severity of hemolytic effects beyond subchronic exposure durations (30,31). In addition, chronic and subchronic studies in rats and mice, as well as reproductive and developmental studies, indicate that the database for EGBE is not particularly limited.

In the default approach, the male rat LOAEL for hepatocellular changes was chosen because it provided the more conservative RfD among the available dose–response data amenable to the default approach. Hepatocellular changes in the male were considered the probable result of adaptation to a subclinical level of hemolysis at this dose. The male rat LOAEL of 55 mg/kg-day from a 91-day continuous drinking water study was used as the unadjusted (because the exposure was continuous) HED. This HED divided by the UFs yielded an RfD of 0.6 mg/kg-day [i.e., $55 \text{ mg/kg-day} \div (UF_H = 10 \times UF_A = 3 \times UF_L = 3)$].

In the other approaches, the RfD is based on a different critical effect. While HCC and hemolysis both result from EGBE exposure, there was more direct scientific support for the dose metric causing hemolysis than HCC. The hemolytic effects were observed more consistently in females than males; thus, a LOAEL of 59 mg/kg-day (in female rats) was chosen as the POD from which C_{max} BAA was to be estimated in blood using the PBPK model developed by Corley et al. (27,29). The model predicts a C_{max} BAA in blood of 103 mM (assuming a 12-hour wake drinking cycle 7 days/wk). Subsequently, the human PBPK model was exercised with varying EGBE concentrations in water to derive an equivalent C_{max} BAA in blood, assuming a water intake of 2 L/day/70-kg person. Regression of the internal dose metric (C_{max} BAA) to the EGBE dose level (mg/kg-day) led to the result that 7.6 mg/kg-day EGBE in drinking water would yield the C_{max} BAA of 103 mM in man, i.e., the equivalent concentration estimated in rat. This HED of 7.6 mg/kg-day is thus the POD from this approach. The HED divided by the UFs yielded an RfD of 0.3 mg/kg-day [7.6 mg/kg-day ÷ (10 × 1 × 3)].

A third approach incorporated BMD analysis for several of the end points used above. MCV was the most sensitive effect, and thus the basis for a BMD_{05} of 49 mg/kg-day as the POD. This POD was divided by the UFs to derive an RfD of 2 mg/kg-day [49 mg/kg-day ÷ (10 × 3 × 1)].

The fourth approach combined the BMD analysis with the PBPK model. The rat PBPK model was exercised to simulate blood concentrations of metabolite resulting from each of the exposure concentrations. BMD analysis was then performed on the estimated C_{max} values for BAA, resulting in a $BMDL_{05}$ of 64 μM. The human PBPK model yielded a corresponding HED of 5.1 mg/kg-day as the POD, assuming that rats and humans receive their entire dose of EGBE from drinking water over a 12-hour period each day. This HED was divided by the UFs to derive an RfD of 0.5 mg/kg-day [5.1 mg/kg-day ÷ (10 × 1 × 1)].

The derivations for the RfC were similar to those for the RfD (Table 2), except for the UFs. In BMD analysis for the RfC, the moderate LOAEL-to-NOAEL UF_L of 3 was retained (i.e., $UF_L = 3$ instead of $UF_L = 1$) because the RfC derivation was based on a critical effect of decreased RBC count, which was deemed more serious than the precursor effect of increased MCV. The default RfC methodology also derives a scientifically based HEC, so the UF_A is 1 for both the nonmodel and the PBPK model approaches. For the combined PBPK/BMD approach, the RfC was 13 mg/m³ [380 mg/m³ ÷ ($UF_H = 10 \times UF_A = 1 \times UF_L = 3$)].

Table 2 Comparing Different Reference Dose and Reference Concentration Derivations

	Default	PBPK	BMD	BMD/PBPK
Reference dose				
$LOAEL_{HEC}$	55 mg/kg-day	7.6 mg/kg-day	Admin. dose	Internal dose
BMD_{05}	–	–	49 mg/kg-day	5.1 mg/kg-day
$UF_{H, A, L}$	10, 3, 3	10, 1, 3	10, 3, 1	10, 1, 1
RfD	0.6 mg/kg-day	0.3 mg/kg-day	2 mg/kg-day	0.5 mg/kg-day
Reference concentration				
$LOAEL_{HEC}$	150 mg/m³	474 mg/m³	Admin. dose	Internal dose
BMD_{05}	–	–	130 mg/m³	380 mg/m³
$UF_{H, A, L}$	10, 1, 3	10, 1, 3	10, 1, 3	10, 1, 3
RfC	5 mg/m³	16 mg/m³	4 mg/m³	13 mg/m³

Abbreviations: RfD, reference dose; RfC, reference concentration; PBPK, physiologically based pharmacokinetic; BMD, benchmark dose; LOAEL, lowest-observed-adverse-effect level; UF, uncertainty factors.

A quantitative assessment for carcinogenicity was not performed because the relevancy of the observed test animal tumors to humans was not clear at the time the assessment was developed.

Evaluation Criteria, Why Used or Not Used

The analysis with just the PBPK model resulted in the most conservative estimate among all of the RfD options, and the least conservative estimate for the RfC. Similarly, BMD analysis alone did not yield a consistent bias in reference values. The BMD/PBPK model approach was considered the most appropriate for the RfD derivation because it was considered to be the method that incorporated the most mechanistic and dose–response information.

Current Status of Model and Assessment

Deisinger and Boatman (32) have further applied the EGBE PBPK model to evaluate whether inhalation exposure in mice to EGBE are likely to generate 2-butoxyacetaldehyde (BAL) concentrations sufficient to cause DNA damage in target tissues. BAL is a transient, labile intermediate in the oxidation of EGBE to BAA. Their results indicate that this is unlikely.

The assessment for EGBE (as of March 2006) is in the final stages of review for an updated characterization of the dose–response for carcinogenic effects.

VINYL CHLORIDE

Weihsueh Chiu

The 2000 U.S. EPA IRIS Assessment for VC (33) made extensive use of PBPK modeling. For noncancer, the model was used for both interspecies extrapolation and route-to-route extrapolation of internal doses for the development of an oral RfD and inhalation RfC from an oral rodent study. For cancer, the model was used for interspecies extrapolation in the development of an oral cancer slope factor and inhalation unit risk from rodent oral and inhalation cancer bioassays, respectively. The PBPK model analysis applied UFs of $UF_{A-PK} = 1$, $UF_{A-PD} = 3$, and $UF_H = 10$ to an HED of 0.09 mg/kg-day to yield an RfD of 0.003 mg/kg-day (0.09 mg/kg-day ÷ 30), about 2.3-fold higher than the value that would have been obtained using the default factors and the NOAEL. There was no default methodology for calculating an RfC from an oral study, so without the PBPK model the RfC would not have been developed. In the derivation of the cancer slope factor for oral exposures, the PBPK model predicted that 1 mg/kg-day in rats corresponds to 0.6 mg/kg-day in humans, a difference of a factor of two from the default approach, which assumes equivalent risks for equivalent mg/kg$^{\frac{3}{4}}$-day (i.e., 1 mg/kg-day in the rat would be assumed equivalent to about 0.3 mg/kg-day in humans). For the inhalation unit risk, the model-derived estimate, even with the addition of adjustments for early exposure, is about 10-fold lower (i.e., less risk) than the estimate based on the administered dose and standard defaults assumptions. The default assumptions wrongly represent absorption and do not account for species differences in metabolic activation and other PK determinants of internal dose. The PBPK model accounts for these PK differences and predicts that rats will have a considerably greater steady-state concentration of the active metabolite of VC than humans, and thereby a greater risk.

Model Structure, Focus, and History

The VC risk assessment implemented a PBPK model developed by Clewell et al. (34,35). This model describes the dosimetry of VC and metabolism by P450 isoforms in the liver, and was an adaptation of a previously developed PBPK model for vinylidene chloride (36). The model has four tissue compartments: a richly perfused tissue compartment for all richly perfused organs except liver, a slowly perfused tissue compartment that includes muscle and skin, a fat compartment, and a liver compartment. The model assumes perfusion-limited kinetics.

The model represents P450 metabolism of VC in the liver with two saturable pathways—one by CYP2E1 and one by an unspecified isoform with higher capacity but lower affinity. Both pathways are assumed to produce chloroethylene oxide, chloroacetaldehyde, or other intermediate reactive products. These reactive metabolites may undergo further metabolism (leading to CO_2), or react with glutathione (GSH) or with other cellular materials, including DNA. Because exposure to VC has been shown to deplete circulating levels of GSH, a simple description of GSH kinetics was also included in the model.

MOA, Choice of Dose Metric, and Specific Application

Reactive metabolites were presumed to be the active toxin leading to both cancer and noncancer effects. Andersen et al. (37) proposed that an appropriate dose metric for reactive metabolites would be the total amount of the reactive metabolite formed divided by the volume of the tissue into which it is produced. For liver toxicity/carcinogenicity, all metabolism is assumed to occur in the liver, so the dose metric for liver toxicity used in the assessment to derive the reference values is milligram metabolite/L of liver (AML).

The U.S. EPA's compiled biological values (38) were used for all physiological parameters except for alveolar ventilation in the human, which was calculated from the standard U.S. EPA value for the ventilation rate in the human, $20\,m^3$/day, assuming a 33% pulmonary dead space. The partition coefficients for Fischer-344 (F344) rats were taken from Gargas et al. (39), and those for Sprague–Dawley rats were taken from Barton et al. (40). The Sprague–Dawley values were also used for modeling of Wistar rats. Blood/air partition coefficients for the other species were obtained from Gargas et al. (39) and the corresponding tissue/blood partition coefficients were estimated by dividing the Sprague–Dawley rat tissue/air partition coefficients by the appropriate blood/air value.

The affinity for the 2E1 pathway (KM1) in the rat, mouse, and hamster was set to 0.1 on the basis of studies of the competitive interactions between CYP2E1 substrates in the rat (41,42). The affinity used for the non-2E1 pathway (KM2) in the mouse and rat was set during the iterative fitting of the rat total metabolism, GSH depletion, and rate of metabolism data, described below. The capacity parameters for the two oxidative pathways (VMAX1C and VMAX2C) in the mouse, rat, and hamster were estimated by fitting the model to data from closed-chamber exposures with each of the species and strains of interest (40,43–45). After the other parameters were scaled from animal weights obtained from individual studies, the model was exercised for optimization to a single pair of values, VMAX1C and VMAX2C, to be used for all of the data on a given sex/strain/species. The parameters for subsequent metabolism are not discussed here, because the dose metrics used in the risk assessment do not depend on them.

For humans, because there was no evidence of high-capacity, low-affinity P450 metabolism for chlorinated ethylenes in the human, VMAX2C in the human was set to zero. The human VMAX1C was set to a value derived from the rhesus monkey (46). Because VMAX1C is an allometrically scaled parameter, it was thus assumed the actual metabolic capacity scaled as body weight to the three-fourth power. This was based on the interspecies scaling of CYP2E1 for other substrates between rodents and humans (37). A value for KM1 was set equal to the rodent value of 0.1 mg/L, based on in vitro data suggesting no significant species differences for kinetic parameters for CYP2E1 substrates (47).

In developing the oral RfD and the inhalation RfC, PBPK model results replaced the PK component of the default interspecies UF (UF$_{A-PK}$); the PD component was retained. For the RfD, the rodent PBPK model was run setting the dose to the NOAEL of 0.13 mg/kg-day [from the lifetime dietary study of Til et al. (48,49)] to yield the value of the internal AML dose metric. The HED of 0.09 mg/kg-day was then derived using the human PBPK model to back estimating from an equivalent value for AML. As information for the oral route of exposure was limited, the conservative assumption of 100% oral absorption over a continuous 24-hour period was made to maximize the formation of the reactive species in both humans and rodents. It should be noted that at these doses, the relationship between exposure and AML was linear in all cases, so all the conversions between species and routes of exposure were linear.

The RfC was based on liver cell polymorphism and cysts observed in the chronic dietary rat study of Til et al. (48,49). Similar to the RfD derivation, a human equivalent concentration (HEC) of $2.5\,\text{mg/m}^3$ was derived by back estimating an equivalent value for AML with the human PBPK model.

For the cancer slope factors, the rodent PBPK model was first used to convert the bioassay administered doses or exposures to internal AML metric. Both oral and inhalation studies were used, and 100% oral absorption was assumed similar to the noncancer derivations. Cancer risk modeling (e.g., $q1^*$ values from linearized multistage model as well as 0.1/LED10 values from BMD modeling) was then performed using the values for AML to derive a cancer slope factor expressed in terms of excess cancer risk per unit AML. The human PBPK model was then used to express this slope factor as an oral slope factor or an inhalation unit risk (noting that at doses where this low-dose extrapolation was being performed, the relationship between exposure and AML was linear). Uncertainty in the PBPK model was determined by conducting a Monte Carlo analysis, in which risk was calculated by sampling distributions for the parameters used in the model, resulting in a distribution of calculated risks. This analysis of the VC model demonstrated that the 95th percentile of the distribution of upper confidence limit (UCL) risks was within 50% of the mean UCL risk. In addition, sensitivity analyses showed that the most sensitive parameters were all of the ones that could be reasonably characterized from experimental data in rodents. However, the sensitivity of the risk predictions to the human values of these parameters indicated that the risk from exposure to VC could vary considerably from individual to individual, depending on the individual's physiology, level of activity, or metabolic capability.

Comparison of Model Application to Default Value

The default methodology for the RfD would have applied a $UF_A = 10$ and $UF_H = 10$ to a NOAEL of $0.13\,\text{mg/kg-day}$ to yield an RfD of $0.0013\,\text{mg/kg-day}$ ($0.13\,\text{mg/kg-day} \div 100$). The PBPK model analysis applied UFs of $UF_{A\text{-PK}} = 1$, $UF_{A\text{-PD}} = 3$, and $UF_H = 10$ to an HED of $0.09\,\text{mg/kg-day}$ to yield an RfD of $0.003\,\text{mg/kg-day}$ ($0.09\,\text{mg/kg-day} \div 30$), about 2.3-fold higher than the default value.

There is no default methodology for calculating an RfC from an oral study, so without the PBPK model no RfC would have been developed.

A default cancer slope factor for oral exposures would have converted equivalent risks for equivalent $\text{mg/kg}^{\frac{3}{4}}$-day, so $1\,\text{mg/kg-day}$ in the rat would be assumed equivalent to about $0.3\,\text{mg/kg-day}$ in humans. The PBPK model predicts $1\,\text{mg/kg-day}$ in rats to correspond to $0.6\,\text{mg/kg-day}$ in humans, a difference of a factor of two from the default.

The derived inhalation unit risk estimate of $4.4\,\text{E-6/}\mu\text{g/m}^3$, even with the addition of adjustments for early exposure ($8.8\,\text{E-6/}\mu\text{g/m}^3$), is about 10-fold lower (i.e., less risk) than the previous U.S. EPA "HEAST" value of 8.4E-5 (50). There are several reasons for this. First, in the earlier estimate, absorption was assumed to equal 50% in the rat versus 100% in humans. Such an assumption is invalid in that virtually all VC is absorbed in both species until a blood concentration determined by the inspired concentration and the blood-to-air partition coefficient is reached. Because the partition coefficient is about twice as large in rats as in humans, arterial blood concentration will be greater in rats than in humans, rather than less. Metabolic activation of VC (VMAX/KM) is about 10 times faster in rats than in humans. Blood flow to the liver is more rapid. After accounting for these and other PK differences, the model predicts that rats will have a considerably greater steady-state concentration of the active metabolite of VC than humans, and thereby a greater risk. Use of administered dose and standard defaults, on the other hand, would result in a prediction of lower risk in rats than in humans.

Evaluation Criteria, Why Used or Not Used

Five different PBPK models were reviewed as part of the IRIS assessment for VC. Many of the models gave similar predictions as to interspecies and route-to-route extrapolation. The Clewell et al. (34) model was considered the most tested against experimental data for both total metabolism and GSH depletion in rats as well as closed-chamber VC exposure data in humans.

Current Status of Model and Assessment

The IRIS assessment for VC was completed in 2000. No further work on the VC PBPK model was identified as of early 2006.

DICHLOROMETHANE

Paul Schlosser

The U.S. EPA revision of the cancer assessment for DCM was one of the first to incorporate PBPK modeling in the derivation of an inhalation unit risk value (1,51–54). The model was not used for other derivations such as an inhalation-to-oral extrapolation. The PBPK model incorporated the relevant information on the PKs and metabolism of DCM, and estimated an internal dose metric of total amount of metabolite produced over time from the GSH–S-transferase (GST) pathway. The unit risk based on the PBPK model results was approximately ninefold lower (i.e., ninefold lower risk per ppm DCM in air) than the previous unit risk estimate, which was based on the administered dose and assumed 100% absorption of the inhaled dose of DCM.

Model Structure, Focus, and History

The DCM inhalation cancer assessment implemented the five-compartment PBPK model of Andersen et al. (55). The model was developed to describe the dosimetry of DCM and its metabolic conversion in the liver and lung, where it can either be conjugated with GSH or be oxidized by cytochrome P450 (MFO), after both inhalation and oral exposure. The model is an elaboration of an earlier model by Ramsey and Andersen (56) for styrene. The DCM model adds a lung tissue compartment in addition to (and downstream from) the gas-exchange compartment, and represents a linear metabolic pathway for GSH conjugation as well as a saturable pathway for oxidation in both the lung tissue and the liver compartments.

Metabolic steps subsequent to GSH conjugation and oxidation were not specifically represented. Reaction products include reactive metabolites (formyl chloride in the oxidative pathway and chloromethyl GSH in the conjugation pathway), which could contribute to the observed toxicity, along with carbon monoxide and carbon dioxide. Dosimetry data on the metabolites were not available, so the model parameters for rodents were estimated by fitting gas uptake data and setting the ratio of lung:liver activity for each pathway based on in vitro activity data. Based on comparisons of the rodent-fitted parameters, the authors concluded that extrapolation to humans should be based on an allometric factor of 0.7. The fully developed model, with no further adjustments to the parameter values, was tested against additional experimental data, including human dosimetry data, and the fits were quite good.

The authors hypothesized that the reactive metabolites would be short-lived and removed by linear rate processes, and hence that the area under the tissue concentration–time curve (AUC_{tissue}) for these metabolites in liver and lung would be proportional to the AUC for their respective rates of production in those tissues. Also, the metabolites were assumed not to distribute outside of the tissue in which they were formed. Three alternative dose metrics were thus considered for assessing the carcinogenicity of DCM: area under the tissue concentration curve (AUC_{tissue}) of DCM itself, time-integrated production of CYP pathway metabolites (AMT-CYP), and time-integrated production of GST pathway metabolites (total amount of metabolite produced, AMT-GST). The model was developed specifically to address rodent-to-human extrapolation of the DCM-induced cancer dose–response based on one of these dose metrics; however, the authors did not know ahead of time which metric would be selected as most applicable to DCM's carcinogenic effects.

MOA, Choice of Dose Metric, and Specific Application

Both oxidative and conjugation pathways produce reactive metabolites from parent DCM. Only a correlation (or lack thereof) between the estimated amount of each metabolite and the observed tumor response was available to suggest a potential causal connection because the mode of action for DCM was and is still not known. The model results were therefore used to compare the different dose metrics with the rodent tumor data to determine which metric was the best predictor of the response.

A clear dose–response was observed for lung and liver tumors in B6C3F1 mice after inhalation exposure to DCM 6 hr/day, 5 days/wk for 2 years (57), while a two-year drinking-water bioassay failed to show a dose-related increase in this same test strain (58). The oxidative metabolic rates were predicted to be fairly similar for inhalation doses of 4000 and 2000 ppm (due to saturation of this pathway) and for drinking water exposure of 250 mg/kg/day, but the tumor incidence in both lung and liver after drinking water exposure was much lower than after inhalation exposure. The AUC for both the DCM and the metabolite production from the conjugation pathway, on the other hand, correlated fairly well with this route-dependent difference, as well as providing a better correlation with the difference between 2000 and 4000 ppm inhalation responses. Thus, the amount of metabolite from the oxidative pathway was not considered a good dose metric. Because it was deemed more likely that the tumors were formed by a reactive metabolite than by DCM itself, the AMT-GST dose metric was selected as the best metric to use in the derivation of an inhalation unit risk.

It should be noted here that while the key metabolic rate used in the assessment was identified with the conjugation pathway, in practical terms that rate represents the linear part of both pathways because the rate constants were informed but not completely identified from the in vitro data on DCM–GSH conjugation. In other words, it was not known how well the estimated rate actually corresponded only with production of the GSH conjugate, rather that the representation described well the linear component of DCM metabolic clearance.

The application of the Andersen et al. (55) model in the risk assessment differed from that suggested by Andersen et al. in three important details: (i) a standard value of $0.0407 \, m^3/day$ for the ventilation rate of mice was used rather than $0.084 \, m^3/day$, which was measured during the gas uptake studies used to parameterize the model by Andersen et al., but presumed not to be representative of bioassay conditions; (ii) the human respiration rate was set at $20 \, m^3/day$ for average daily activity rather than $12.5 \, m^3/day$, which is a resting level; (iii) and a correction factor of 12.7 for potential differences in PD sensitivity, based on surface-area scaling (body weight raised to the two-third power), was used. The first two changes would clearly reduce the level of AMT-GST associated with tumors in mice and raise the level predicted in humans, respectively. The third point was a matter of some debate because the surface-area correction factor had not been specifically identified by the U.S. EPA as an adjustment for PD versus PK sensitivity prior to this assessment. If that correction had been for PK differences, then the use of the PBPK model would have obviated it.

After implementing these changes in the mouse inhalation rate, the predicted mouse AMT-GST for liver and lung were multiplied by 5/7 to obtain a weekly average metric (since the bioassay exposures were 5 days/wk), and the resulting values were used in place of the applied dose in calculating q1* values (unit risks; upper bounds on the linear component of the linearized-multistage risk model) for the mouse liver and lung tumor responses. The human PBPK model was then used to calculate the value of AMT-GST in liver and lung resulting from a 1 ppm DCM exposure, with the recognition that the exposure to AMT-GST relationship was linear at these low levels. The resulting values were multiplied by the 12.7 correction factor to account for the sensitivity differences. Multiplying the mouse-derived q1* values by those values of AMT-GST then yielded unit risk values (per ppm) for human lung and liver tumors, respectively. The result is the same as if the mouse internal q1* had been multiplied by 12.7 to adjust for sensitivity, and then that human equivalent q1* was multiplied by the human AMT-GST value per ppm. Finally, because the lung and liver tumors were considered to arise independently, the resulting inhalation unit risks were added to obtain an overall cancer unit risk for humans.

Comparison of Model Application to Default Value

The previous default approach to the unit risk assumed that 100% of the inhaled DCM dose would be absorbed, resulting in a unit risk approximately nine times higher than that obtained with the use of the PBPK model. The difference would have been greater had other aspects of the PBPK model derivation not been included. For example, the PBPK model estimate of AMT-GST was corrected for interspecies differences in sensitivity by using a surface-area correction factor. Had the surface-area correction factor been ascribed to PK differences, rather than PD differences, it would not have been included (because the model accounted for the PK differences) and the PBPK model-based unit risk would have been 111-fold lower than the previous default. Had the ventilation rates proposed by Andersen et al. (55) been retained (i.e., 0.084 and 12.5 m^3/day for mice and humans, respectively), the model-based unit risk would have been lower still, at 167-fold.

Evaluation Criteria, Why Used or Not Used

The model of Andersen et al. (55) was used in the IRIS assessment for VC because it incorporated what was then known about animal and human PKs and metabolism of VC and reduced some of the uncertainty inherent in the default approach to interspecies extrapolation of the dose–response relationship. The Andersen et al. (55) PBPK model also performed well when tested against experimental animal and human dosimetry data and proved useful in resolving which among the internal dose metrics correlated best to the observed tumor response (i.e., the amount of metabolite produced by the GST pathway).

The assessment noted that a great deal of uncertainty remained specifically regarding the PKs, PDs, and mechanisms of carcinogenicity for DCM. Here too, the PBPK model proves more useful than the default approach because the deficits in the model (e.g., poor fits, need to optimize sensitive parameters, or parameter identifiability issues) helps guide further research. The results from some of this research are now available, and are discussed in the next section.

Current Status of Model and Assessment

Andersen et al. have continued to develop the DCM PBPK model for mice, first by extending the model to describe the kinetics of carbon monoxide and carboxyhemoglobin in blood (55), and then by addressing time-related changes such as the enzyme induction of both the oxidative and the GSH conjugation pathways (59). Recently, other researchers have used the DCM PBPK model to characterize human interindividual variability relative to known polymorphism in GSH–S-transferase theta 1—first by El-Masri et al. (60), then by Johanson et al. who made use of distinct human data sets (61,62), and then by Sweeney et al. (63). A PBPK model estimate of human variability could be used in future revisions of the assessment to replace the PK part (UF_{H-PK}) of the default intraspecies UF (UF_H).

A reestimation of human metabolic constants has also been conducted by ENVIRON using Bayesian analysis, assigned distributions for all physiological variables, and the extant human data (64). A caveat for this analysis is that it included a term for DCM metabolism in rapidly perfused tissues other than the liver and lung in humans, which lowered to some extent the target tissue (liver and lung) dose metrics in humans. However, this metabolic term was not included in the mouse model, which would lead to an overestimate of the liver dosimetry in mice and hence an underestimate of the risk. While the mean posterior value for the ratio of extrahepatic and extrapulmonary metabolic constants (VMAX values) in humans was only 2% (due to the fact that blood flow tends to limit hepatic metabolism) the proportion of metabolic rate in richly perfused tissues versus the liver is likely to be higher than 2%.

Subsequent to the ENVIRON report, a scientific peer review panel was convened in mid-2005, to review the "state of the science" for DCM and to critically evaluate the new information and analyses for use in assessing potential human cancer risks from DCM exposure (65). Currently (mid-2006), the U.S. EPA is considering these new data and PBPK model analyses in an update of the IRIS assessment for DCM.

XYLENE

Robert S. DeWoskin

An *m*-xylene PBPK model was used in a 2003 IRIS assessment for xylenes (1) to derive a HEC from an inhalation exposure, but only for comparison (and further support) of the RfC derived using the standard RfC methodology for a category 3 gas. The PBPK model–derived RfC was 0.16 mg/m^3, compared to the standard RfC approach value of 0.1 mg/m^3. One reason for the similarity is that, in this example with xylene, there were very little differences between the default and the PBPK modeling approach with respect to the UFs.

The *m*-xylene PBPK model did not simulate absorption from an oral exposure, so the model was not used to derive an RfD. The carcinogenic potential of xylenes could also not be assessed because the data were inadequate to characterize a dose–response.

Model Structure, Focus, and History

A PBPK model was developed to simulate the PKs of *m*-xylene in male Sprague–Dawley rats (66–69) and humans (70–72) following an inhalation exposure. The model has also been used to simulate the kinetics of inhaled mixtures of xylenes and other aromatic solvents (66–68,70–73). The model has no terms to represent oral absorption of *m*-xylene, so it cannot be used for a route-to-route extrapolation. The inhalation PBPK model was developed from inhalation exposure data for both rats (68,73) and humans (71,73).

The *m*-xylene PBPK model is a classic perfusion-limited four-compartment PBPK model (liver, fat, other slowly perfused tissues, other richly perfused tissues) with absorption of inhaled *m*-xylene represented by a gas-exchange lung compartment. Metabolism of *m*-xylene is modeled as only occurring in the liver (as a surrogate for total body metabolism) and as following Michaelis–Menten kinetics. Rat metabolic parameters were derived from closed chamber inhalation studies, and tissue partition coefficients were derived from gas-phase vial equilibration data. The human maximum velocity metabolic parameter was estimated by scaling the rat value on the basis of body weight to the three-fourth power, and the human blood:air partition coefficient was determined via vial equilibration. All other metabolic and physiochemical parameters were considered to be species-invariant.

The data used to test the *m*-xylene model included Sprague–Dawley male rat data from closed chamber inhalation studies (for metabolic parameters), and urinary metabolites, blood and tissue levels of *m*-xylene alone or in combination exposures with toluene and ethylbenzene (68,73). The human data (from male volunteers) consisted of *m*-xylene venous blood concentrations from exposure to xylene alone or to xylene in combination with toluene and ethylbenzene (71,73).

MOA, Choice of Dose Metric, and Specific Application

As of 2003, there were no human oral subchronic or chronic exposure studies to determine the toxicity of xylenes. Available animal studies did not identify a single sensitive health effect other than changes in body weight. Although changes in body weight have been consistently reported in several male rat studies, no mode of action has been proposed. The *m*-xylene PBPK model does not represent absorption from an oral exposure, so the PBPK model was not used in the derivation of the RfD or the oral slope factor. The RfD for xylenes was derived with a NOAEL (from a two-year study in rats) for change in body weight, using the standard default UFs.

For inhalation exposures, the weight of evidence from limited human data and more extensive animal data indicate neurological impairment, and developmental effects are the most sensitive adverse effects following repeated inhalation exposure to xylenes. Reversible symptoms of neurological impairment and irritation of the eyes and throat are well-known acute effects from inhaled xylene or other aromatic solvents. In general, acute effects are expected to result from reversible molecular interactions of the parent compound (i.e., not a metabolite) with target tissue membranes, e.g., neuronal membranes, and are most pronounced at high exposure levels in excess of 1000 ppm.

The standard methodology (74) was used to derive the RfC. The available human data were insufficient for derivation of an RfC, and chronic animal inhalation data were lacking, so a subchronic study by Korsak et al. (75) was selected as the principal study. A NOAEL of 50 ppm and a LOAEL of 100 ppm were identified for decreased rotarod performance (impaired motor coordination). This neurologic test was administered 24 hours after termination of the exposure period, when xylenes would be expected to have been eliminated from the body.

The NOAEL of 50 ppm ($217\,mg/m^3$) was duration adjusted to $39\,mg/m^3$ (5/7 days, 6/24 hours). Xylene is considered a category 3 gas because of its low water solubility, its potential for accumulation in blood during exposure, and the most sensitive effect is an extrarespiratory effect. For a category 3 gas, the $NOAEL_{[HEC]}$ is calculated by multiplying the duration-adjusted $NOAEL_{(ADJ)}$ by a ratio of the blood:gas partition coefficients for the rat over the blood:gas partition coefficient for the human.

$$NOAEL_{(HEC)} = NOAEL_{(ADJ)} \times [(Hb/g)A/[(Hb/g)H]] \qquad (1)$$

where Hb/g = blood/gas partition coefficient for the species in question, animal (A) or human (H).

Tardif et al. (71) reported an (Hb/g) H of 26.4 for *m*-xylene, and an earlier study from the same group (68) reported an (Hb/g) A of 46.0 for *m*-xylene in the rat. Thus, the ratio of the animal-to-human blood:gas partition coefficients is 1.7 ($1.7 = 46.0/26.4$). However, when (Hb/g) A>(Hb/g) H, a value of 1 is used for the ratio (74). The result is a $NOAEL_{(HEC)} = NOAEL_{(ADJ)}$ or $39 \, mg/m^3$.

The $NOAEL_{[HEC]}$ of $39 \, mg/m^3$ was divided by a total UF of 300 for an RfC of $0.1 \, mg/m^3$. The UFs were a $UF_{A-PD} = 3$ for animal-to-human extrapolation using standard RfC methodology for the dosimetric adjustment (i.e., $UF_{A-PK} = 1$), $UF_H = 10$ for intrahuman variability, $UF_S = 3$ for extrapolation from subchronic to chronic duration, and $UF_D = 3$ for deficiencies in the database.

A UF of 10 was applied for intraspecies uncertainty to account for human variability and sensitive populations. The degree of human variance in abilities to absorb or dispose of xylenes is unknown, as is the degree of human variance in responding to xylene's neurotoxicity. Results from developmental toxicity studies of rats exposed by inhalation during gestation indicate that adverse developmental effects occur only at doses higher than chronic doses producing the critical effects observed in adult male rats in the principal and supporting studies, suggesting that the developing fetus is not at special risk from low-level exposure to xylenes. However, as with oral exposure, the effects of inhaled xylenes in other potentially sensitive populations such as newborns or young children or animals have not been assessed.

A UF of three was applied for extrapolation from subchronic to chronic duration. A factor of 10 was not used because the changes in rotarod performance did not increase with time from 1 to 3 months, and they were similar to those described in a separate study of 6 months duration (76).

A UF of three was applied for uncertainties in the database. The inhalation database includes some human studies, subchronic studies in rats and dogs, neurotoxicity studies, a one-generation reproductive toxicity study, developmental toxicity studies, and developmental neurotoxicity studies. Although the available developmental toxicity studies are confounded by a lack of litter incidence reporting, the data reported for fetal incidences did not indicate effects at levels lower than the level found to induce neurologic impairment in several end points in male rats. The database is lacking a two-generation reproductive toxicity study.

In the PBPK model analysis, three different dose metrics were evaluated to estimate the HEC based on the animal dose–response data:

1. An overall time-weighted-average (TWA) blood concentration ($0.198 \, mg/L$; averaged over 1-hour intervals across 13 weeks)
2. The maximum (MAX) blood concentration attained on any given day during exposure ($1.09 \, mg/L$; which appeared to be essentially a constant over 13 weeks

3. The midpoint (MID) between the maximum (1.09 mg/L) and the minimum (0/mg/L) concentration on any given day during exposure (0.55 mg/L)

The model predicted air concentrations that would produce the target steady-state concentrations in human blood for the three different dose metrics resulting from a continuous inhalation exposure as follows: 10.5 ppm (46 mg/m^3) for the TWA surrogate, 49.8 ppm (216 mg/m^3) for the MAX, and 27.4 ppm (106 mg/m^3) for the MID.

The rat PBPK model also predicted that blood concentrations were essentially zero when the critical effects on rotarod performance were measured (24 hours after cessation of exposure). This result supported the idea that the observed effects were not dependent on the concurrent presence of xylenes in the blood, and that there may be a persistent neurological effect.

Brain concentrations would have been a better dose surrogate to use in the model, but the model did not have a compartment that represented the brain (i.e., most likely due to lack of supporting data).

The TWA dose metric was considered the most relevant description of the exposure experienced by the rats in the study compared to the other two metrics (MID or MAX), especially since the effects were measured after *m*-xylene had completely cleared from the blood. Based on the TWA metric and the rat data from Korsak et al. (75) the model predicted an HEC of 46.5 mg/m^3. Applying UFs totaling 300 to the model-derived HEC of 46.5 mg/m^3 resulted in an RfC of 0.16 mg/m^3, very similar to the standard RfC approach value of 0.1 mg/m^3. The UFs were a $UF_{A-PD} = 3$ for animal-to-human extrapolation using the PBPK model for a dosimetric adjustment (i.e., $UF_{A-PK} = 1$), and comparable $UF_H = 10$, $UF_S = 3$, and $UF_D = 3$ as in the standard approach.

Comparison of Model Application to Default Value

The PBPK model–derived RfC of 0.16 mg/m^3 is similar, although slightly higher than the standard RfC approach value of 0.1 mg/m^3. The difference may be due to the rounding of the animal-to-human blood:gas partition coefficients from a value of 1.7 to 1 in the standard RfC methodology, whereas the PBPK model runs did not modify these coefficients. Or there may be other PK differences that are represented in the model that would account for the higher RfC. The assessment does not discuss what those might be.

One reason for the similarity is that, in this example with xylene, there were very little differences between the default and the PBPK modeling approach with respect to the UFs. The standard approach for a category three gas includes a dosimetric adjustment to derive the HEC that replaces the default UF_{A-PK} factor, similar to the replacement of this factor when using a PBPK model. All of the other default UFs remained the same for both the approaches.

Evaluation Criteria, Why Used or Not Used

The *m*-xylene PBPK model was considered to be sufficiently well-developed and tested to be described in detail in the assessment, to support the RfC value derived with the default approach, and to evaluate different dose metrics for concordance with the response and thus provide some insight into a potential mode of action for xylene-induced neurotoxicity. The assessment does not explicitly state why the PBPK model–derived RfC was not used in lieu of the one developed with the default approach; however, because it was slightly higher (i.e., less risk), the lower value might have been selected as more conservative. Because the model did not represent oral absorption, it was not applicable to the RfD derivation. The absence of adequate data to assess xylene's carcinogenic potential precluded the use of the model to derive unit risk.

Current Status of the Model and the Assessment

The xylene PBPK model continues to have applications in the derivation of safe levels for acute inhalation exposures at rest or with an active workload (77), to evaluate the interaction effects of xylene in mixtures (78) or to evaluate the dermal bioavailability of aqueous xylene in rats and human volunteers.

As of mid-2006, there have been no revisions to the IRIS assessment for xylene from 2003, nor any on-going activities to update the 2003 assessment.

BORON AND COMPOUNDS

Carolyn Smallwood, Jeff Swartout, Linda Teushler, and John C. Lipscomb

A BMD analysis was used to develop a POD for developmental toxicity of boric acid (BA) in rats. PBPK modeling was not employed, but chemical-specific PK data and compound-related physiologic data were used to develop data-derived UFs for boron and compounds (borates, BA), based substantially on methodology proposed by the International Programme on Chemical Safety (IPCS) (see Chapter 2). Renal clearance measures from studies with pregnant rats and pregnant humans served as the basis for nondefault intraspecies PK adjustments (UF_{A-PK}) and measures of the variability of glomerular filtration in pregnant humans served as the basis for nondefault values for intraspecies PK adjustments (UF_{H-PK}). The approach employed relied on scientifically valid data, and the UF was reduced from a default value of 100 to 66.

Boron (B) is a commercially important mineral, frequently marketed as BA (H_3BO_3). Production of boron accounts for 2,700,000 tons annually, based on figures from 1994 (79). Human exposures are anticipated to be via the oral route, resulting from dietary consumption primarily from fruits and vegetables.

BA has been evaluated for decades, and the study data are rich. For risk assessment the RfD was derived using a $BMDL_{05}$ for decreased fetal body weight in rats. Because the critical effect was a developmental effect, no duration adjustment was necessary; therefore, a UF for subchronic-to-chronic duration (UF_S) was not required, additionally no database UF (UF_D) was necessary because of the richness of the available database. A UF for extrapolation from a LOAEL to a NOAEL was also not necessary because BMD modeling was used to determine the POD. Thus, only the values for the interspecies extrapolation UF (UF_A) and human variability UF (UF_H) remained.

Each of the UF_A and UF_H factors has a default value of 10; the presently available RfC methodology (74) indicates that UF_A may be divided into components addressing TK and TD, each component valued at one-half order of magnitude ($10^{0.5}$), mathematically employed as 3.16. At present, there is no formal U.S. EPA policy on subdividing UF_H into TK and TD components, but the WHO's IPCS has developed guidance [IPCS, (80) finalized in 2005] that informs this process and recommends an even subdivision of UF_H into TK and TD components (each valued at one-half order of magnitude). Because of the availability of relevant data, based in part on this guidance, and following intense review, the U.S. EPA (81) divided UF_H into TK and TD components for the Boron and Compounds Risk Assessment.

Following chronic and developmental exposures, BA demonstrates testicular effects and fetal growth effect, respectively. Consideration of the design of the experiments demonstrating testicular effects in dogs produced some concerns about the age dependence of the effect, due to testicular effects in three of four control animals. Additional concerns for the dog studies included the small number of test animals per dose group, shared controls for different studies, and the NOAEL and LOAEL were taken from two different studies of different duration. The reported NOAEL for that study was, however, 8.8 mg B/kg-day. Developmental toxicity studies were conducted in rats exposed to BA in the diet, under adequate experimental design, and were well-reported (82,83). These data demonstrated effects over a broad range of concentrations; decreases in fetal body weight were deemed the most sensitive and a BMD analysis was undertaken (84) using combined data from both studies. From this analysis, the $BMDL_{05}$ was chosen as the POD. This value was 10.3 mg B/kg-day.

Several pieces of information and relevant and quantitative data sets were available, which were relevant to species extrapolation. BA is well-absorbed from oral exposures (85) freely soluble at physiologic pH. Borate compounds and BA in the body exist primarily as undissociated BA that distributes freely among body tissues and is not protein bound in blood. Indeed, it has been demonstrated (86) that boron concentrations in multiple tissues were approximately equal, bone and adipose being outliers. This may be explained by boron's free distribution with body water, serving as the basis for a lower boron concentration in adipose tissue; boron is sequestered in bone, to some degree. Because of these attributes, it was concluded that the placental transfer of boron would be similar in exposed rats and in humans: the placenta in neither would serve as a barrier to fetal exposures. It was assumed that the fetal tissues were the targets for the observed adverse effects, with decreased fetal body weight being more sensitive than the observed skeletal variations.

Because available data demonstrated that boron (i) is freely absorbed via the oral route, (ii) is excreted unchanged, (iii) distributes with total body water, and (iv) is eliminated (greater than 90%) in the urine, additional investigations were aimed at determining whether and how species differences in urinary elimination could be used to inform an other-than-default approach to extrapolating boron internal exposures (TK) between rats and humans. The available evidence demonstrated similarity in boron's urinary clearance in pregnant rats (87) and pregnant humans consuming boron as part of their diet (88). Mean values for urinary clearance of boron were calculated and served as the basis for quantification of interspecies differences in TK and subsequent development of the TK component of UF_A. In this instance, clearance was combined with information on the relative bioavailability in rats and humans and study-specific body weights to develop measures of boron clearance. Renal clearance of boron in pregnant rats was 1.0 mL/min, and body weight was 0.38 kg; renal clearance in pregnant humans was 66.1 mL/min and body weight was 67.6 kg. Boron clearance values were approximately 3.3 mL/min-kg and 0.98 mL/min-kg in rats and humans, respectively. This difference was modified to take into account relative bioavailability of an oral dose (0.95 and 0.92 in rats and humans, respectively). Clearance was used to modify the internal exposures from absorbed doses in pregnant rats and humans, yielding a factor of 3.3. This replaced the default UF value of 3.16, and represents a corresponding reduction in the uncertainty in species extrapolation. With respect to TD, some initial data indicate an effect of boron on the *hox* gene family that control somatic development. Even in light of the conservation of this gene among mammalian species, it was decided that there was insufficient evidence to implicate these genes in the mechanism of decreased fetal body weight in mammalian species; the default UF value of 3.16 was retained for interspecies differences in TD.

Finally, the available data were examined relative to developing a quantitative, nondefault value to replace UF_H. The nature of the distribution of boron throughout the body and renal elimination mechanisms were again examined. Similarities between boron clearance in humans (88–90) and glomerular filtration rates (GFRs) in humans were evident. The low molecular size and uncharged nature of circulating BA reinforced passive, rather than active renal mechanisms in determining renal clearance (81), a supposition supported by the effectiveness of hemodialysis in removing BA from human patients. Although some data were available on a relatively small number of pregnant human subjects (88) the data were very limited (nonrecorded dietary exposures). It was determined that this data set was insufficient to determine human variability in boron clearance with any degree of certainty. However, data were available on variability of GFR in pregnant humans from three different studies (91–93). These data sets were available as means and standard deviations, and so were used as such. The sigma method has been previously applied to data on GFR to estimate human variability on boron clearance (94). That approach determines variability as the difference between the mean measured value and the value at a certain number of standard deviations from the mean. While Dourson et al. (94) defined variability as the difference quantified at two standard deviations below the general population mean, the U.S. EPA's boron risk assessment (81) employed the value at three standard deviations below the mean to quantify variability in GFR among pregnant humans. The resulting sigma values were 1.54, 1.97, and 2.29 for data from Dunlop (91), Krutzen et al. (93), and Sturgiss et al. (92), respectively. The mean of these values is 1.93, which was rounded to a value of 2.0 to account for some uncertainty in the method employed. This factor (the GFR value at 3 SD below the population mean) was also supported by the relationship between GFR and serum creatinine levels; excessive creatinine levels are indicators of renal impairment and represent a risk to human health. The data-derived value for replacement of the TK portion of UF_H was 2.0. Replacement of the TD component was not considered; a value of 3.16 was retained for the default UF.

The RfD of 0.2 mg/kg-day was determined as $BMDL_{05}/UF$, where $BMDL_{05} = 10.3$ mg/kg-day and $UF = (3.3 \times 3.16) \times (2.0 \times 3.16) = 66$.

THE IEUBK MODEL

Allan Marcus

The IEUBK model for estimating lead (Pb) in children's blood is an example of a complex model that accounts for both exposure and PKs. The IEUBK model allows the user to estimate, for a hypothetical child or population of children, a plausible distribution of PbB concentrations centered on a geometric mean PbB concentration. The geometric mean PbB is predicted from available information about the child's or children's exposure to Pb. From this distribution, the model estimates the risk (i.e., probability) that a child's or a population of children's PbB concentration will exceed a certain level of concern (typically 10 µg/dL). The IEUBK model has been most often used to evaluate soil lead remediation goals for Superfund Comprehensive Environmental Response Compensation Liability Act (CERCLA) sites.

Model Focus and History

The metal Pb (atomic number 82) was known to be toxic to humans at least 2000 years ago (see Ref. 95 for a concise historical summary and Ref. 96 for details). In the late 19th century, the fetus and young children were identified as the most susceptible subpopulations for general nonoccupational exposure because Pb neurotoxicity targets their rapidly developing brains and central nervous systems. The most common sources of Pb exposure in young children within the last century were (i) contaminated house dust and soil from deposition of deteriorating Pb-based paint, (ii) deposition of airborne particles from burning leaded gasoline, (iii) industrial Pb product processing such as making Pb batteries or tetra-ethyl Pb, and (iv) Pb contamination in drinking water supplies and in food. Because Pb is so ubiquitous—and so useful—the safety margins between identified adverse effect levels and Pb levels in highly exposed subpopulations of children are often rather small, and the need for more precise quantification of Pb risk is particularly urgent.

The IEUBK model for Pb in children's blood was conceived in 1988, developed by the U.S. EPA and its contractors in the 1990s (97,98), and released in 1994 after review by the U.S. EPA's Science Advisory Board (SAB).

The model utilizes four interrelated modules (exposure, uptake, biokinetic, and probability distribution) to estimate PbB levels in children exposed to Pb-contaminated media. The Windows version of the IEUBK model (IEUBKwin) allows the user to estimate, for a hypothetical child or population of children, a plausible distribution of PbB concentrations centered on a geometric mean PbB concentration. The geometric mean PbB is predicted from available information about the child's or the children's exposure to Pb. From this distribution, the model estimates the risk (i.e., probability) that a child's or a population of children's PbB concentration will exceed a certain level of concern (typically 10 µg/dL).

IEUBKwin is a tool for making rapid calculations and recalculations of an extremely complex set of equations that includes many exposure, uptake, and biokinetic parameters. It has been recommended as a risk assessment tool to support the implementation of the July 14, 1994, Office of Solid Waste and Emergency Response (OSWER) Directive entitled Revised Interim Soil Lead Guidance for Comprehensive Environmental

Response Compensation Liability Act (CERCLA) Sites and Resource Conservation and Recovery Act (RCRA) Corrective Action Facilities, as well as the subsequent August 1998 OSWER Directive entitled Clarification to the 1994 Revised Interim Soil Lead Guidance for CERCLA Sites and RCRA Corrective Action Facilities. The development of the IEUBKwin model included independent validation and verification and a limited peer review of the software and documentation (99–102).

A greatly enhanced version of the IEUBK software, the All Ages Lead Model (AALM) (103) is being prepared by the U.S. EPA's NCEA to replace both the IEUBK model and the adult Pb model, which was developed as a multimedia dose–response model without a PK component.

Overview of the IEUBK Model Structure and Parameters

The focus of the IEUBK model for Pb in children is the prediction of blood Pb concentrations in young children exposed to Pb from several sources and by several routes. The model incorporates a four-step process that mathematically and statistically links environmental Pb exposure to PbB concentrations for a population of children (0–84 months of age). Figure 4 is a schematic showing the sources of exposure to environmental Pb and the absorption and processing of Pb by the human body [IEUBK Guidance Manual (97)].

Each of the four model components reflects a different aspect of the overall biologic process.

Exposure Component

Exposure can be thought of as the contact of a chemical or other agent with the absorption or exchange boundaries of an organism, such as the gut, lungs, and skin. Quantitation of a child's exposure to Pb (μg/day) requires estimation of the concentration of Pb in the environmental media that the child contacts (usually μg/g, μg/m^3, or μg/L), multiplied by a term to describe the amount of contact the child has with the medium (usually g/day, m^3/day, or L/day), and a term for the duration of that contact (usually days). The results from the exposure component of the IEUBK model are estimated intake rates for the quantities of Pb inhaled or ingested from environmental media. The media addressed by the IEUBK model include soil, house dust, drinking water, air, and food. Paint is usually addressed in terms of its contribution to the measured concentration of Pb in soil or house dust. While ingestion of chips of Pb-based paint can cause serious and sometimes fatal Pb poisoning, the IEUBK model does not include paint ingestion or other acute exposes because the model is programmed for calculating hypothetical risks from more typical subchronic or chronic exposures.

Uptake Component

The uptake component models the process by which Pb intake (Pb that has entered the child's body through ingestion or inhalation) is transferred to

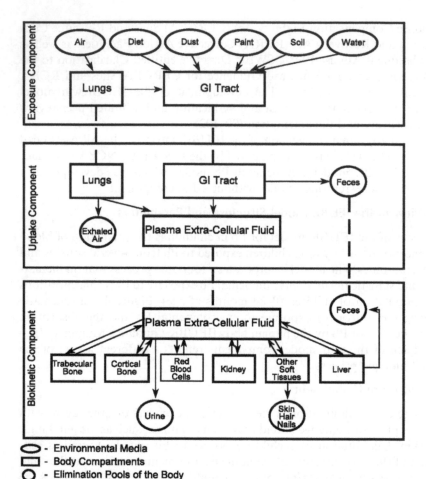

Figure 4 The integrated exposure, uptake, and biokinetic model. *Abbreviation*: GI, gastrointestinal.

the blood plasma. Uptake (μg/day) is the quantity of Pb absorbed per unit time from portals of entry (gut, lung) into the systemic circulation of blood. Only a fraction of the Pb entering the body through the respiratory or GI tracts is actually absorbed into the systemic circulation. This absorption fraction is, by convention, termed bioavailability and provides the most convenient parameterization of the uptake process. The IEUBK model addresses the different bioavailabilities of Pb from different environmental media and provides for a partial saturation of absorption at high levels of Pb intake.

Biokinetic Component

The biokinetic component of the IEUBK model is a mathematical expression of the movement of absorbed Pb throughout the body over time by physiologic or biochemical processes. The biokinetic component converts the total Pb uptake rate from the uptake component into an input to the central plasma-extracellular fluid compartment. Transfer coefficients are used to model movement of Pb between the internal compartments and to the excretion pathways. The quantities are combined with the total Pb uptake rate to continuously recalculate the Pb masses in each of the body compartments and especially the changing concentration of Pb in blood. The PK parameters were derived from a wide variety of sources, including human clinical and epidemiologic studies and experimental studies in many relevant animal models, including large nonhuman primates (monkeys, baboons) and other large mammalian species (dogs, miniature pigs) with relevant structural or functional physiological characteristics. Because Pb is a bone-seeking element, a child with high Pb exposure will carry a large internal source of Pb exposure in bone for decades.

Variability

An important goal of the IEUBK model is to address variability in PbB concentrations among exposed children. Children having contact with the same concentrations of environmental Pb can develop very different blood Pb concentrations due to differences in behavior, household characteristics, and individual patterns of Pb uptake and biokinetics. The IEUBK model uses a lognormal probability distribution to characterize this variability. The combined output of the exposure, uptake, and biokinetic components based on deterministic submodels provides a central estimate of PbB concentration, assumed to be the geometric mean of the blood distribution for children with a specified environmental Pb exposure scenario, specified uptake and biokinetic parameters, at a specified age. The IEUBK model user is permitted to specify the variability of the assumed lognormal PbB distribution conditional on the calculated geometric mean by assigning a value to the geometric standard deviation (GSD). The recommended default value for this parameter 1.6 was derived from empirical studies with young children where both blood and environmental Pb concentrations were measured.

The model includes variability and heterogeneity explicitly by allowing time-varying input values and by assuming that within each subpopulation there was a lognormal distribution of PbB concentrations with the model-predicted PbB as the geometric mean and with a user-specified GSD. A number of investigators have subsequently performed Monte Carlo simulations of various types using the IEUBK batch mode input capability. The stochastic output has been variously interpreted as either (i) the probability that a child residing at a particular site (residence, yard, playground, etc.) would have a PbB exceeding the health-specified level of concern, or (ii) the proportion of a specified population of children of a certain age, all exposed to the same environment, having an elevated PbB.

The IEUBK model includes certain kinetic nonlinearities: (i) a capacity limitation on the amount of Pb in the RBCs, affecting the plasma-to-red-blood-cell transfer coefficient, and (ii) absorption from the gut lumen that has both saturable and nonsaturable absorption of Pb in oral intake. These are documented in a U.S. EPA Parameters and Equations document (98). Many of these parameters will be updated in the AALM.

All of the exposure and uptake parameters can be adjusted by the user to match actual or potential site-specific properties. The user cannot adjust the biokinetic parameters. The option of adjusting PK parameters will be included in the AALM.

Risk Assessment Applications

The IEUBK model has been most often used to evaluate soil Pb remediation goals for Superfund (CERCLA) sites. The model may be used to evaluate prospective risk to children who may reside in a community in the future if no remedial action is taken by running the model in batch mode with current soil and dust Pb concentrations and determining the probability that a child of a certain age will have a PbB exceeding some level of concern for adverse health effects. The current value of the PbB level of concern is 10 µg/dL. The aggregate or individual site-specific risks of the no-action alternative may then be compared with the aggregate or individual risk for children where soil Pb concentrations have been reduced below a certain criterion level.

There is no requirement that a single soil Pb remediation goal be used at all sites when site-specific information suggests that an adequate margin of safety can be achieved by setting a different goal, i.e., by remediating fewer residential sites or community areas. Situations in which a different goal (higher or lower remediation criterion) may be appropriate include finding that the bioavailability of Pb in soil is higher or lower than usual due to particle size and solubility, mineral matrix embedding, and other factors. Typical soil Pb remediation goals are often in the range of 400 to 500 ppm Pb in soil, but other values have sometimes been used. These cases are usually reviewed by the U.S. EPA's Lead Technical Review Work Group.

Superfund sites where the IEUBK model has been used to inform a risk assessment include many Pb smelter and mining sites in the intermountain west: Bunker Hill/Silver Valley, Idaho; Midvale, Utah; Sandy, Utah; Bingham Creek, Utah; Leadville, Colorado, Telluride, Colorado; Butte, Montana; East Helena, Montana. Other sites including secondary Pb smelters and Pb battery recycling facilities are Granite City, Illinois, and Palmerton, Pennsylvania.

The IEUBK model has also been used for some drinking water assessments, including evaluation of potential risk from Pb leaching out of water distribution systems in Washington, D.C., and Pb leaching out of newly installed brass faucets over a period of some months.

Current Status of the IEUBK Model

The IEUBK model is maintained by the U.S. EPA's NCEA office. The AALM successor model was reviewed by the U.S. EPA's SAB in November 2005. It is expected that the new AALM will become available for public use in 2006. The proposal to bundle a parallel PBPK model with the AALM is currently under consideration.

ACKNOWLEDGMENTS

Examples of PK/PBPK models used in risk assessment: Allan Marcus (Lead); Chad Thompson (EGBE); Carolyn Smallwood, Jeff C. Swartout, Linda K. Teuschler and John C. Lipscomb (Boron); Paul Schlosser (DCM); Rob DeWoskin (xylenes); Weihsueh Chiu (VC).

REFERENCES

1. U.S. EPA, Integrated Risk Information System (IRIS). Washington, D.C.: U.S. Environmental Protection Agency, National Center for Environmental Assessment, 2005. Available at http://www.epa.gov/iris.
2. Krishnan K, Andersen ME. Physiologically based pharmacokinetic modeling in toxicology. In: Hayes AW, ed. Principles and methods of toxicology. Philadelphia: Taylor & Francis, 2001:193–241.
3. Gentry PR, Covington TR, Andersen ME, et al. Application of a physiologically based pharmacokinetic model for isopropanol in the derivation of a reference dose and reference concentration. Regul Toxicol Pharmacol 2002; 36: 51–68.
4. Dorne JL. Impact of inter-individual differences in drug metabolism and pharmacokinetics on safety evaluation. Fundam Clin Pharmacol 2004; 18:609–620.
5. Ginsberg G, Hattis D, Sonawane B. Incorporating pharmacokinetic differences between children and adults in assessing children's risks to environmental toxicants. Toxicol Appl Pharmacol 2004; 198:164–183.
6. Hattis D, Banati P, Goble R. Distributions of individual susceptibility among humans for toxic effects. How much protection does the traditional tenfold factor provide for what fraction of which kinds of chemicals and effects? Ann N Y Acad Sci 1999; 895:286–316.
7. Krishnan K, Johanson G. Physiologically-based pharmacokinetic and toxicokinetic models in cancer risk assessment. J Environ Sci Health C Environ Carcinog Ecotoxical Rev 2005; 23:31–53.

8. Clark LH, Setzer RW, Barton HA. Framework for evaluation of physiologically-based pharmacokinetic models for use in safety or risk assessment. Risk Anal 2004; 24:1697–1717.
9. U.S. EPA, Toxicological Review of Chloroform, EPA/635/R-01/001, U.S. Environmental Protection Agency, Integrated Risk Information System, Washington, D.C., 2001. Available at http://www.epa.gov/iriswebp/iris/index.html.
10. Heywood R, Sortwell RJ, Noel PR, et al. Safety evaluation of toothpaste containing chloroform: III. Long-term study in beagle dogs. J Environ Pathol Toxicol 1979; 2:835–851.
11. Corley RA, Mendrela AL, Smith FA, et al. Development of a physiologically based pharmacokinetic model for chloroform. Toxicol Appl Pharmacol 1990; 103:512–527.
12. Reitz RH, Mendrela AL, Corley RA, et al. Estimating the risk of liver cancer associated with human exposures to chloroform using physiologically-based pharmacokinetic modeling. Toxicol Appl Pharmacol 1990; 105:443–459.
13. U.S. EPA, Toxicological Review of Methyl Ethyl Ketone, EPA/635/R-03/009, U.S. Environmental Protection Agency, Integrated Risk Information System, Washington, D.C., 2003. Available at http://www.epa.gov/iriswebp/iris/index.html.
14. Liira J, Johanson G, Riihimaki V. Dose-dependent kinetics of inhaled methyl ethyl ketone in man. Toxicol Lett 1990b; 50:195–201.
15. Leung H-W. Use of physiologically based pharmacokinetic models to establish biological exposure indices. Am Ind Hyg Assoc J 1992; 53:369–374.
16. Dietz FK, Rodriguez-Giaxola M, Traiger GJ, et al. Pharmacokinetics of 2-butanol and its metabolites in the rat. J Pharmacokinet Biopharm 1981; 9:553–576.
17. Thrall KD, Soelberg JJ, Weitz KK, et al. Development of a physiologically based pharmacokinetic model for methyl ethyl ketone in F344 rats. J Toxicol Environ Health A 2002; 65:881–896.
18. Traiger GJ, Bruckner JV. The participation of 2-butanone in 2-butanol-induced potentiation of carbon tetrachloride hepatotoxicity. J Pharmacol Exp Ther 1976; 196:493–500.
19. Dietz FK, Traiger GJ. Potentiation of CCl4 of hepatotoxicity in rats by a metabolite of 2-butanone: 2,3-butanediol. Toxicology 1979; 14:209–215.
20. U.S. EPA, Toxicological Review of Acetone, EPA/635/R-03/004, U.S. Environmental Protection Agency, Integrated Risk Information System, Washington, D.C., 2003. Available at http://www.epa.gov/iriswebp/iris/index.html.
21. Clewell HJ III, Gentry PR, Gearhart JM, et al. Development of a physiologically based pharmacokinetic model of isopropanol and its metabolite acetone. Toxicol Sci 2001; 63:160–172.
22. Kumagai S, Matsunaga I. Physiologically based pharmacokinetic model for acetone. Occup Environ Med 1995; 52:344–352.
23. NTP (National Toxicology Program), Toxicity Studies of Acetone (CAS No. 67–64-1) in F344/N Rats and B6C3F1 Mice (drinking water studies), NTP TOX 3. Research Triangle Park, NC: U.S. Department of Health and Human Services, Public Health Service, National Institutes of Health, 1991 [NIH Publication No. 91–3122].

24. Dietz DD, Leninger JR, Rauckman EJ, et al. Toxicity studies of acetone administered in the drinking water of rodents. Fundam Appl Toxicol 1991; 17: 347–360.
25. Gentry PR, Covington TR, Clewell HJ III, Anderson ME. Application of a physiologically based pharmacokinetic model for reference dose and reference concentration estimation for acetone. J Toxicol Environ Health A 2003; 66:2209–2225.
26. U.S. EPA, Toxicological Review of Ethylene Glycol Monobutyl Ether (EGBE), U.S. Environmental Protection Agency, Integrated Risk Information System, Washington, D.C., 1999.
27. Corley RA, Bormett GA, Ghanayem BI. Physiologically-based pharmacokinetics of 2-butoxyethanol and its major metabolite, 2-butoxyacetic acid, in rats and humans. Toxicol Appl Pharmacol 1994; 129:61–79.
28. Johanson G, Wallen M, Nordquist MB. Elimination kinetics of 2-butoxyethanol in the perfused rat liver—dose dependence and effect of ethanol. Toxicol Appl Pharmacol 1986; 83:315–320.
29. Corley RA, Markham DA, Banks C, et al. Physiologically based pharmacokinetics and the dermal absorption of 2-butoxyethanol vapor by humans. Fundam Appl Toxicol 1997; 39:120–130.
30. NTP (National Toxicology Program), Technical Report on Toxicity Studies of Ethylene Glycol Ethers 2-Methoxyethanol, 2-Ethoxyethanol, 2-Butoxyethanol Administered in Drinking Water to F344/N Rats and B6C3F1 Mice, NTP No. 26. Research Triangle Park, NC: U.S. Department of Health and Human Services, Public Health Service, National Institutes of Health, 1993 [NIH Publication No. 93-3349].
31. NTP (National Toxicology Program), NTP Technical Report on the Toxicology and Carcinogenesis Studies of 2-Butoxyethanol (CAS No. 111–76-2) in F344/N Rats and B6C3F1 Mice (inhalation studies), NTP TR 484. Research Triangle Park, NC: U.S. Department of Health and Human Services, Public Health Service, National Institutes of Health, 1998 [NIH Draft Publication No. 98-3974].
32. Deisinger PJ, Boatman RJ. In vivo metabolism and kinetics of ethylene glycol monobutyl ether and its metabolites, 2-butoxyacetaldehyde and 2-butoxyacetic acid, as measured in blood, liver and forestomach of mice. Xenobiotica 2004; 34(7):675–685.
33. U.S. EPA, Toxicological Review of Vinyl Chloride, EPA/635/R-00/004. Washington, D.C.: U.S. Environmental Protection Agency, Integrated Risk Information System, 2000.
34. Clewell HJ III, Gentry PR, Gearhart JM, et al. The development and validation of a physiologically based pharmacokinetic model for vinyl chloride and its application in a carcinogenic risk assessment for vinyl chloride. Washington, D.C.: ICF Kaiser for the U.S. Environmental Protection Agency, Office of Health and Environmental Assessment, and the Occupational Safety and Health Administration, Directorate of Health and Standards Programs, 1995.
35. Clewell HJ III, Covington TR, Crump KS, et al. The application of a physiologically based pharmacokinetic model for vinyl chloride in a noncancer risk assessment. Washington, D.C.: ICF Kaiser/Clement Associates under EPA

contract number 68-D2-0129 for the U.S. Environmental Protection Agency, National Center for Environmental Assessment, 1995.

36. D'Souza RW, Andersen ME. Physiologically based pharmacokinetic model for vinylidene chloride. Toxicol Appl Pharmacol 1988; 95:230–240.

37. Andersen ME, Clewell HJ, Gargas ML, et al. Physiologically based pharmacokinetics and risk assessment process for methylene chloride. Toxicol Appl Pharmacol 1987; 87:185–205.

38. U.S. EPA. Recommendations for and Documentation of Biological Values for Use in Risk Assessment, EPA/600/6-87/008. NTIS PB88–179874/AS. Washington, D.C.: U.S. Environmental Protection Agency, 1988.

39. Gargas ML, Burgess RJ, Voisard DE, et al. Partition coefficients of low-molecular-weight volatile chemicals in various liquids and tissues. Toxicol Appl Pharmacol 1989; 98:87–99.

40. Barton HA, Creech JA, Godin CS, et al. Chloroethylene mixtures: pharmacokinetic modeling and in vitro metabolism of vinyl chloride, trichloroethylene, and trans-1,2-dichloroethylene in the rat. Toxicol Appl Pharmacol 1995; 130:237–247.

41. Andersen ME, Gargas ML, Clewell HJ, et al. Quantitative evaluation of the metabolic interactions between trichloroethylene and 1,1-dichloroethylene in vivo using gas uptake methods. Toxicol Appl Pharmacol 1987; 89:149–157.

42. Bolt HM, Lai RJ, Kappus H, et al. Pharmacokinetics of vinyl chloride in the rat. Toxicology 1977; 7:179–188.

43. Clement International, Development, and Validation of Methods for Applying Pharmacokinetic Data in Risk Assessment. Final Report. Vol. V: Vinyl Chloride, AAMRL-TR-90–072. Dayton, OH: Department of the Air Force, Armstrong Aerospace Medical Research Laboratory, Wright-Patterson Air Force Base, 1990.

44. Gargas ML, Clewell HJ III, Andersen ME. Gas uptake inhalation techniques and the rates of metabolism of chloromethanes, chloroethanes, and chloroethylenes in the rat. Inhal Toxicol 1990; 2:295–319.

45. Buchter A, Filser JG, Peter H, et al. Pharmacokinetics of vinyl chloride in the Rhesus monkey. Toxicol Lett 1980; 6:33–36.

46. Kedderis GL, Carfagna MA, Held SD, et al. Kinetic analysis of furan biotransformation by F-344 rats in vivo and in vitro. Toxicol Appl Pharmacol 1993; 123:274–282.

47. Til HP, Immel HR, Feron FJ. Lifespan Oral Carcinogenicity Study of Vinyl Chloride in Rats, TNO Report no. V 83.285/291099. Zeist, The Netherlands: Netherlands Organization for Applied Scientific Research, Division for Nutrition and Food Research TNO, 1983.

48. Til HP, Feron VJ, Immel HR. Lifetime (149-week) oral carcinogenicity study of vinyl chloride in rats. Food Chem Toxicol 1991; 29:713–718.

49. U.S. EPA. Health Effects Assessment Summary Tables (HEAST), EPA/540/R-97/036, Washington, D.C.: U.S. Environmental Protection Agency, Office of Solid Waste and Emergency Response, 1997.

50. U.S. EPA Health Effects Assessment Summary Tables (HEAST). EPA/540R-97/036. U.S. Environmental Protection Agency, Office of Solid Waste and Emergency Response, Washington, D.C., 1997.

51. U. S. EPA, Health Assessment Document for Dichloromethane (Methylene Chloride). Final Report, EPA/600/8-82/004F, U.S. Environmental Protection Agency, Office of Health and Environmental Assessment, Washington, D.C., 1985.

52. U.S. EPA, Addendum to the Health Assessment Document for Dichloromethane (methylene chloride). Updated Carcinogenicity Assessment, EPA/600/8-82/004FF, Carcinogen Assessment Group, OHEA for the U.S. Environmental Protection Agency, Washington, D.C., 1985.

53. U.S. EPA, Update to the Health Assessment Document and Addendum for Dichloromethane (Methylene Chloride): Pharmacokinetics, Mechanism of Action and Epidemiology. Review Draft, EPA/600/8-87/030A, U.S. Environmental Protection Agency, Office of Health and Environmental Assessment, Washington, D.C., 1987.

54. U.S. EPA, Technical Analysis of New Methods and Data Regarding Dichloromethane Hazard Assessments. Review Draft, EPA/600/8-87/029A, U.S. Environmental Protection Agency, Office of Health and Environmental Assessment, Washington, D.C., 1987.

55. Andersen ME, Clewell HJ, Gargas ML, et al. Physiologically based pharmacokinetic modeling with dichloromethane, its metabolite, carbon monoxide, and blood carboxyhemoglobin in rats and humans. Toxicol Appl Pharmacol 1991; 108:14–27.

56. Ramsey JC, Andersen ME. A physiologically based description of the inhalation pharmacokinetics of styrene in rats and humans. Toxicol Appl Pharmacol 1984; 73:159–175.

57. NTP (National Toxicology Program), Toxicology and Carcinogenesis Studies of Dichloromethane (methylene chloride)(CAS No. 75–09-2) in F344/N Rats and B6C3F1 Mice (inhalation studies), NTP-TR-306. Research Triangle Park, NC: U.S. Department of Health and Human Services, Public Health Service, National Institutes of Health, 1986 [NIH Publication No. 86–2562].

58. NCA (National Coffee Association), Twenty-four Month Oncogenicity Study of Methylene Chloride in Mice. Final Report, Hazleton Laboratories, America, Inc., Vienna, VA, 1983.

59. Thomas RS, Yang RS, Morgan DG, et al. PBPK modeling/Monte Carlo simulation of methylene chloride kinetic changes in mice in relation to age and acute, subchronic, and chronic inhalation exposure. Environ Health Perspect 1996; 104:858–865.

60. El-Masri HA, Bell DA, Portier CJ. Effects of glutathione transferase theta polymorphism on the risk estimates of dichloromethane to humans. Toxicol Appl Pharmacol 1999; 158:221–230.

61. Jonsson F, Johanson G. A Bayesian analysis of the influence of GSTT1 polymorphism on the cancer risk estimate for dichloromethane. Toxicol Appl Pharmacol 2001; 174:99–112.

62. Jonsson F, Bois F, Johanson G. Physiologically based pharmacokinetic modeling of inhalation exposure of humans to dichloromethane during moderate to heavy exercise. Toxicol Sci 2001; 59:209–218.

63. Sweeney LM, Kirman CR, Morgott DA, et al. Estimation of interindividual variation in oxidative metabolism of dichloromethane in human volunteers. Toxicol Lett 2004; 154:201–216.

64. ENVIRON Health Sciences Institute. Development of Population Cancer Risk Estimates for Environmental Exposure to Methylene Chloride (DCM) Using a Physiologically Based Pharmacokinetic Model. Ruston, LA: ENVIRON Health Sciences Institute, 2005.
65. Starr TB, Matanoski G, Anders MW, et al. Workshop overview: reassessment of the cancer risk of dichloromethane in humans. Toxicol Sci 2006; 91:20–28.
66. Tardif R, Laparé S, Plaa GL, et al. Effect of simultaneous exposure to toluene and xylene on their respective biological exposure indices in humans. Int Arch Occup Environ Health 1991; 63:279–284.
67. Tardif R, Plaa GL, Brodeur J. Influence of various mixtures of inhaled toluene and xylene on the biological monitoring of exposure to these solvents in rats. Can J Physiol Pharmacol 1992; 70:385–393.
68. Tardif R, Lapare S, Krishnan K, et al. Physiologically based modeling of the toxicokinetic interaction between toluene and m-xylene in the rat. Toxicol Appl Pharmacol 1993; 120:266–273.
69. Kaneko T, Horicuhi J, Sato A. Development of a physiologically based pharmacokinetic model of organic solvent in rats. Pharmacol Res 2000; 42: 465–470.
70. Tardif R, Laparé S, Krishnan K, et al. A descriptive and mechanistic study of the interaction between toluene and xylene in humans. Int Arch Occup Environ Health 1993; 65:S135–S137.
71. Tardif R, Laparé S, Charest-Tardif G, et al. Physiologically-based pharmaco-kinetic modeling of a mixture of toluene and xylene in humans. Risk Anal 1995; 15:335–342.
72. Haddad S, Tardif R, Viau C, et al. A modeling approach to account for toxi-cokinetic interactions in the calculation of biological hazard index for chemical mixtures. Toxicol Lett 1999; 108:303–308.
73. Tardif R, Charest-Tardif G, Brodeur J, et al. Physiologically based pharmaco-kinetic modeling of a ternary mixture of alkyl benzenes in rats and humans. Toxicol Appl Pharmacol 1997; 144:120–134.
74. U.S. EPA, Methods for Derivation of Inhalation Reference Concentrations and Application of Inhalation Dosimetry, EPA/600/8-90/066F, U.S. Environmental Protection Agency, Office of Research and Development, Washington, D.C., 1994. Available at http://nepis.epa.gov/pubtitleORD.htm.
75. Korsak Z, Wisniewska-Knypl J, Swiercz R. Toxic effects of subchronic combined exposure to n-butyl alcohol and m-xylene in rats. Int J Occup Med Environ Health 1994; 7:155–166.
76. Korsak Z, Sokal JA, Górny R. Toxic effects of combined exposure to toluene and m-xylene in animals. III. Subchronic inhalation study. Pol J Occup Med Environ Health 1992; 5:27–33.
77. Dennison JE, Bigelow PL, Mumtaz MM, et al. Evaluation of potential toxicity from co-exposure to three CNS depressants (toluene, ethylbenzene, and xylene) under resting and working conditions using PBPK modeling. J Occup Environ Hyg 2005; 2:127–135.
78. Dennison JE, Andersen ME, Yang RS. Characterization of the pharmacoki-netics of gasoline using PBPK modeling with a complex mixtures chemical lumping approach. Inhal Toxicol 2003; 15:961–986.

79. WHO (World Health Organization). Environmental Health Criteria 204: Boron. Geneva, Switzerland: World Health Organization, 1998.
80. IPCS (International Programme on Chemical Safety), Guidance Document for the Use of Data in Development of Chemical-specific Adjustment Factors (CSAF) for Interspecies Differences and Human Variability in Dose/concentration Response Assessment, World Health Organization, Geneva, Switzerland, 2001. Available at www.who.int/entity/ipcs/publications/methods/harmonization/en/csafs_guidance_doc.pdf.
81. U.S. EPA, Integrated Risk Information System (IRIS), Online Toxicological Review of Boron and Compounds (CASRN 7440–42-8), entered August 2004, U.S. Environmental Protection Agency, Integrated Risk Information System, Washington, D.C., 2006. Available at http://www.epa.gov/iris/subst/0410.htm.
82. Heindel JJ, Price CJ, Field EA, et al. Developmental toxicity of boric acid in mice and rats. Fundam Appl Toxicol 1992; 18:266–277.
83. Price CJ, Strong PL, Marr MC, et al. Developmental toxicity NOAEL and postnatal recovery in rats fed boric acid during gestation. Fundam Appl Toxicol 1996; 32:179–193.
84. Allen BC, Strong PL, Price CJ, et al. Benchmark dose analysis of developmental toxicity in rats exposed to boric acid. Fund Appl Toxicol 1996; 32:194–204.
85. Schou JS, Jansen JA, Aggerbeck B. Human pharmacokinetics and safety of boric acid. Arch Toxicol Suppl 1984; 7:232–235.
86. Ku WW, Chapin RE, Moseman RF, et al. Tissue disposition of boron in male Fischer rats. Toxicol Appl Pharmacol 1991; 111:145–151.
87. Vaziri ND, Oveisi F, Culver BD, et al. The effect of pregnancy on renal clearance of boron in rats given boric acid orally. Toxicol Sci 2001; 60:257–263.
88. Pahl MV, Culver BD, Strong PL, et al. The effect of pregnancy on renal clearance of boron in humans: a study based on normal dietary intake of boron. Toxicol Sci 2001; 60:252–256.
89. Jansen JA, Schou JS, Aggerback A. Gastrointestinal absorption and in vitro release of boric acid from water emulsifying ointments. Food Chem Toxicol 1984; 22:49–53.
90. Jansen JA, Andersen J, Schou JS. Boric acid single dose pharmacokinetics after intravenous administration to man. Arch Toxicol 1984; 55:64–67.
91. Dunlop W. Serial changes in renal haemodynamics during normal human pregnancy. Br J Obstet Gynecol 1981; 88:1–9.
92. Sturgiss SN, Wilkinson R, Davison JM. Renal reserve during human pregnancy. Am J Physiol 1996; 271:F16–F20.
93. Krutzén F, Olofsson P, Beck SE, et al. Glomerular filtration rate in pregnancy; a study in normal subjects and in patients with hypertension, preeclampsia and diabetes. Scand J Clin Lab Invest 1992; 52:387–392.
94. Dourson M, Maier A, Meek B, et al. Boron tolerable intake re-evaluation of toxicokinetics for data derived uncertainty factors. Biol Trace Elem Res 1998; 66:453–463.
95. Needleman H. Lead poisoning. Ann Rev Med 2004; 55:209–222.
96. Nriagu Jerome O. Lead and Lead Poisoning in Antiquity. John Wiley & Sons. New York, NY, U.S.A. 1983.

97. U.S. EPA, Guidance Manual for the Integrated Exposure Uptake Biokinetic Model for Lead in Children, EPA/540/R-93/081, PB93–963510, U.S. Environmental Protection Agency, Washington, D.C., 1994.
98. U.S. EPA, Technical Support Document: Parameters and Equations Used in the Integrated Exposure, Uptake, and Biokinetic Model for Lead in Children (v0.99d), EPA/540/R-94/040, NTIS PB94–963505, U.S. Environmental Protection Agency, Washington, D.C., 1994. Available at http://www.epa.gov/superfund/programs/lead/products.htm.
99. Hogan K, Marcus A, Smith R, et al. Integrated exposure uptake model for lead in children: empirical comparisons with epidemiologic data. Environ Health Perspect 1998; 106:1557–1567.
100. Marcus AH, Elias RW. Some useful statistical methods for model validation. Environ Health Perspect 1998; 106:1541–1550.
101. Mickle MH. Structure, use, and validation of the IEUBK model. Environ Health Perspect 1998; 106:1531–1534.
102. White PD, Van Leeuwen P, Davis BD, et al. The conceptual structure of the integrated exposure uptake biokinetic model for lead in children. Environ Health Perspect 1998; 106:1495–1503.
103. U.S. EPA, Guidance Manual for the All Ages Lead Model (AALM). Draft Version 1.05, EPA/600/C-05/013, U.S. Environmental Protection Agency, Washington, D.C., 2005. Available at http://cfpub.epa.gov/ncea/cfm/recordisplay.cfm?deid=139314.

Index

Absorption, distribution, metabolism, elimination (ADME), kinetic process study, 96
Acyl glucuronides and sulfonates, 118
Adenocarcinomas, 43
Agency for Toxic Substances and Disease Registry (ATSDR), 3
Alpha2u-globulin nephropathy, in epithelial cells, 56
Alveolar ventilation rate, 274, 277
Aqueous xylene, bioavailability of, 330
Arterial blood concentration, 274, 323

Bayesian analysis, 280, 327
  to statistical inference, 272
Benchmark dose (BMD), 3, 11, 306
Biological fluids, metabolites in, 123
Biopharmaceutical factors, 96
Blood–air partition coefficient, 132, 274
Blood partition coefficients (P_{tb}), 77, 275, 316
Boron's urinary clearance, 334
1,3-Butadiene (BD)
  disposition models of, 63
  oxidation of, in vitro kinetics of, 51
  in rats, metabolism and pharmacokinetics of, 51

2-Butoxyacetaldehyde (BAL) concentrations, 319

Cancer
  bioassays, 41
  human relevance of, 39
  and noncancer effects, hazard characterization for, 28
  precursor lesions for, 41
  toxicity, 2
Carboxyhemoglobin, kinetics of, 327
Carcinogenesis, 20
Carcinogenic agents, 47
Carcinogenicity, 53
  dose selection
    bioassays studies, 101
    systemic exposure margins in, 101
  of epoxides and epoxide-forming chemicals, 53
  mechanism of, 327
Cardiac output, 277
Center for Drug Evaluation and Research (CDER), 287
Chee's culture medium, 112
Chemical-specific adjustment factors (CSAFs), 1, 27, 225, 252
  development of, 17, 28–29
  for interindividual variability, 34

[Chemical-specific adjustment factors
(CSAFs)]
mode of action (MOA), 28, 39
for toxicodynamic components, 33
for toxicokinetic components, 31
Chemical-specific/compound-related
composite factor
for human variability, 35
for interspecies differences, 35
Chemical-specific/compound-related
toxicokinetic adjustment
factors, 31
Chemical-specific toxicodynamic
adjustment factors, 33
Chloroform
carcinogenic effects of, 311
inhalation exposures to, 39
oxidation rates of, 39
Chloroform hepatotoxicity, 20
Chloroform-induced tumors,
in animals, 39
Chloroform metabolism, effects of, 311
C-labeled compounds, mass balance
studies with, 119
Clarithromycin, cytochrome (CYP)
inhibitor, 111
Cytochrome (CYP), 106
assessment, 107
conditions, 107
determination, by liquid
chromatography–mass
spectrometry (LS–MS), 107
induction, 106, 111
gelled collagen layer effect
on, 112
inhibition, 106, 107
isoforms, 107
metabolic-based drug–drug
interactions for, 106
recombinant systems, 107–108
metabolism evaluation, 108
Cytochrome P450 (CYP) enzymes, 6,
187, 197, 324
catalyzed reactions, 196
dependent mixed-function oxidase
system, 195
electron transfer mechanism in, 196

[Cytochrome P450 (CYP) enzymes]
ferrous, 196
furan bioactivation in, 203
glucuronic acid, conjugated with, 187
of hemeprotein oxidoreductases, 195
in hepatocytes, 196
human hepatic, distribution effects
of, 205
induction, 197
in inner mitochondrial membrane, 196
monooxygenase, toxicants and
solvents for, 192
polymorphisms in, 197
stoichiometry, 195
in tissues, 196
trichloroethylene studies and
oxidation mediated by, 205
Cytomatrix, 187
Cytosolic fractions, of rats and mice, 51

Dermal absorption, 234
Dichloromethane, 323
1,2,3,4-Diepoxybutane (DEB), 51
7,12-Dimethylbenz(a)anthracene, 199
Dioxin-induced enzymes, dose–response
of, 53
DNA damage, 319
DNA-reactive carcinogens, 5
DNA-reactive metabolites, dosimetry
of, 51
Dose–limiting toxicity (DLT), 97,98
adverse effects, 97
dose range-finding studies, 98
dose selection based on, 100
Dose metric measurements, 125
Dose–response, 124
analyses of, 49, 306
assessment, 2
measure of, 29
Dosimetry, evaluation of, 21
Doxin-induced receptor-mediated
processes, pharmacodynamic
models for, 51
Drug
absorption, 96
saturable, 98

[Drug]
area under the curve (AUC) for, 98
candidate, cytochrome induction by, 107–108, 111
elimination, 106
metabolism
in animals and humans, 104
assessment strategy, 114
development, 113, 118–119
plasma, 97
safety assessment (SA) studies, 96
species selection for, 117
uptake in tissues, 103
Drug–drug interactions, 105
absorption-mediated, 105, 106
distribution-mediated, 105, 106
metabolic-mediated, 105, 106
assessment, 106
clinical implications of, 106
for cytochrome P450 (CYP), 106
Drug development, 95, 285
bioanalytical techniques of, 286
biomarkers of, 285
body surface area (BSA), 288
carcinogenicity, 287, 290
cessation of dose escalation, 298
chronic toxicology, 287
drug applications, 289–290
evaluation, 95
classical pharmacokinetics, 95
toxicokinetic investigations in, 96
Food and Drug Administration (FDA), 287
human efficacy studies of, 288
human equivalent dose (HED), 293
human preliminary safety studies of, 288
International Conference on Harmonization (ICH) guidances, 287
investigational new drug (IND) applications, 286
bioavailability, 286
biodistribution, 286
health care professionals, 290
microdosing studies of, 286
requirements for, 290

[Drug development]
maximum plasma levels, 298
maximum tolerated dose (MTD), 287–288
metabolism, 286
pharmacokinetic (PK) properties, 286
human bioavailability, 298
metabolism, 298
toxicity mechanism, 298
pharmacologic effects of, 286
pharmacology and toxicology activity, 291–292
clinical pharmacokinetic applications, 291
elimination rate parameters of, 292
maximum concentration, 292
on physiologic function, 291
volume of distribution, 292
plasma area under curve (AUC) values, 287
toxicology studies, 289
postmarketing surveillance, 290
process method of, 286
safety signals, 290
species evaluations, 292–293
adverse effects of, 292
allometric scaling of, 293
calculation and conversion factors, 294
dose-limiting toxicity for, 292
drug elimination rate of, 296
life-threatening toxicity, 293
oral bioavailability of, 294
pharmaceutical ingredient of, 294
pharmacokinetic measurements of, 293
pharmacologic activity of, 293
steady-state concentration, 292
ventricular arrhythmias, 294

Endoplasmic-reticulum-bound epoxide, 52
Environmental agents, absorption of, 234

Environmental toxicants
  data for, 233
  studies with, 244
Enzyme-catalyzed biotransformation
      reactions, 187
  apparent velocity of, 188
Enzyme inhibitor, 191
  competitive inhibitors, 191–192
    steady-state velocity
        equation, 191
  noncompetitive inhibitors, 192–193
    devastating effects, 194
    rapid equilibrium velocity
        equation, 192
    reversible, 193
    steady-state velocity equation
        for, 193
Enzyme, inhibitor, and substrate (ESI),
      ternary complex, 192
Enzymes, 116, 186
  catalysis, 187
  characteristics, 191
  in chemical involvement
    biotransformation, 194
    detoxication, 194
    toxication, 194
    xenobiotic metabolism, 195
  cytochrome P450, 196
  function, 186
  glutathione S-transferases (GST)
        family of, 198
  induction, 197
    by drugs, 198
    by xenobiotics, 198
  inhibition types, 191
    diagnostic patterns for, 192
  inhibitors
    endogenous compounds, 191
    xenobiotics, 191
  in vitro biotransformation,
        191, 199
    design, 200
    extrapolation of kinetic
          parameters, 201
    factor in, 200
    kinetic prediction of, 200–201
    problems, 202

[Enzymes]
    uridinediphospho (UDP)-
        glucuronosyltransferase
        in, 202
  kinetics, 187
  in liver cytosol, 116
  in mammalian cells, 186
  metabolic, inhibition of, 6
  ping-pong mechanism of, 194
  soluble, 187
  structure, 186
  sulfotransferase, 199
  xenobiotic-metabolizing, 191
Enzyme–substrate complex
      (ES), 189
  concentration, 190
  dissociation constant of, 189
  equilibrium expression, 190
  steady-state level of, 190
Epoxide hydrolase (EH), 51, 239
3,4-Epoxy-1,2-butandediol (EBD), 51
  liver steady-state concentrations
        of, 52
1,2-Epoxy-3-butene (EB), metabolic
      elimination of, 51
Erythrocytes
  cellular molecules in, 316
  sensitivity of, 19
Erythromycin, 111
  with astemizole, 111
  cytochrome (CYP) inhibitor, 111
  with pimozide, 111
  with terfenadine, 111
Ethylene glycol butyl ether (EGBE), 19,
      310, 315
Extrahepatic and extrapulmonary
      metabolic constants, ratio
      of, 327
Extrarespiratory effect, sensitive
      effect, 330

Fetal growth effect, 333
Fluid dynamic modeling, respiratory
      tract morphology with, 235
Food and Drug Administration
      (FDA), 101

Food Quality Protection Act
    (FQPA), 260
Furan, 204
  bioactivation, 204
  oral bolus administration of, 204
  steady-state blood concentrations
      of, 204

Geometric standard deviation
    (GSD), 75
  adverse severity level, 80
  human interindividual variability,
      75, 89
  non-immune systemic adverse
      effect, 80
  percentile drug, 76
  population estimation of, 79
  predicted sampling-error variance
      in, 82
Glomerular filration rate, 19, 335
  variability of, 332
Glucuronosyl transferases, 55, 198
  endoplasmic reticulum, 198
Glutathione, 198, 239
  degradation, 198
  reserves, depletion of, 130
Glutathione-S-transferase (GST),
    51, 239
  enzymes family, 198

Half-life ratios span, adult and child, 240
Hazard quotient (HQ), 4
Hemolysis, 19
Hemolytic anemia, 19
Henri–Michaelis–Menten equation, 188
Hepatic blood flow, 274
  and V_{max}/k_m, 271, 274
Hepatic extraction, rate of, 235
Hepatic function, immaturity of in vivo
    data, 237
Hepatic microsomal protein,
    of humans, 237
Hepatocarcinogenicity, 58
Hepatocellular cytoplasmic changes
    (HCCs), 316

Hepatocytes, 200
  human, 112–113
    cryopreserved versus
        primary, 112
    culture conditions for, 112
    medium, 112
    parameter, 112
    uses, 112
  isolatation of, 200
  proliferation
    Kupffer cells in, 60
    by peroxisome proliferators, 60
Hepatotoxicity, chloroform, 20
Homozygous genotype, 222
Hormonal cycle, 136
Hormone disruption, 55
Human epidemiological study, 126
Human equivalent concentration
    (HEC), 276, 322
  for inhalation exposure, 10
Human equivalent dose (HED), 10, 276
  development of, 304
  human route extrapolation of, 276
Human health risks
  from animal and mechanistic
      data, 48
  biologically based approaches
      for, 49
  mechanistic information for, 47
Human interindividual variability,
    characterization of, 327
Human internal dose, estimate of, 308
Human metabolic constants, 43
  re-estimate of, 327
Human population, heterogencity
    of, 211
Human respiration rate, 324
Human tissues
  and blood, lipid, and water contents
      of, 176
  neutral lipids and water in, 172
Human variability
  degree of, 330
  experimental animal to, 10
  factor for, 30
  in toxicokinetics/toxicodynamics, 18
Hyperplasia, regenerative, 2

Inhalation cancer bioassays, 319
Integrated Risk Information System
    (IRIS) program, 302
Interindividual variability, 9, 233, 272
International Agency for Research on
    Cancer (IARC), 101
International Conference on
    Harmonization (ICH), 101
International Programme on Chemical
    Safety (IPCS), 7, 28, 69, 251
  assumptions of, 72
  chemical-specific adjustment factors
      (CSAFs), 72
  chemical-specific information, 69
  for human interindividual
      variability, 70
    dynamic factors, 73
    kinetic factors, 73
    uncertainty factors, 79
  maximum systemic concentration, 73
  in Monte Carlo simulations, 70
  uncertainty effects, 85
  pharmacodynamic (PD) variability
      database, 70
    drug testing, 78
    epidemiological observations, 74
    human interindividual variability
        components, 85
    individual susceptibility, 80
    lognormal distribution of, 85
    physiological parameters, 82
    quantitative definition of, 88
    straw man simulations of, 81
    uncertainty distributions
        for, 84
  pharmacokinetic (PK) variability
      database, 70
    chemical-to-chemical differences
        in, 81
    data-derived uncertainty factors
        for, 84
    decision theory, 81
    elimination half-life, 79
    human interindividual variability
        components, 85
    individual susceptibility, 80
    lognormal distribution of, 85

[International Programme on Chemical
    Safety (IPCS)
  pharmacokinetic (PK) variability
      database]
    maximum systemic
        concentration, 75
    physiological parameters, 82
    quantitative definition of, 88
    straw man simulations of, 81
    uncertainty distributions for, 84
  referred doses (RfDs), 84
    decision theory, 81
    definition of, 84
    health protection, 70
    mild adverse effects, 78
    multiple regression analysis, 91
    multiplicative reduction in, 90
    pharmacodynamic (PD)
        variability, 70
    pharmacokinetic (PK)
        variability, 70
    straw man procedure, 76, 88
  risk management criteria, 71
Interspecies extrapolation, 132
  Monte Carlo approach, 134
Intravenous bolus dose, administration
    of, 37
In vivo kinetic studies, in animals and
    humans, 38
Isopropanol, 215

Kidney metabolic rate constants, 43
Kidneys
  blood flow to, 43
  chronic cytotoxic/proliferative
      response, 39
  metabolism in, 43
  microsomal fraction of, 43

Limit dose (LD), 97
  saturable, 98
Lipid-soluble chemicals, retention
    of, 236
Lipophilic toxicants, retention of, 240
Liquid chromatography–mass
    spectrometry (LS–MS), 107
Liver cell polymorphism, 321

Liver dosimetry, estimation of, 327
Liver tissue subvolumes, 43
Liver tumor induction, 59
  mechanisms of, 59
  by peroxisome proliferators, 59
Lognormal (LN) sampling
    distribution, 273
Lowest-observed-adverse-effect level
    (LOAEL), 3, 16, 74
  extrapolation, 11
Lysosomal proteolytic enzymes, 57

Macromolecular adduct formation, 125
Mammalian cells, enzymes in, 186
Margin of exposure (MOE) approach, 5
Margin of safety (MOS) approach,
    advantage of, 5
Markov chain Monte Carlo
    integration, 273
Mass flow controller (MFC), 153
Maximal feasible dose (MFD),
    saturable, 98
Maximum recommended human dose
    (MRHD), 101
Maximum tolerated dose (MTD), 97
Mean corpuscular volume (MCV), 316
Metabolism, 108, 185
  chemical, 185
  evaluation, 108
  maximum rate of, 270
  parameters, V_{max}, 280
  role in toxicological response, 49
Metabolites, 103
  dominant, 104
  of droloxifene, 119
  human, 104
  identification, advantages, 113
  in human microsomes, 115
  quantitative assessment of, 103–105
  in rat liver microsomes and plasma,
      115, 116
  separation, 119
  toxic, 104
Methemoglobinemia, 9
Methyl ethyl ketone (MEK), 310
Methyl t-butyl ether (MTBE), 57

Michaelis–Menten kinetics, 270,
    312, 328
Michaelis–Menten metabolism,
    in liver, 271
Michaelis–Menten saturation
    kinetics, 312
Microsomes and cytosol, 200, 201
  for enzymes studies, xenobiotic
      metabolism, 200
  extrapolation, in vivo, 201
Monte Carlo analysis, 20, 160,
    322, 339
  use of, 308
Moolgavkar–Venzon–Knudson (MVK)
    model, 20
mRNA, 61
  degradation, 61
    by proteolytic enzymes, 61
    by ribonucleases, 61
m-Xylene, tissue levels of, 328

National Academy of Sciences
    (NAS) risk assessment
    paradigm, 2
National Center for Environmental
    Assessment (NCEA), 308
National Health Nutrition Examination
    Survey (NHANES), 277
National Toxicology Program (NTP),
    101, 314
N-demethylation, 115
Nephrotoxicity, mechanism of,
    in rats, 57
Neurological impairment, reversible
    symptoms of, 329
Neurologic test, 329
Neurotoxic effects, of solvents, 129
Nicotinamide adenine dinucleotide
    phosphate (NADPH), 60, 195
Noncancer effects, incidence of, 302
Nonreactive gases, dose of, 235
No-observable-adverse-effects level
    (NOAEL), 1, 16, 70, 292
  allometric scaling, 297
  dose-limiting effect, 296
  extrapolation method, 11, 296

[No-observable-adverse-effects level
    (NOAEL)]
  nonclinical studies, 297
  subchronic/chronic data
      projection, 74

O-demethylation, 115
Olefins, epoxidized by cytochrome
    (CYP), 196
Oral cancer slope factor, development
    of, 319
Organism, pharmacokinetic model of,
    types, 202
Oxidative metabolic rates, 325

Paraoxon, 224
Paraoxonase (PON1), 224
Parathion, 224
Partition coefficient
  allometric scaling, 155
  blood and air ratio, 169
    blood water content on, 172
    neutral lipid content on, 172
  chemical-specific, 150
  chemical structural features, 168
  control and reproducibility
      concentrations, 151
  on electrical potentials, 152
  Free-Wilson approach, 169
  freezing effects on tissue quality, 152
  in silico approaches for, 168
  in vitro filtration method, 151
  interindividual differences in, 169, 178
  mechanistic in silico approaches
      for, 169
  for metabolic parameters, 143
  metabolism, 152
  physiochemical properties, 149
  physiological parameter in PBPK
      model, 149
  for physiologically based
      pharmacokinetic (PBPK)
      models, 167
  tissue and air ratio, 170
    neutral lipid content on, 174

[Partition coefficient]
    tissue water content on, 174
    tissue and blood ratio, 145, 173
    for volatile and semivolatile
        compounds, 150
    of volatile organic compounds
        (VOCs), 169
    volatilization, 151
    water-soluble compounds, 151
Pediatric pharmacokinetic data, 240
Pediatric trials, with zidovudine, in
    children, 234
Perchloroethylene, 277
  exhaled air indication, 279
  inhalation exposure, 278
  pharmacokinetics analysis, 280
  venous blood concentrations, 279
Peroxisome proliferation, 58
  dose–response model for, 59
  in rats, 59
Peroxisome proliferator-activated
    receptor (PPARa), 59
  liver carcinogenesis, 59
  mechanism of, 59
  role of, 61
Pharmaceutical testing, 260
Pharmacodynamic component,
    magnitude of, 136
Pharmacodynamic modeling, tissue
    responses, 50
Pharmacodynamics, 167
  composite factor of, 167
  variability factor, 167
Pharmacokinetic database, 237
Pharmacokinetic models, 202
  organism, 202
  types, 202
    data based, 202
    physiologically based, 202
Pharmacokinetic and pharmacodynamic
    measures, 50
  derivation of, 50
  from toxicology and mechanistic
      information, 50
Pharmacokinetic–pharmacodynamic
    (PKPD) modeling, 136
  iterative process, 61–62

[Pharmacokinetic–pharmacodynamic
    (PKPD) modeling]
pharmaceutical agents, 30
Pharmacokinetics, in children, 231
Physiologically based pharmacokinetic
    (PBPK) models, 1, 142, 49, 167,
    185, 211
  advantages of, 142
  anatomical parameters, 146
  animal component of, 42
  application, 276
  application of statistical methods
      to, 269
  approaches to, 271
    discomfort of, 272
  arteriovenous substrate
      concentration, 159
  bioactivation, 30
  biotransformation, 159
  chemical-independent data, 270
  chemical-specific data, 271
  chemical-specific parameters, 146
    calculating bag concentrations, 151
    components of, 149
    partition coefficients, 149
  classical compartmental kinetic
      models, 144
  detoxification processes, 30
  diffusion limitation, 162
  disadvantages of, 142
  flow-limited diffusion, 143
  human component of, 43
  human variability, 33
  identifiability, 270
  of internal dose
    by Monte Carlo simulation
        methodology, 168
  impact of, 212–213
  in vitro activity of, 212
  in vitro extrapolation, 186
  in vivo, 273
  in vivo dosimetry, 212
  in vivo exrapolation, 186
  local sensitivity analysis, 271
  metabolism
    chemical, 158
    enzyme-limited clearance, 159

[Physiologically based pharmacokinetic
    (PBPK) models]
  flow-limited clearance, 159
  gas uptake method, 152
  in vitro methods, 152
  maximum hepatic metabolic
      rate, 158
  Michaelis–Menten equation, 153
    for particular volatile
        compound, 153
    saturable process, 153
    venous equilibration, 153
  metabolites elimination, 159
  model development and validation
    airborne chemical absorption, 156
    dermal absorption, 156
    gastrointestinal (GI) transit
        times, 157
    metabolism rate, 158
    oral uptake method, 157
    routes of administration for, 156
    steady-state equation, 156
    systemic circulation, 157
  model development process, 142
  model verification process, 161
  Monte Carlo analysis
    consequential parameters, 160
    dose metric simulation, 161
    nonlinear metabolic
        mechanisms, 160
    parameter variability effects, 160
    probability distribution
        functions, 160
    sensitivity coefficients, 160
  noncompartmental organ model, 144
  nonlinearity of, 271
  nonstatistical approach, 272–273
  partition coefficients, 143, 168
  pharmacodynamic (PD)
      models, 144
  physiochemical and biochemical
      parameters, 144
  physiological parameters, 146, 147
    age-related growth, 148
    allometric scaling of, 148
    in drug metabolism, 149
    growth equations for, 149

[Physiologically based pharmacokinetic
(PBPK) models]
for perinatal and lifetime
models, 148
point estimate predictions, 270
Ramsey–Andersen model, 143
risk assessment applications, 163, 269
sampling distribution, 273
sensitivity of, 270
sinusoidal perfusion model, 144
sodium-iodide symporter (NIS), 163
species extrapolation, 39
statistical approach, 273
statistical issues in, 277
target site dosimetries, 142
for 2,3,7,8-tetrachlorodibenzo-
p-dioxin (TCDD), 54
tissue blood flow, 162
tissue dose metrics, of parent
compound, 50
translocation and sequestration, 157
uses of, 269
versus kinetic models, 142
volatile organic chemical, 144
volatile organic solvents, 159
Piperonyl butoxide, 6
PK and PBPK models in IRIS program,
use of, 302, 304
Placental transfer
rate of, 236
Plasma-based toxicokinetic
measurements, 32
Plasma concentration measurements, 29
Plasma drug concentrations, 96
Plasma-extracellular fluid
compartment, 339
Plasma GSH levels, 239
Plasma-to-red-blood-cell transfer
coefficient, 340
Point of departure (POD), 303
Polymorphisms, 217
cancer risk of, 218
case of warfarin, 219
cytochrome P450 isozymes, 219
effects of, 218
overall population, 222
tissue levels of, 218

[Polymorphisms]
toxicology field of, 218
(R)-warfarin CYP2C19, 219
(S)-warfarin CYP2C9, 219
Population variability, 272
Portal of entry (POE), 125
Potential human cancer risks, 327
Preeclampsia, 19
Premature neonates, skin of, 235
Proteins, 186
Proteolytic degradation, 56
rate of, 56
by lysosomal enzymes, 56
Pulmonary region
deposition in, 235

Quantitative structure–activity
relationship (QSAR), 227
Quantitative structure–property
relationships (QSPRs), 20

Ramsey–Andersen model, 143, 145
Receptor-mediated carcinogens, 54
Receptor-mediated mechanisms, of
toxicity, 53
Red blood cell (RBC), 316
Reference dose (RfD), 3
Reference dose values, quantitative
relationship for, 36
Reference values, derivation of, 302, 314
Regenerative hyperplasia, 311
Regional deposited dose ratio
(RDDRr), 13
Regional gas dose ratio (RGDRr), 13
Renal adenomas, 43
Renal clearance measurement, 330
Renal elimination, development of, 239
Renal elimination mechanisms, 335
Renal tubular tumor induction, 41
Respiratory distress syndrome, 239
Respiratory tract regions, 14
Risk assessment
applications of, 340
pharmacokinetic requirement for, 125
Route-to-route extrapolation, 274

[Route-to-route extrapolation]
parameters and, 277
predictions for, 277
for vinyl chloride, 274

Salmonella typhimurium, 51
Sensitive populations
epidemiology studies, 260
mechanistic information, 259–264
high-quality human toxicology, 260
risk of, 259
principles of, 251
reference concentration (RfC), 252
reference dose (RfD), 252
risk assessment, 264
subthreshold doses in, 251
uncertainty factors of, 254–259
Sensitivity coefficient, 161
Serum creatinine levels, 335
Sinusoidal perfusion model, 144
liver metabolism, 159
Somatic development, control of, 334
Sprague–Dawley male rat data, 328
Sprague–Dawley values, 274
tissue/air partition coefficients, 274
Steady-state route-to-route
relationship, 275
Subcellular fractions, 200
Subchronic to chronic
extrapolation, 10–11
Surface-area correction factor, 325

Target organ toxicity (TOT), 95
dose, 97
Tedlar® bags, 151
Terfenadine, 106
2,3,7,8-Tetrachlorodibenz-p-dioxin
(TCDD), 53, 176, 215
biological effects of, 54
concentration, adipose tissue and
blood ratio of, 175
females, 216
gestation model of, 216
growth of, 216
males, 216

[2,3,7,8-Tetrachlorodibenz-p-dioxin
(TCDD)]
maternal levels of, 216
toxicity and carcinogenicity of, 54
Thyroid growth
dose–response for, 56
Thyroid hormone disruption
aryl hydrocarbon (Ah) receptor
dependent, 55
Thyroid hormones, regulation of, 55
Thyroid stimulating hormone (TSH),
inhibition of, 55
Thyroid tumors, 56
Thyrotropin-releasing hormone
(TRH), 55
Time-weighted-average (TWA) blood
concentration, 330
Tissue, drug concentrations, 96
Tissue and blood partition coefficients
for acetate esters, 177
for alcohols, 177
estimation of, 180
for ketones, 177
and n-octanol and water partition
coefficient, 177
in rat, 176
Tissue dosimetry
exploration of, 132
models, mechanistic information
for, 47
Tissue/organ delivery
human variability in, 30
Tissue partition coefficient, 328
in animals, 32
Tissues, of animals and humans, 34
Tolerable daily intake (TDI), 3
Toxic agents, 47
Toxicants
detoxication of, 187
inhibition, 192
Toxic effects, 136
in humans or animal models, 48
of metabolic products, 48
Toxic moiety, 126
Toxic response
components of, 33
and external dose, 29

[Toxic response]
  in vivo, 35
Toxicity, 185
  dose selection for study of, 97
    dose-limiting toxicity (DLT), 97
    good laboratory practice
        (GLP), 98
    limit dose (LD), 97
    maximal absorbable dose, 97
    maximal feasible dose (MFD), 98
    maximum tolerated dose
        (MTD), 97
    perspective, 97
    target organ toxicity (TOT)
        dose, 97
  evaluations, 95
  in drug, 95
  major organs, 98
  repeated dose studies of, 98
    dose selection for, 100
    high dose, 98
    low dose, 99
    middle dose, 99
Toxicodynamic model, 20
Toxicodynamics (TD) fractions, 303
Toxicokinetic and toxicodynamic
    components, 28
  in dose/concentration–response
      assessment, 31
  of interspecies differences and human
      variability, 28
  subfactor for, 30
Toxicokinetics, 6, 17, 96
  in animals and humans, 38
  application, 96
    evaluation in pharmaceutical
        industry, 96
  on chemical concentrations, 30
  fractions, 303
  nonlinearity in, 96
  in rats and humans, 41
  role in, 21
  tissue drug, 102
    distribution of, 103
    in target organs, 103
Toxicology program, 97
Trichloroacetic acid (TCA), 279

[Trichloroacetic acid (TCA)]
  in blood, 279
  concentration of, 215
  and urine, 279
1,1,1-Trichloroethane, 170
Trichloroethylene, 205
Trimethylpentane-induced nephropathy
    in rats, 58
Trimethylpentane (TMP), 57
Tritium, 116
  in vitro with rat and human
      microsomes, 116
Truncated lognormal population
    distributions, 278
Tumors
  in animals, 39
  liver and kidney, in mice, 39
  mode of induction of, 39
  weight of evidence for, 40–41

U.S. Environmental Protection Agency,
    1, 234, 251, 302
U.S. Food Quality Protection Act
    (FQPA), 6
Uncertainty factors (UFs), 1, 259
  dose–response analysis, 254
  Food Quality Protection Act
      (FQPA), 260
  human interindividual
      variability, 256
  in vivo methods, 259
  inter-human variability
      extrapolation, 252
  interspecies extrapolation, 252
  lowest-observed-adverse-effect level
      (LOAEL), 252
  mechanistic toxicology, 254
  no-observed-adverse-effect level
      (NOAEL), 252
    benchmark dose (BMD), 256
    bioassays, 259
    database incompleteness, 264
    epidemiology, 259
    frequency histograms of, 258
    geographic correlation
        studies, 260

[Uncertainty factors (UFs)
no-observed-adverse-effect level
(NOAEL)]
health statistics, 260
high-quality human toxicology, 259
reproductive and developmental
toxicity, 258
trimodal distribution, 256
perspective of toxicodynamics, 256
perspective of toxicokinetics, 256
reference dose(RfD), 253
in risk assessment, use of, 6–9
Upper confidence limit (UCL) risks
distribution of, 322
Uridine diphosphate (UDP), 55, 198
Urinary metabolite levels, 125

Ventilation rate
in human, 274
Vinyl chloride, 274, 319

V_{max}/K_m ratio, 270
hepatic blood flow and, 271
V_{max}, 270
Volatile organic chemical
(VOC), 274

Warfarin
distribution for, 223
Water-soluble gases, 235
William's E culture medium, 112
World Health Organization
(WHO), 7

Xenobiotic biotransformation, 186
Xenobiotic metabolism, 195
enzymes in, 195
extrapolation of, 132
Xylene, carcinogenic potential of, 328